# THE SYMMETRY OF SAILING

*The Physics of Sailing for Yachtsmen*

# THE SYMMETRY OF SAILING

## *The Physics of Sailing for Yachtsmen*

Ross Garrett

*Drawings by Dave Wilkie*

SHERIDAN HOUSE

This edition first
published 1996 by
Sheridan House Inc.
145 Palisade Street
Dobbs Ferry, NY 10522

Copyright © Ross Garrett 1987
First published by
Adlard Coles Ltd 1987

*Library of Congress Cataloging-in-Publication Data*

Garrett, Ross.
    The symmetry of sailing: the physics of sailing for yachtsmen /
Ross Garrett; drawings by Dave Wilkie.
        p.   cm.
    Originally published: London: Adlard Coles, 1987.
    Includes bibliographical references (p.   ) and index.
    ISBN 1–57409–000–3
    1. Sailing.   I. Title.
VK543.G37   1996
623.88'223—dc20                                              95–46519
                                                                 CIP

Printed in Great Britain

ISBN 1–57409–000–3

Felix qui potuit rerum cognoscere causas.
*Happy is he who gets to know the reasons for things.*

MOTTO OF CHURCHILL COLLEGE, CAMBRIDGE

For Jennifer, Catherine, Patsy and Brian

# CONTENTS

# PREFACE

The purpose of this book is to explain to the 'average yachtsman' how a yacht works. One doesn't usually think of a sail boat as something that works, but if you have done a reasonable amount of sailing you will not need convincing that there are many subtleties in the often strange behaviour of the sailing yacht.

The book is based on a course of lectures given over the last ten years at the Centre for Continuing Education of the University of Auckland, New Zealand. The men and women for whom the course was designed were of widely varying occupations with rarely any special educational background in physics or mathematics. Although the scientific background and abilities of the audience were widely divergent the success of the lectures was undoubtedly due to a single unifying influence: all were keen and experienced sailors, readily able to relate the phenomena with which they were so familiar with an explanation of why and how those phenomena occurred.

Many sailors, though admirably able to sail their boats against wind and waves, are not at all clear just why this is even possible. Why, for instance, is it essential to make leeway in order to go to windward? What causes downwind rhythmic rolling and how can it be prevented? When is it quicker to tack downwind? Not too difficult to figure out if your boat has instruments, but understanding the reasons also leads to a method of improving your overall speed on a shy reach. Balance can be a mysterious business; it is generally grossly oversimplified in books for the yachtsman. An indication of its complexity and variability is the use of 'lead' by yacht designers. Although this is a convenient and sensible approach to a difficult aspect of yacht design, it is also a measure of the degree to which balance is not understood. A knowledge of some of the factors that contribute to balance will give a better appreciation of the designer's dilemma as well as add to your enjoyment and appreciation of your own boat.

Thus the thrust of this book is understanding the 'whys' rather than telling you the 'what'. Understanding such things is a subtle business. When a professor gives an explanation of a new phenomenon to his class he bases it on a logical sequence of other phenomena with which he thinks the students are already familiar. The student who says, 'Oh, now I understand' is seeing the new and hitherto not understood phenomenon as a combination of ideas with which he is already familiar. The key to success then is in making sure that one's audience is familiar with the basic set of ideas familiar to the 'average yachtsman'. Part II is a shorter explanation of the same phenomena in more formal and mathematical language for those familiar with a slightly different set of basic ideas.

I think it is not an exaggeration to say that in recent times all sports have been affected to an increasing extent by scientific investigation. Better knowledge of human physiology has made it possible for athletes to train and prepare themselves in ways which result in greater accomplishments. Those sports which involve the use of a 'weapon' or 'tool' have been transformed even more by science. Thanks to glass-reinforced plastic the pole vaulter has attained new heights. New materials have given greater freedom in the design of tennis racquets so that their shape is no longer limited by strength of materials and the centre of percussion can be near the centre of the head giving better feel and control. New materials and fabrication methods have revolutionised skiing.

Obviously motor racing and powerboat racing depend heavily upon scientific research.

The interaction of the sport of yachting with scientific research is in a league of its own. First, the yachtsman's 'weapon', his boat, is infinitely more complex than a tennis racquet or a pole vaulter's pole. It is also less well understood in a scientific sense than a racing car, even though the latter may *seem* a lot more complicated. Second, a knowledge of the physical processes in sailing is far more important to the competing yachtsman than is an equivalent knowledge in most other sports. Understanding why your boat behaves the way it does will increase both your appreciation of and ability in your chosen sport.

This is not a book which intends to tell you directly how to win races. The aim is to tell you how you can figure out how to win races or make efficient passages with your particular type of boat. Add to this constant practice to the exclusion of most other activities, the drive to defeat all opposition, and you are well on the way to getting the gold, if this is your desire.

Another objective of this book is to make available the quite considerable body of information on sailing theory which has appeared in the scientific literature since the mid-thirties. Much of this has crept into the popular literature but usually disguised in a mélange of sound scientific statements plus the pet opinions of a gold-winning helmsman. This is not meant to imply that such opinions are necessarily wrong but simply that their validity or generality may not have been subject to the same rigorous investigation.

For instance, many authors of sailing books feel the urge at some point to give a diagram showing the air flow around a headsail and main. These diagrams and the conclusions drawn from them are almost always incorrect. One can only assume that the author had a preconceived notion and drew his picture to correspond, or that he repeated earlier errors. Such diagrams look very scientific to the reader yet are not based on measurements. All the diagrams appearing in this book are drawn from measured data. Though these measurements have often been made under ideal conditions the extent of their validity will always be made clear (this applies particularly to the discussion in chapter 4 on sail interaction).

The words 'drag' and 'lift', borrowed from aeronautics, refer to key concepts in understanding sailing. After first introducing an unusual view of sailing in chapter 1, aspects of sailing which depend primarily on drag, namely downwind sailing, are discussed in chapter 2. Both lift and drag, or more specifically their relationship, form the kernel of understanding windward sailing in chapter 3.

Sailing is rarely smooth; waves produce pitching, waves and wind may produce rolling, hull motion adds to the hydrodynamic resistance, and an oscillating sail produces a very peculiar driving force even in a steady wind: these are among the topics discussed in chapter 6. Unfortunately not all questions can be answered as we are now well into unexplored territory near the frontiers of knowledge.

Chapter 5 is hull design for sailors. It is not intended for designers but for those who want to appreciate the design on which they are sailing.

An attraction of sailing is that one gets a little closer to nature. The forces that produce this environment are discussed in chapter 11, where some of the questions you must have asked yourself about waves will be answered.

Part II of the book is for those who would like to be filled in on some of the technical details. It follows the same order as Part I but is not meant to stand on its own; it must be read in conjunction with the first part.

There is some variability in the level of presentation between the chapters. Chapter 6, by nature of its subject, became more technical than anticipated.

It is inevitable that a book so full of facts as this one will contain errors. I would be grateful, therefore, to receive readers' comments.

Acknowledgement should be made to the several hundred sailors who attended my lectures over the past few years, for it was their response that fuelled the enthusiasm needed to drop all leisure activities for a year in order to write. As my own field of research is in an area of physics completely unrelated to sailing, I have relied very heavily on material published in a wide range of scientific journals. I have made no attempt to make formal reference to all these works as that seemed inappropriate in a book of this kind. Nevertheless I am indebted to their authors, the real innovators. However some references are given in the Bibliography so that the serious student with access to a university library may easily follow up the original work.

Final acknowledgements are due to Dave Wilkie whose unique blend of artistic ability and technical know-how made possible the production of more than two hundred diagrams in a space of time rather shorter than the optimum.

Ross Garrett
Physics Department
University of Auckland
New Zealand

# PART I

Chapter One

# THE SYMMETRY OF SAILING

*I have become an avid symmetry fan, addicted beyond cure, utterly convinced of the fertility of symmetry in scientific study and research as a unifying, clarifying and simplifying factor.*

JOE ROSEN, *Symmetry Discovered*

There is an old sailing conundrum that goes something like this: two sailors with a competitive spirit desire to race against each other. There is, however, a difficulty: they have only one boat and to complicate matters the only available sailing water is a river that flows at a constant 10 knots. They decide therefore to sail separately, timing their runs over a 1 mile stretch of water and sailing downstream. When our first sailor does his measured mile there is a 10 knot wind blowing directly downstream. By the time the boat is brought back to the starting line ready for our second sailor to demonstrate his prowess the wind has fallen away to a flat calm. Undaunted, he proceeds with the race. Now our two sailors are in truth both very clever and able yachtsmen who, if racing normally in one-design boats, would be neck-and-neck around the course. Given these facts the outcome of the race is completely predictable – one of them wins hands down.

If you are not sure who the winner is and why the outcome is so certain then you owe yourself a different view of sailing. The view is a beautiful one. Not the sort of beauty that might be associated with a work of art or a fine sunset, but more akin to the kind of beauty the ancient Greeks saw in the symmetry of the circle or the intriguing properties of the golden

rectangle. It is a scientific beauty. As you sail your boat to windward, whether it be a 40 ft classic of yesteryear or a praam dinghy with a daggerboard, there is acting on your vessel a perfect symmetry of forces. Whether you sail well or poorly, Nature sees to it that the forces between wind and boat are always precisely balanced by the forces between water and boat. This symmetry of sailing is happening all the time. You cannot see it, but knowing that it is ever present and understanding why will surely add to your appreciation of one of the hidden beauties of the sport.

What makes a boat go to windward? To say that it is the result of the wind reacting with the sails is only half the answer. In fact it is *precisely* only half the answer. To see why it is possible to make such an uncompromising statement we have to take a look at sailing from a rather unconventional point of view. Remember how the ancients thought that the sun and the moon and all the rest of the heavenly bodies rotated about the earth as a centre? Stupid weren't they! Stand on the earth at night and look out and it is easy to believe why they thought as they did. Stand on the deck of your boat when sailing to windward and feel the wind and see the sails and you too can easily fall into the trap of the ancients. The forces of Nature have no regard for a mere human's point of view. To understand them one must use a little cunning.

Imagine that your ship, instead of floating on water, is floating on air, a hot-air balloon perhaps with crew in the basket below all floating in the air above the sea. Your conveyance is totally immersed in the air, so where the air goes you go. If you brought along with you a wind indicator it would read zero.

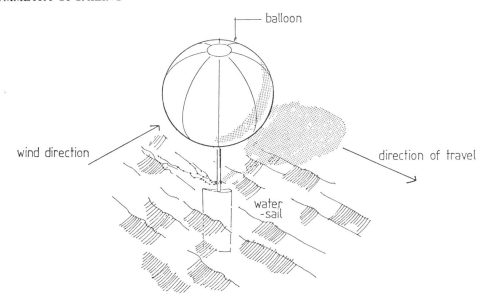

*Fig. 1.1* The pilot of a craft floating on air would consider the apparent water flow beneath him a usable source of energy to push him through the air. By lowering a 'water-sail' into the water and modifying the shape of his balloon he could even fly 'to waterward' at a reasonable speed through the air. If the words 'water' and 'air' are interchanged we find ourselves describing a sailing boat. The symmetry of sailing lies in the fact that if we view our vessel in a mirror which turns air into water and water into air, it is still able to move over the same range of directions. When viewed in this way it is clear that the underwater body of a yacht contributes just as much to its windward ability as the cut of its sails.

You are in a windless world and seemingly motionless. Then you look at the sea below and notice that it is flowing past you at a good 10 knots. Being a resourceful person and becoming bored of being motionless in the air you decide to exploit the motion of the water and lower a water-sail into the sea (fig. 1.1). Careful design and appropriate orientation of the water-sail allows you to set a course through the air, and for the first time the wind indicator starts to register and with the wind in your hair you feel the exhilaration of motion. This newly found power is tempered by the fact that you cannot move through the air in a direction less than about 45° to the direction of water flow, and even to get to this requires a few radical changes to the shape of the balloon. Nevertheless you are pleased that you are now able to fly through the air to 'waterward'. Naturally on days when the water flow is fast you can make better time through the air, but when the water is motionless you are becalmed.

This flying thing and a sail boat are of course propelled by exactly the same means: it's just that when you are floating on water you think of the water as being fixed and when you float on air you think of the air as being fixed. So the balloonist claims that it is the flow of water past his water-sail that propels him to waterward, whereas the sailor claims that it is the flow of air past his boat's sail that moves him to windward. The truth of the matter is of course that the flow of water and the flow of air are equally important in making a boat go to windward. Now if both air and water are flowing at 10 knots from the north then the balloonist will look down at the sea and find no movement past his water-sail. The sailor whose ship is floating in a 10 knot stream will find no apparent wind if it is blowing at 10 knots in the same direction as the stream. They both consider themselves becalmed. We can see from this that what counts is not how fast the air or water is moving but what their *relative* speeds are. The boat, being partly immersed in each medium, is able to turn this relative motion into a motion of its own in almost any direction. How is this accomplished?

That the flow of air around sails can be such as to produce a side force is a familiar concept to all sailors. (The details of how this comes about are given in chapter 3.) Fig. 1.2 will remind you of how the side force is related to the wind direction and the sail orientation. It is intuitively obvious that when sailing to windward the force developed by the sail pushes the boat sideways more than forwards. But boats sail more forwards than sideways. If you push something you expect it to go where you push it. Boats seem to be an exception, but the laws of Nature admit of no exceptions. Obviously there is more going on than just wind on sails.

Since a boat does not go the way the wind pushes, it would seem that the water forces adding to the wind forces result in a forward force. Sounds like sense, but it isn't true for a boat sailing at a steady speed to windward. In fact under these conditions the wind force and the water force are exactly equal in magnitude and exactly opposed in direction. It is just like the two teams in a tug-of-war: there is a lot of grunting and straining going on but nobody is going anywhere. How do I then reconcile my statement that the air and water forces are equal and opposite with the fact that the boat is certainly moving forward?

This is where the symmetry of sailing really unfolds. Remember that the force on the sail is due to the flow of air which is approaching the sail at about 30° to the boom. If there is no relative motion of air and sail there is no force. If a boat is motionless in the water there can be no forces on its keel or hull (we are not concerned with buoyancy or gravity). The water force is only present when the boat is moving. So the

water force and the wind force can be equal and opposite because these forces exist only when the boat is moving through both air and water.

Now you will be asking, why should these forces be exactly equal and opposite? It was all very well for the ancient Greeks to feel that the existence of symmetry was alone sufficient explanation for a phenomenon. Perhaps human nature has changed since then, but even the most philosophical cruising yachtsman (racing people don't get time to philosophise) would like a better explanation.

To see how it is possible for the wind force and the water force to oppose each other like a tug-of-war let's conduct a 'thought' experiment! Imagine that there is a good steady breeze of about 10 to 15 knots at about 40° to the port bow, genoa and main are set and correctly sheeted, but any movement of the boat through the water is prevented by mooring lines fore and aft. Suddenly and simultaneously both mooring lines are cut. We ask, what is the motion of the boat through the water in the next few seconds? Obviously there is an acceleration period whose duration depends mainly on the displacement of the boat. At the end of this period the boat will be sailing on port tack at a steady speed determined by wind strength and its design. It is the detail of what happens in between that interests us.

The sequence of diagrams in fig. 1.3 shows the events in the important few seconds of acceleration. The instant after the mooring ropes are cut there is only one force acting on the boat, that developed by

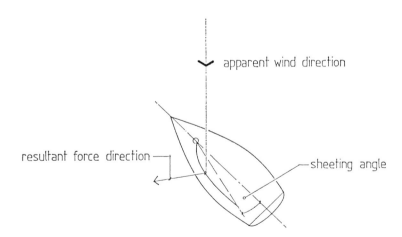

*Fig. 1.2* As every sailor knows, the flow of air around an efficient sail produces a force, not in the direction of the wind, but almost at right angles to it. For the discussion in this chapter this fact must be accepted; an explanation of how it comes about is given in chapter 3.

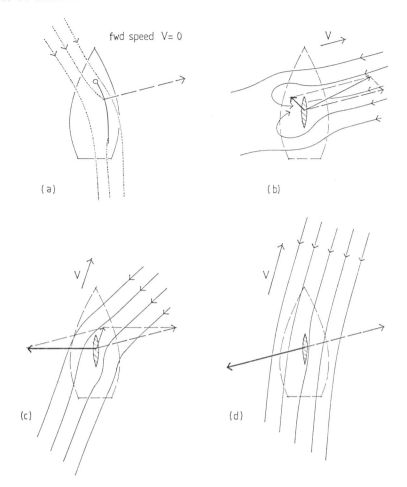

*Fig. 1.3* The sequence of events immediately after a closehauled yacht, which is immobilised by bow and stern lines, is suddenly released. The dotted lines and dashed arrows refer to wind flow around the sail and wind forces, whereas solid lines refer to water flow around the keel and water forces. (a) is at the instant of release when the boat speed V = 0. Only wind forces are present and their resultant is shown by the dashed arrow perpendicular to the sail. For clarity the wind flow is not shown in subsequent diagrams. A further simplification is that the wind force is shown fixed even though the boat starts moving through the air. In (b), (c) and (d) the magnitude and direction of the boat speed is indicated by the arrow V. The total force on the boat is the resultant of the water force and the wind force. As the boat speeds up this resultant swings around from pointing dead to leeward to pointing forward, to disappearing altogether as in (d) where the air and water forces exactly balance. This balance results from the forward motion of the boat which then continues at a steady unaccelerated speed. The principles involved in this diagram are discussed in greater detail in chapter 3.

the wind. But this force is mainly sideways, so the boat must be pushed sideways. Great eddies will be seen in the water as the flat faces of keel and rudder are forced sideways through the water. Immediately this movement starts water forces are produced; the keel is acting just like a sail sheeted in too hard on a broad reach. Although it is inefficient there is a small forward component in this force produced by the keel which, when added to that produced by the sail, gives a forward push to the boat as well as opposing the side force of the wind. The boat is now moving forwards as well as sideways so the angle of attack of water on

4

the keel is less. It is now a more efficient hydrodynamic surface, just as increasing the boom angle on a broad reach lessens the heeling force and increases the forward force. As long as the wind and water forces are at an angle to one another there will be a net forward force which will keep the boat accelerating. But the faster the boat goes the more the keel force turns to directly oppose the sail force. Finally, when they are both equal and opposite there is no net force on the boat at all. However this condition requires that the boat be moving in order to produce the forces. It is a bit like arranging a tug-of-war with a bent rope, the bend being where it goes around a railway car on a horizontal track. As the opposing teams pull, the car accelerates until the rope straightens out after which the car continues at a constant speed (if well oiled) with no further acceleration.

Now just as the sail will luff if the wind is dead ahead, so the keel will luff if the water flow is from dead ahead. The keel and underwater body of the boat, being symmetrical, must move so that the water flow comes in at a few degrees from the leeward side if a side force is to be generated. This is the leeway angle. Without leeway a boat cannot sail to windward for then there would be no side force to counteract that of the wind and the boat would simply drift downwind. It seems almost ironical that this 'defect' of the sailing craft when going to windward is in fact what makes it possible to go to windward at all.

The symmetry of sailing thus refers to a symmetry between air and water forces. It is not only while sailing to windward but on all points of sail that these two forces are always equal and opposite if the boat is moving at a steady speed. It is analogous to the equal and opposite reflection you see of yourself in the mirror.

This symmetry goes far deeper than the simple discussion so far might lead one to believe. It will be shown in chapter 3, for instance, that the pointing ability of a yacht depends on a property of the rig known as the drag angle; also on the same property of the underwater body and to exactly the same extent. To be more specific, the optimum course to windward is such that the sum of the rig drag angle and the underwater drag angle is equal to the angle between the apparent wind and the boat's course. This means that both the rig and the hull contribute equally to a boat's ability to point.

The symmetry further manifests itself in that the physics of water flow and air flow are identical if one takes account of the differences in density and viscosity. It also turns out that the quantity that counts in many aspects of fluid flow is the ratio of viscosity to density. Remarkably, this ratio for water differs only by a factor of 15 from that for air, although the density ratio is about 1000.

Throughout the chapters that follow this idea of symmetry will be a recurring theme. Although symmetry implies similarity between air and water flow, there are of course obvious differences. A difference in speed of the air and water flow is, for instance, vital to enable a yacht to sail in a direction dictated by the crew rather than one dictated by drift. This was seen in our example of the hot-air balloon. If you think about it a little you will see that it is not really necessary to have two different media to make 'sailing' possible. If the pilot of the hot-air balloon were able to lower a sail into a region of air moving at a different speed from that of the balloon itself, then he could guide his craft just as easily as if he used the sea. So all that is really required is a variation of the flow speed with height, a velocity gradient. In fact such a variation always exists in air flow over the earth because of a greater viscous drag near the surface. That ancient friend of the sailor, the albatross, can fly against the wind without ever flapping his wings. Cunning use of the velocity gradient allows the bird to use the same physical principles in getting to windward as does the sailor. Clearly the albatross and the sailor have a lot more in common than even old Sam Coleridge ever envisaged.

This chapter began with a problem. You have probably figured out by now who won. In case you haven't the argument goes as follows. The first sailor drifts downstream at 10 knots. But this is just the true wind speed so the apparent wind speed is zero. He is truly becalmed; his sails don't fill, he is not moving with respect to the water and has no steerage way. So he makes rather uncontrolled progress downstream at 10 knots. For the second sailor there is no true wind, so in drifting downstream at 10 knots he feels an apparent wind of 10 knots coming upstream. In order to reach his destination downstream he must tack against this apparent headwind. The second sailor thus makes good a speed to windward with respect to the water. Thus his overall speed with respect to the ground is the 10 knot drift speed plus his speed made good to windward. Even if he sails poorly he can always beat his opponent.

Sailing is made possible by the *relative* movement of air and water. For the first sailor there is no such relative movement and so he loses the race.

# Chapter Two

# DOWNWIND SAILING

---

*It is a deplorable fact that no theory of drag yet exists which even approximately does justice to the experimental results.*

PRANDTL AND TIENTJENS, *Fundamentals of Aero and Hydrodynamics*, 1934

---

## §2.1   Form of the Drag Equation

The key concept associated with downwind sailing is fluid drag. We would like to minimise the hydrodynamic drag of the hull through the water and maximise the aerodynamic drag of the sails.

Drag between a solid and a fluid is a lot more complex than the more familiar frictional drag between two solids. (In this context fluid refers to either air or water.) Familiarity with the characteristics of a phenomenon greatly aids our comprehension of its causes. By its nature we are more familiar with frictional drag between solids than we are with fluid drag. We test a surface by passing our fingertips over it. This is mainly solid-to-solid friction and does not necessarily tell us what we want to know about solid-to-fluid friction. For instance a coating of wax will make a surface feel slippery but will make absolutely no difference to the fluid drag, as we shall see.

Drag is defined as the net force in the direction of the undisturbed flow.

Fluid drag is not brought about by a single mechanism but is the combination of several effects. These can be divided into two main categories: surface friction drag and normal pressure drag. Before looking at the detailed mechanisms of these

drag forces let us look at it in a purely phenomenological way and answer the question: what quantities determine the magnitude of the drag?

Physicists have a procedure known as dimensional analysis which makes it possible to predict how quantities such as velocity, length, density etc will enter into the drag. (Mathematical details of this procedure are given in Part II.) Essentially, for an object whose size is characterised by a length L (the waterline length for a hull), moving at a speed V in a fluid of density $\varrho$ and viscosity $\mu$, we can predict, without doing an experiment, that the drag depends upon these quantities in the following way:

$$\text{Drag} = \text{Constant} \times \left[ \begin{array}{c} \text{Some function of} \\ (\varrho LV/\mu) \text{ and } (V/\sqrt{Lg}) \end{array} \right] \\ \times \varrho \times L^2 \times V^2 \qquad \textbf{(1)}$$

You may think the formula looks complicated and so it should, for drag is a complex phenomenon. The procedure of dimensional analysis has only told us what quantities matter in determining drag, but at least we know which way to direct our attention.

What does equation (1) means physically? The unknown constant in front means that we cannot by this means determine the absolute value of the drag, only relative values as the waterline length L or the speed V are changed.

The quantities in parentheses inside the square brackets require some explanation. They are dimensionless quantities – pure numbers – and do not depend upon the system of units used. Whether the individual quantities that make up these numbers are measured in the Imperial units of feet, pounds and seconds or the International System using metres, kilogrammes and seconds makes no difference to these dimensionless numbers. This independence from an arbitrary man-made construct such as a system of units suggests that these quantities have a fundamental significance in describing the phenomena of nature. The fact that both these quantities are named after famous researchers in fluid dynamics and naval architecture testifies to their usefulness.

$(\varrho LV)/\mu$ means: the product of fluid density, length and velocity divided by viscosity of the fluid

A small object moving through a fluid will thus have a low Reynolds' number whereas a large fast object will have a high one. It is called Reynolds' number after the English engineer Osborne Reynolds (1842–1912), Fellow of the Royal Society and Professor of Engineering at Manchester University, and given the symbol $R_n$. Quite clearly Reynolds' number has to do with the speed of fluid flow or the speed of a hull through the water and also its length. In many applications Reynolds' number can be thought of as simply a measure of speed, but it has a much deeper significance than just that.

It is shown in Part II, §7.1 that Reynolds' number is a measure of the ratio of inertia forces to viscous forces in the fluid. Inertia forces have to do with the difficulty of moving a heavy object. It requires more force to get a garden roller up to walking speed in ten seconds, say, than it does to get a lawnmower up to the same speed in the same time. As a boat moves along it pushes aside a volume of water determined by its displacement and the force associated with moving this volume of water is the inertia force. There is also a viscous drag associated with the same volume of water (see below). So Reynolds' number measures the relative importance of these:

$$R_n = \frac{\text{inertia forces}}{\text{viscous forces}}$$

If $R_n$ is low, as it is for a small object at low velocity, then the drag due to the viscosity of the fluid is more important than the drag associated with pushing the fluid aside. If $R_n$ is large it means that the viscous part of the drag is less important.

The second quantity in parentheses, $V/\sqrt{Lg}$, is called the Froude number. It is a fixed number if the ratio of the speed to the square root of the waterline length is fixed (since the acceleration of gravity,

represented by g, certainly does not change significantly). More will be said later about the important relationship between speed and waterline length, but for the moment we are interested in the fact that the Froude number, $F_n$, may also be thought of as the ratio of inertia forces to gravitational forces:

$$F_n = \frac{\text{inertia forces}}{\text{gravitational forces}}$$

Waves are discussed in §11.3 where you will see that the speed of water waves depends on gravity, so it is not surprising to find that the Froude number is associated with that part of hull drag which is due to the energy needed to create the waves in the boat's wake.

Referring back to equation (1), the statement inside the brackets can now be written 'some function of $R_n$ and $F_n$'. This is the mathematician's euphemism for, 'I have no idea in what mathematical way drag depends on $R_n$ and $F_n$ except that I know it does'.

Equation (1) presents a somewhat more positive aspect when we look at the last three terms, which are the product of fluid density $\varrho$, the square of the speed and the square of a length. It tells us, for instance, that if $F_n$ and $R_n$ are fixed and if the speed and size of the object are fixed, the drag force is directly proportional to the density of the fluid. Since water is about a thousand times more dense than air, it means that for the same conditions the drag force in water will be a thousand times that in air. 'Nothing new', you are saying, of course the drag in water is greater. But our drag equation is a quantitative statement: the drag is exactly proportional to the density, not its square or some other function of it.

Similarly, we see that it depends on the square of the velocity if all else is kept fixed. So if we double the speed the drag will go up four times, if we triple the speed nine times, and so on.

The factor $L^2$ in equation (1) is to be interpreted as the square of an appropriate linear dimension. This sounds a bit vague: it could be the waterline length, the draft or the beam; we could use any one. All it does is to change the things we don't know such as the constant or the functional dependence on $F_n$ and $R_n$. A sensible interpretation of $L^2$ is that it represents the wetted surface area of the hull, which of course is measured in square metres or square feet.

We can now write the drag equation (1) in a more succinct form:

$$D = Kf(R_n, F_n)\varrho V^2 S.$$

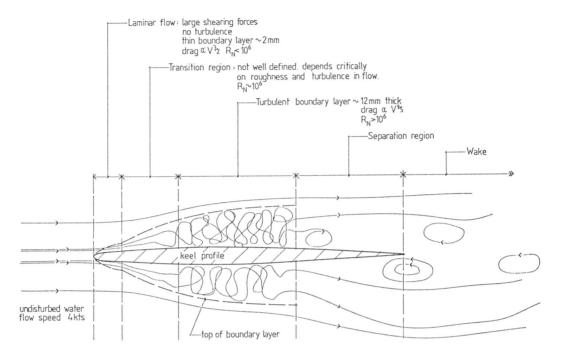

Laminar flow: large shearing forces
no turbulence
thin boundary layer ~2mm
drag $\alpha\ V^{3/2}\ R_N < 10^6$

Transition region: not well defined. depends critically
on roughness and turbulence in flow.
$R_N \sim 10^6$

Turbulent boundary layer ~12mm thick
drag $\alpha\ V^{9/5}$
$R_N > 10^6$

Separation region

Wake

keel profile

undisturbed water
flow speed 4kts

top of boundary layer

*Fig. 2.1* The general nature of the flow which is always found when fluid moves past an obstacle. The relative size of the various flow regimes depends upon the shape of the object and the undisturbed flow speed. Figures given here are for water flow at 4 knots past a fin keel. The dotted outline shows the relative thickness of the two boundary layers although the absolute scale is magnified somewhat for clarity. The separation region will be discussed in chapter 3 because of its greater importance for lift than for drag. The same general flow patterns also occur around a mainsail or genoa but because of the sail curvature there will no longer be this port-starboard symmetry.

D is the drag force, K is the unknown constant, $f(R_n, F_n)$ means some function of $R_n$ and $F_n$. S is the wetted surface area.

To determine the constant K and the functional dependence on Reynolds' and Froude's numbers we have to make measurements, either of a model in a towing tank or of the full size boat. (For scaling techniques and the problems associated with model measurements see chapter 5.)

The fundamental drag equation (1) tells us that drag has several more or less independent physical origins. Clearly it depends upon the wetted surface area, as every sailor knows. Coupled with this is a dependence on the viscosity of sea water through Reynolds' number. The presence of Froude's number tells us that wave-making also contributes to resistance. Because of these different physical origins it is possible, to a close approximation, to regard them as occurring independently of each other. This is a simplification because instead of one complex phenomenon we can now break drag down into separate simpler phenomena which can be studied individually.

## §2.2   Types of Drag

Total drag consists of:

    (i)  Surface friction drag
    (ii)  Normal pressure drag.

The latter consists of three distinct parts:

    boundary layer pressure drag
    vortex drag
    wave drag

The way in which these various forms of drag manifest themselves is best seen by looking carefully at the fluid flow close to the surface of a keel. This flow will be close to the ideal, unlike that around the mainsail for instance, which is disturbed by the mast. What we would like to study here is the flow close to the surface because it is in this region that the drag-producing effects mainly operate.

The nature of the flow at a particular point on the keel depends on the history of events leading up to that point. Air or water flowing uniformly in the absence of any obstruction will continue to flow without turbulence, but in the presence of some object like a sail or keel the nature of the flow will change as one moves from the leading edge to the trailing edge. The change does not occur immediately on encountering the obstruction but in three fairly discrete steps as we move back along the surface. This is best described with the aid of a diagram and fig. 2.1 is a sort of microscopic view of the cross-section of a keel and the water flow around it. It is a strange sort of microscope, however, since only the direction transverse to the foil is magnified. Furthermore it allows us to view something which is intrinsically invisible, the motion of small parcels of air or water within the overall flow. What an incredibly different sport sailing would be if we could see the movement of the air. Imagine, for instance, if air was coloured and the colour depended upon its speed relative to the sea so that every complicated nuance of air flow could be readily observed by the crew. In such a Technicolor world of fluid flow we would undoubtedly all be better sailors and there would be little need for a book like this. But the flow cannot be visualised, except in a very crude manner with wool telltales, so the best one can do is read what follows, study the diagrams which are based upon half a century of scientific investigation, and when you are out sailing try and imagine that Technicolor vision which will help your sailing enormously.

Fig. 2.1. shows the extent of a flow regime known as the boundary layer, a region near the surface where the flow velocity is changing rapidly with distance out from the surface. Within the boundary layer the flow speed varies from zero at the surface to that of the external flow. In this chapter we will be looking only at the flow within the boundary layer since it is here that the effects of viscosity are manifested and where most of the drag has its origins. The character of the 'external' flow will be the subject of chapter 3.

There are three kinds of boundary layer: a laminar one, a turbulent one and a transition region.

## §2.3 Laminar Boundary Layer

To understand the formation of the boundary layer and also the fundamental origin of nearly all drag forces one must understand clearly what is meant by viscosity. As with all quantities in physics, viscosity is defined in terms of how it is measured: take two freely rotating concentric cylinders and place between them the fluid of unknown viscosity (fig. 2.2). If the inner cylinder is rotated it is evident that the outer cylinder will have a tendency to turn in the same direction and this tendency will be greater for a more viscous fluid.

Unfortunately this eminently practical apparatus with cylindrical geometry leads to some mathematically technical difficulties which are not very interesting. So it is better to consider the impractical device of fig. 2.3 and define viscosity in terms of it. Those who are skilled in the art of mathematical jiggery-pokery can easily translate the

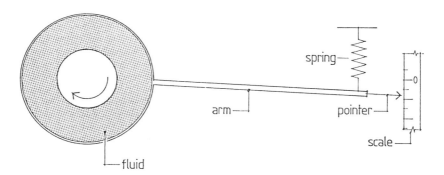

*Fig. 2.2* One way of measuring viscosity. The inner cylinder is rotated which gives the outer cylinder a tendency to rotate. This is measured by the torque, which is the force extending the spring multiplied by the distance from the point of attachment of the spring to the common centre of rotation of the cylinders.

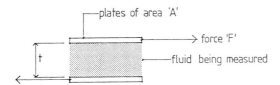

*Fig. 2.3* Viscosity is defined in terms of this 'thought experiment' in which two smooth flat plates of area A and separation t have a fluid filling the space between them. It is found experimentally that the drag force F depends on the relative velocity of the plates, their area, the thickness t of fluid as well as the viscosity. This contrasts with the very different behaviour of friction between solid surfaces.

results from the practical apparatus of fig. 2.2 into those of the 'ideal' apparatus of fig. 2.3, in which two smooth plates have a fluid between them. To move the plates at a constant speed requires a force F which is found by experiment to depend on the parameters of the apparatus: it increases in proportion to the area and to the velocity, and is inversely proportional to the thickness of fluid t. This may be written more succinctly as:

$$\text{Force F is proportional to } \frac{\text{area} \times \text{velocity}}{\text{thickness}}$$

So far we have not mentioned the fluid between the plates. Clearly F will be greater for a high viscosity than a low one so we simply multiply by another factor called viscosity and write:

$$\text{Force} = \frac{\text{viscosity} \times \text{area} \times \text{velocity}}{\text{fluid thickness}}$$

or in symbols,

$$F = \eta \times \frac{vA}{t}$$

where the Greek letter $\eta$ (eta) is called the coefficient of viscosity. This formula defines viscosity because it tells us how to measure it. If v, A, t and F are all measured, or rather the equivalent quantities in the practical apparatus of fig. 2.2, we can determine $\eta$. Thus $\eta$ is a measure of the resistance to sliding of one surface over another when they are separated by a thickness of fluid.

In order to see how this relates to the boundary layer and fluid drag in general there is one more important ingredient that must be added: intermolecular forces. If it comes as a surprise that

the forces between individual molecules play a major role in determining how fast your boat goes, then read on.

## §2.4   Intermolecular Forces

The ultimate origin of fluid viscosity lies in the forces between molecules. Such forces exist between all molecules of whatever kind, so that identical molecules in the fluid attract one another just as they are also attracted by different molecules of a solid immersed in the fluid. The force is always an attractive one and is brought about by a distortion of the electrically charged cloud of electrons surrounding the atoms of the fluid molecules when in the presence of another similar electron cloud. In short, viscosity has its origins in electrostatic forces known as van der Waals forces. They have a very short range: the force is appreciable for molecules that are close together but negligible when their separation is more than a few molecular diameters. For example, if the distance between two molecules is halved the force of attraction becomes $2^7$ or 128 times as great. Thus for all practical purposes a molecule feels an attraction only for those others immediately surrounding it. This is a very important fact, for as we shall see it explains an often misunderstood aspect of skin friction drag.

It is now easy to see what must happen very close to the surface of a hull or sail when the water or air is flowing past without turbulence. The layer of fluid molecules next to the solid interact strongly with its molecules and also with the adjacent fluid molecules, but not to any significant extent with any others. So this first layer moves hardly at all with respect to the solid (fig. 2.4). The second layer, being held by the first one, moves slowly with respect to it but therefore a bit faster with respect to the solid. As one continues out through millions of such layers a point is reached where the speed of flow is the same as that of the external flow at some distance from the surface.

It is therefore clear that the velocity of flow increases smoothly as one moves away from the surface. Thus the frictional drag of the laminar boundary layer is brought about by the forces that have to be applied to make these layers move with respect to one another against the intermolecular attraction.

Now with this picture of the origin of 'skin friction drag' we can make an important prediction: for a fixed speed of flow the drag depends only on the viscosity of the *fluid*, not on the material of the

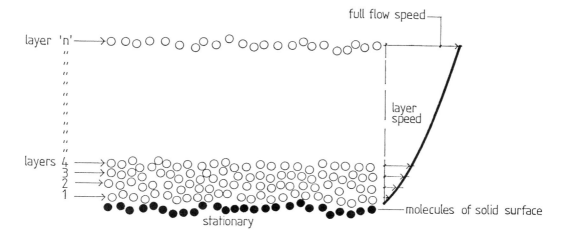

*Fig. 2.4* Skin friction drag has its origins in the short range forces between molecules. As the fluid flows past the stationary solid only those molecules in the layer of fluid adjacent to the solid interact with the molecules of the solid. This layer is held by intermolecular forces and can move past the solid only at an imperceptible speed. Layer 2 is not affected directly by the presence of the solid because of the short range of the molecular forces, but is held back by layer 1 which is moving very slowly. As we move outwards from layer to layer the speed with respect to the solid gradually increases until at layer n, where n is a few million, the flow speed is close to that of the bulk of the external flow.

surface. Although the molecular attraction between surface molecules and fluid molecules may be different from that between pairs of fluid molecules, the solid affects only layer 1 in fig. 2.4 and there are millions more layers involving only fluid pairs contributing to the total drag. Thus the material of the surface can have no measurable influence.

This may come as a surprise for we have all seen dinghy sailors polish their boats with silicone wax prior to a race and test the surface by running a dry finger over it to convince themselves of the low friction.

Such a procedure is false for two reasons. First, the nature of the surface material cannot affect the skin drag, as already explained. Second, the surface was tested by measuring the drag between two solids, the hull and the finger. Friction between solids behaves in a completely different way to that of a fluid: it is greatly affected by lubrication and will seem small for a waxed surface. Of course it is possible that the layer of wax has filled small cracks and scratches thereby making the surface smoother and reducing skin friction, but the fact that it feels slippery to the touch does not guarantee a low fluid drag.

The correctness of these statements has been tested at the National Physical Laboratory in England and because of its relevance to this section the report, which appeared in *The Naval Architect* for April 1974, is reproduced in full below.

## The Skin Friction Bogey

J. A. H. PAFFETT, RCNC, FRINA

In the ordinary merchant ship by far the greater part of the resistance to motion ahead is provided by 'skin friction' – the sum total of tangential forces acting on the bottom plating. These forces arise from the sliding motion of the water relative to the ship and the viscosity of the water.

It is every naval architect's dream to reduce the skin friction to zero or thereabouts. A great deal can be done simply by achieving mechanical smoothness, and the combination of flush-welded construction, shot-blasting and modern paint finishes has reduced ship skin friction to 80% or so of that common before the war. How much further can one get?

One approach is to insert a film of air between ship and water. This is far from easy, as many aspiring inventors have found. Air ejected under the bottom usually insists on rolling around as discrete bubbles embedded in the water, doing nothing to reduce friction. The only successful approach to air lubrication to date employs a bottom specially formed to retain one huge bubble, and an enormous flux of air to replace the continuous leakage round the edges; the result is the Hovercraft.

Another technique is suggested by the properties of certain long-chain molecules. These substances, by a mechanism not fully understood, reduce the skin friction shear stress for a given turbulent flow velocity when they are dissolved or suspended in water. They are effective in remarkably small concentrations, sometimes only a few parts per million, and they have been exploited in pipe-flow installations, for example, in increasing the water delivery of pumps used in fire-fighting. Why should we not, therefore, feed suitable long-chain molecules into the water near the bows and let them do their work while sliding aft? It has been tried of course. (Canham, Catchpole and Long – Trans. RINA 1971.) Some marginal drag reduction is undoubtedly possible, at the expense of fouling the ocean with large quantities of chemical whose cost would heavily outweigh the cost of fuel saved, even at current prices. Nevertheless, it is not inconceivable that some ingenious inventor may find a way of tacking long-chain molecules by their ends to the bottom plating in such a way that they do their curious job of friction reduction while not getting washed away downstream. (One wonders whether Nature did not get there first when she evolved hairy and scaly animals.)

Then there may be other mechanisms awaiting exploitation. Is some sort of super-smoothness achievable in formulating the paint itself? Here some hopes may have been raised recently by the literature of one paint company. Their product '. . . is a revolutionary type of bottom treatment which increases the speed of sailing boats and motor boats. . . . The dolphin has an ingenious way of reducing this friction. It simply uses water as a lubricant. In the outer surface of its skin it stores water, thereby creating a water-to-water boundary layer that reduces the friction. . . . Hydron High Speed works the same way thanks to this water-to-water-effect.'

This concept of the dolphin's functioning was new to NPL. Nevertheless, the claim made on behalf of the product appeared to be worthy of

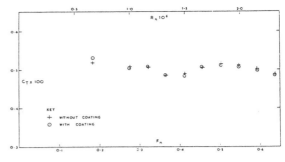

MEASURED RESISTANCE RESULTS FOR A COATED FLAT PLATE

examination and amenable to quantitative assessment. Tests were therefore carried out with a plank-type test model, 1.14 m long, 0.76 m draught and 25.4 mm thick, with tapered entry and run and having turbulence studs at the fore end. The surface was of highly polished polyurethane varnish. Runs were carried out in the No. 2 towing tank at NPL over a range of speeds. On completion the plank was raised, dried and coated with two coats of the product, which was allowed to dry overnight. The plank was then immersed for an hour, at the end of which the resistance experiments were repeated. The results are plotted in the accompanying graph, from which naval architects can draw their own conclusions. In fairness to the manufacturers of Hydron, it should be pointed out that these tests related only to the application of the product to an already smooth surface. The effect of Hydron on a bad surface, or upon toxic release from antifouling compositions, was not examined.

There still seems to be plenty of scope for those friction-reducing inventions.

*(Reprinted from* The Naval Architect, *Journal of RINA London, April 1974, by permission of the Royal Institution of Naval Architects)*

It should be clear from this discussion that the relative speed of flow very close to the hull or sail must be zero but increases smoothly as we go further away. Put another way, when a hull moves through the water it drags a layer of water with it, but not all parts of this layer move as fast as the hull. The variation in velocity is shown in fig. 2.5.

The force needed to slide these layers past one another is the origin of the drag. Remember this is only one part of the overall drag, that part produced by laminar flow. The region of laminar flow is

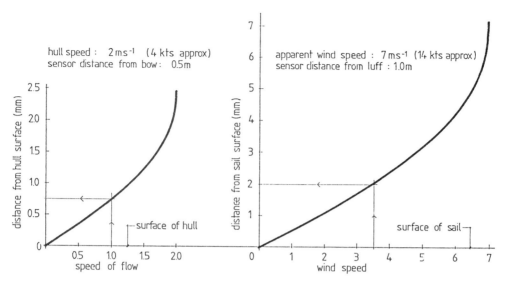

Laminar flow characteristics

*Fig. 2.5* These diagrams show how the relative velocity of flow varies with distance from a surface, where laminar flow conditions prevail. The water or air very close to the surface does not move with respect to the surface because of strong molecular attraction. Layers of fluid slide with respect to one another according to their viscosity, so the speed of flow increases with distance from the surface. The region over which this velocity change occurs is called the boundary layer. In (a) the boundary layer half a metre from the bow is shown for a hull speed of 4 knots in a smooth non-turbulent sea. Notice that the boundary layer is there about 2.5 mm thick as beyond that the flow speed is unchanging. Note also that at 0.75 mm from the hull the speed of flow is only about half that of the hull with respect to the main body of water. Thus the boat is dragging along with it a sort of envelope of fluid which affects the dynamics of its motion (see chapter 6). A similar diagram (b) gives some idea of the boundary layer flow in a wind speed of 14 knots at a point 1 metre from the luff in non-turbulent air. Here the air flow reaches half of its maximum at 2 mm from the surface.

typically quite small (fig. 2.1). What we would like to know is: how large is this region and how much does it contribute to the total drag?

## §2.5  Reynolds' Number

The man who first investigated the characteristics of laminar flow was Osborne Reynolds in 1883, using an apparatus in which the flow of a stream of coloured fluid was observed. By varying the speed of flow and the viscosity of the fluid he found experimentally a result of far-reaching importance. What he did might nowadays be more easily visualised in the following way. Light a cigarette but don't smoke it; place it on an ashtray in a completely still room with no air currents. The smoke from the cigarette will rise in a thin straight column for perhaps 10 cm and then quite suddenly and at a well defined position it will appear to spread out. The nature of the flow changes at this point from an orderly state to a turbulent one. It's as

though the flow were happy to go along in an orderly fashion for a while but then got bored and became disorderly, like a column of school children who when they pass the teacher remain quiet and in formation but when far enough away find it more agreeable to talk and move about. More will be said later about the onset of turbulence, which is such an important part of the fluid flows which power all boats.

For the moment we are concerned only with the fact discovered by Reynolds that the onset of turbulence in an otherwise non-turbulent flow depends on three quantities: the speed of flow, the viscosity (actually the kinematic viscosity) of the fluid, and the 'length of the flow' which refers to the distance along the flow direction from the leading edge of the sail or keel. It is assumed that the air is moving without turbulence, a pretty crude approximation of course, and it is simply the introduction of a sail into this stable state that produces turbulence in the flow. This doesn't happen immediately but only after the air has been in contact with the sail over a distance which we will call L.

Reynolds' discovery was that turbulence always occurs when the product of L and the speed divided by the kinematic viscosity has a numerical value equal to about one million, or:

$$(LV)/v = 10^6.$$

This is just the Reynolds' number introduced in §2.1 except that here I have used the Greek letter $v$ (nu) instead of $\mu/\varrho$. Rearranging this formula allows us to determine the length L of the laminar flow region:

$$L = (10^6 v)/V.$$

V is the velocity and $v$ is the kinematic viscosity. For air and water $v$ has the values: $v$ air = $1.5 \times 10^{-5}$ m²/sec and $v$ water = $1.0 \times 10^{-6}$ m²/sec. Note that these numbers do not differ a great deal which might seem strange until we realise that $v$ is the actual viscosity $\mu$, divided by the density, and the density of water is about a thousand times that of air. (m²/sec can also be written $m^2s^{-1}$, as in some of the diagrams.)

## §2.6 Laminar Flow Region

We have two results for the maximum length of the laminar flow region, depending on whether we are looking at sails or hull.

| Water | | Air |
|---|---|---|
| $L\text{(metres)} = \dfrac{1}{V\text{(m per sec)}}$ | | $L\text{(metres)} = \dfrac{15}{V\text{(m per sec)}}$ |
| OR | | |
| $L\text{(metres)} = \dfrac{2}{V\text{(knots)}}$ | | $L\text{(metres)} = \dfrac{30}{V\text{(knots)}}$ |

For a very smooth keel moving through non-turbulent water at 4 knots downwind (i.e. without leeway), we might expect laminar flow over the region from the leading edge to 0.5 m downstream. In practice it will always be considerably less than this since this calculation is for a hydrodynamically smooth flat plate moving in a turbulence-free fluid. Air flow conditions around a headsail are even less ideal so that in practice this laminar region could be shorter by a factor of 10.

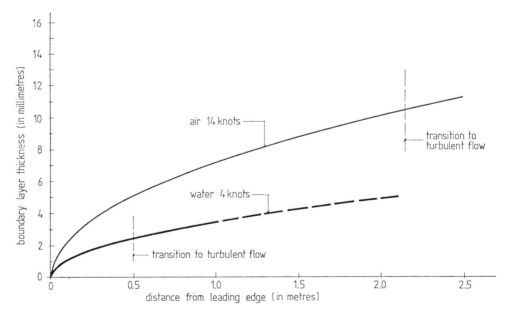

Fig. 2.6 Increase of thickness of boundary layers with distance from the leading edge of the flow. To give a feeling for the typical thickness of air and water laminar boundary layers, these are plotted on the same scale and have been calculated for our standard situation of 14 knots apparent wind with a hull speed of 4 knots. The broken lines showing the transition regions give an idea of the maximum length over which laminar flow can be expected under these conditions.

Because of the physically simple circumstances of laminar flow scientists are able to calculate quite precisely the drag due to this part of the flow. The first step is to calculate the velocity distribution through the boundary layer, already shown in fig. 2.5, and then the rate of growth of the thickness of the laminar boundary layer. Fig. 2.6 shows this for the same speeds as used in fig. 2.5.

Since the frictional drag between two layers of water is (see §2.3):

$$\text{Drag} = \frac{\text{viscosity} \times \text{area} \times \text{velocity difference}}{\text{thickness of layer}}$$

it is clear that the friction drag over a small area located 0.5 m from the bow can be calculated from fig. 2.5a by adding up the separate drags between all the layers in the boundary layer. Since the thickness of the boundary layer varies with distance from the leading edge so that the relative velocity between the layers is different, one must carry out this calculation for a large number of small regions from the leading edge to as far back as laminar flow conditions prevail. All this sounds tedious but fortunately there exist mathematical methods for simplifying this particular problem.

Before presenting the results it is worthwhile thinking about what we might expect. As mentioned above, the drag between layers of water is proportional to the velocity difference between those layers. We can determine this from fig. 2.5 if for instance we take layers 0.10 mm thick and plot the velocity differences (fig. 2.7). Since the drag is proportional to the velocity difference, it follows that the part of the boundary contributing most to the drag is that near the hull. By adding up all the speed differences in fig. 2.7 between each of the 0.1 mm thick layers from the surface out to 2.45 mm, you get the total speed difference of 2 m/sec between the hull and the sea.

Thus we expect the local drag, that due to a small

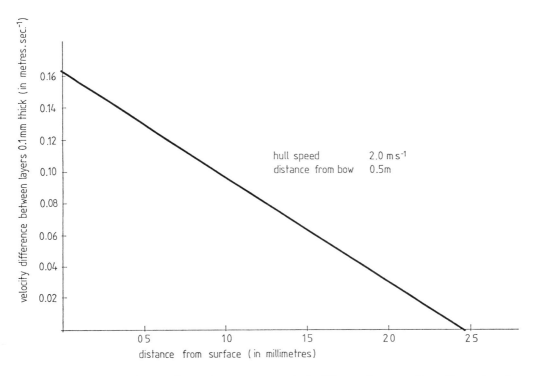

hull speed          2.0 m s⁻¹
distance from bow   0.5m

*Fig. 2.7* The velocity difference between layers of water 0.10 mm thick for the conditions of fig. 2.5a. Since the drag depends on the relative velocity between layers it is clear that most comes from near the surface where this velocity difference is greatest. The boundary layer ends where the velocity difference between adjacent layers drops to zero. The diagram shows that at 0.1 mm from the hull surface the speed of flow with respect to the hull is 0.163 ms⁻¹ or about 0.32 knots, whereas at 1 mm the speed difference between adjacent layers 0.1 mm apart is about 0.096 ms⁻¹.

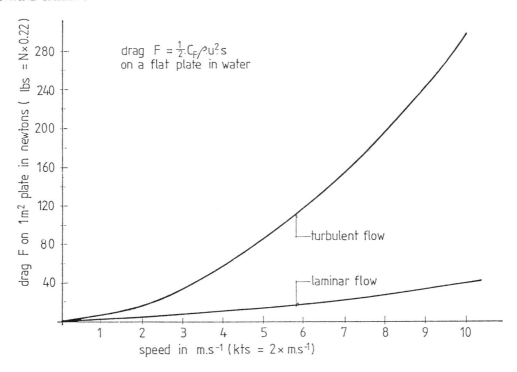

drag $F = \frac{1}{2} \cdot C_F \cdot \rho \cdot u^2 \cdot s$
on a flat plate in water

turbulent flow

laminar flow

speed in m.s$^{-1}$ (kts = 2 × m.s$^{-1}$)

*Fig. 2.8* The surface friction drag on a flat plate. The lower curve gives the drag if laminar flow is maintained over its entire surface (virtually impossible in actual practical sailing) and the upper curve is the drag if the flow is turbulent over the whole surface. In conditions of smooth air and water some laminar flow may be present especially around the keel. The big difference in drag clearly makes it desirable to maximise the region of laminar flow if at all possible.

area of hull surface where the boundary layer thickness is $\delta$, to depend on the speed change between layers near the surface and depend inversely on the boundary layer thickness. This latter is because a thin boundary layer means a large speed difference between layers since these differences must still total 2 m/sec and there are fewer 0.1 mm layers in a thinner boundary layer. Another way of putting this is to say that the shear forces in the fluid are stronger where the boundary layer is thinner and hence the drag per unit area greater. Thus as we proceed back from the leading edge of the flow the drag contribution from an area of say 1 cm$^2$ becomes less as the boundary layer increases in thickness, provided that the flow remains laminar. To get the total flow these contributions are mathematically added and we find that the drag force in purely laminar flow is:

Laminar drag force is proportional to

$$(U^{3/2} \times \text{wetted area})$$

Here U is the hull speed or apparent wind speed for a sail. Of course it also depends on the viscosity, but this is not a variable over which we have much control.

Fig. 2.8 shows the drag on a flat plate a metre square at various speeds, in conditions where laminar flow exists over the whole plate.

To summarise laminar flow: in conditions where there is little or no turbulence in the air or water and where the surfaces are hydrodynamically smooth (see §2.8) laminar flow may exist up to a maximum distance of around 0.5 m back from the leading edge in the case of a fin keel and 2 m in the case of a headsail equipped with a luff foil to assure smooth flow. The drag increases in proportion to the 3/2 power of the speed.

## §2.7 Transition to Turbulence

Fig. 2.1 shows the location of the transition region between the laminar boundary layer and the

16

turbulent boundary layer. As has already been said, fluids which are initially non-turbulent soon get tired of this state of orderliness and turbulence sets in. It is part of the grand scheme of things in Nature, where it is known that the total amount of disorder in the universe is constantly increasing. One can force a degree of orderliness into one small region, like one's workbench, but without a conscious effort to the contrary the natural tendency is always toward disorder.

The quote from Prandtl and Tietjens at the beginning of this chapter refers to the turbulent part of the flow which usually predominates, because as we have seen the laminar part is fully understood and predictable. Although this quotation dates from 1934 it was unfortunately still true in 1984. In fact it is true to say that more is known about the structure of the atomic nucleus than about how turbulence develops in a cubic centimetre of the air that we breathe. It is even difficult to say precisely what we mean by turbulence, although a very recent development in mathematics known as 'strange attractors' could be the key to better understanding.

Nevertheless a naive physical picture which is often used to describe how turbulence starts in a laminar boundary layer is worth discussing. Imagine you are located somewhere in a boundary layer and moving along with the fluid. If the layer of water above you is moving to the right, the one below will seem to be moving to your left, because of the velocity gradient (fig. 2.9). Thus from this point of view we have adjacent layers moving in opposite directions, just the condition associated with rotational motion. However, a motion which appears rotational for a moving observer has an up-and-down oscillatory character for a fixed observer. This is the beginning of turbulent motion, as shown in (d).

This turbulent motion has a characteristic size or amplitude which depends upon the viscosity and the velocity gradient (see Part II, §7.3). A velocity gradient is the change in velocity with distance out from the surface. Since such a change only occurs within the boundary layer it is evident that turbulence also has its origin within this layer. From the conditions of fig. 2.5a, for a hull moving at 4 knots and at a position of 0.5 m from the bow, we find that close to the surface the characteristic size of the turbulence is about 0.025 mm. As the discussion in Part II implies, the energy in these turbulences is used up against the viscous forces which ultimately have their origin in the forces between molecules. A standard view of turbulence is that there is a whole

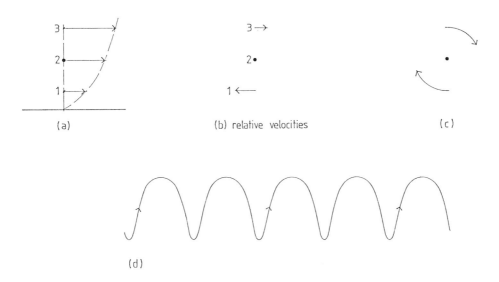

(a)

(b) relative velocities

(c)

(d)

*Fig. 2.9* Graph (a) depicts three layers of fluid in a boundary layer moving with a velocity gradient. If an observer moves along with the fluid particle labelled 2 so that he regards 2 as fixed, then from his point of view 3 will be moving to the right and 1 to the left (b). Such a system of contrary velocities on each side of a diameter is just the situation found in rotational flow (c). It is not difficult to imagine therefore that relative to 2 the motion of the adjacent fluid could become circulatory. The fixed external observer will not see circles, however, but an up-and-down motion (d) which results from adding the velocity 2 in (a) to the circular motion in (c).

hierarchy of sizes of motion in which energy is injected into the system with a large scale of motion and cascades successively into smaller and smaller structures until it reaches the characteristic molecular size at which it is dissipated by viscosity.

It is interesting to note that this idea was first envisaged in 1929 by the English physicist and meteorologist Lewis F. Richardson (1881–1953). He developed the use of the method of finite differences for solving the differential equations for weather prediction, but at that time the absence of computers meant that his equations could not be solved quickly enough to be of any practical value. He also pioneered the mathematical investigation of the causes of war and had a probably better known nephew, the actor Sir Ralph Richardson.

In line with his unusual life was his method of proposing his theory of turbulence. It was in the form of a rhyme: 'Big whirls have little whirls which feed on their velocity. / Little whirls have lesser whirls, and so on to viscosity – in the molecular sense.' This is probably a parody of some lines from a poem by Jonathan Swift, from his book *On Poetry a Rapsody*.

Even the use of the word 'turbulence' to describe this characteristic of fluid flow has not been prevalent very long. When Reynolds was doing his classic experiments he used the adjective 'sinuous'. The term 'turbulent flow' was introduced by Lord Kelvin in 1887.

So far in this discussion nothing has been said about what turbulence *is*. Most of us have a picture in our mind's eye, but where does it come from? Turbulence in water can be seen to some extent but not in detail, and air turbulence is quite invisible.

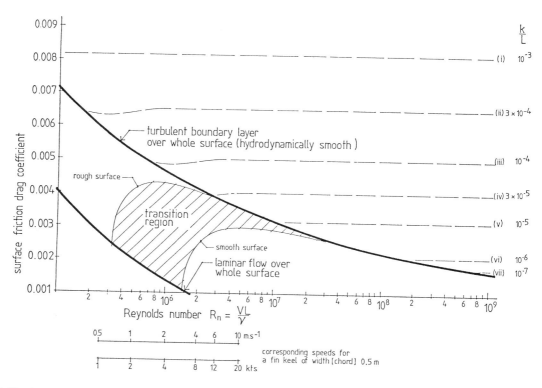

*Fig. 2.10* This diagram summarises the results of many measurements on flat plates under idealised conditions. The vertical axis is proportional to the surface friction drag, the actual drag force being $1/2\varrho U^2 C_f S$ ($\varrho$ = the density of fluid, $U$ = speed, $S$ = surface area). The horizontal axis is Reynolds' number which, for a given fluid, is proportional to the product of speed and length of the flow. The large increase in drag after transition from laminar to turbulent flow is clearly shown, and also the effect of surface roughness on the friction drag. The scales at the bottom allow the corresponding Reynolds' number to be determined for a fin 0.5 m wide moving through water at the speed indicated.

Although turbulence is not understood in a scientific sense a great deal is known about its characteristics, certainly enough for the yachtsman to make decisions as to whether turbulence in a particular situation might be reduced, thereby reducing drag. One is reminded of the statement attributed to the physicist Oliver Heaviside: 'Shall I refuse my dinner because I do not fully understand the process of digestion?' Therefore let me describe the dinner and I am sure you will not find it too indigestible.

Measurements have been made on flat plates in wind tunnels or water channels where the turbulence in the incident flow can be reduced to very low levels. If you feel that this is a far cry from real life and the flow around your genoa or fin keel, you are right but it is only by studying simple 'ideal' systems that we can find out what causes produce what effects. The standard diagram which appears in all books in one form or another is shown in fig. 2.10. It is very important for surface friction drag and will be referred to again later. For the moment we are interested in the transition from laminar flow to turbulent flow, shown schematically by a shaded curve. As the resistance that results from laminar flow is much less than that from turbulent flow, we would like to keep the flow laminar over as much of the surface as possible. So how does transition start?

At some point and at some time in an otherwise laminar flow a tiny spot of turbulence will appear, as at S in fig. 2.11. The spot is initially small and of irregular shape, and grows while maintaining its shape as it is washed along with the boundary layer fluid. Since the spots grow as they move, the downstream area is covered by more turbulence than the upstream parts.

These turbulent sources can be increased in density and frequency by increasing the flow speed, by increasing the turbulence in the flow before it reaches the plate, or a continuous source can be produced by placing a small protuberance on the surface. The onset of turbulence is thus random in nature and turbulent spots can occur anywhere within the laminar boundary layer.

Fig. 2.12 illustrates a phenomenon predicted in 1933 by H. Schlichting in Göttingen but not confirmed experimentally until 1941 in the United States. If the incident flow is almost completely free of turbulence, the region in the laminar flow just ahead of the turbulent region is found to contain oscillations in velocity, shown in (a). Initially the traces are regular, but as the distance from the

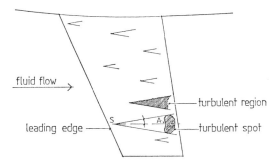

Fig. 2.11 Anywhere in a laminar flow spots of turbulence appear to arise randomly and spontaneously. The turbulent spot grows while maintaining its original shape as it is washed along.

leading edge of the flat plate is increased so increasing the distance Reynolds' number, they become more random as turbulence sets in. These measurements confirmed Schlichting's theory that under certain conditions velocity oscillations in the laminar flow would be amplified as they moved downstream. The initial velocity oscillations are very minute and it is the amplified oscillations that are seen in (a). Thus laminar flow even under ideal conditions contains velocity oscillations of many frequencies. Some of these are preferentially amplified in the fluid and eventually give rise to turbulence, but most of these small disturbances in the laminar flow are damped and disappear.

Fig. 2.12b shows the effect of pressure change on the development of the oscillations. Oscillations which are present in the top three traces are damped out as the pressure falls, only to return again as the pressure increases.

It was further found that by feeding vibrations of the correct frequency into the boundary layer oscillations could be induced and hence turbulence started sooner than otherwise.

In summary, the characteristics of the transition region are as follows.

(a) Sources of turbulence appear randomly in time and position as spots in the laminar flow which grow as they move downstream.
(b) The frequency of appearance of these source spots depends on the turbulence in the incident flow.
(c) Transition can be triggered by a particle of dust which acts as a continuous source spot.

19

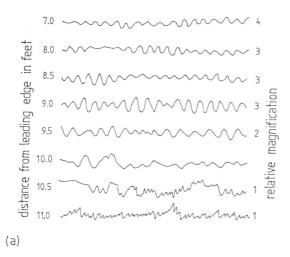

(a)

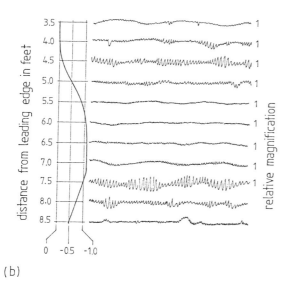

(b)

*Fig. 2.12* Even under the most ideal conditions the laminar flow contains minute velocity oscillations of various frequencies. Depending on the Reynolds' number, some of these frequencies are amplified in the fluid and are shown in (a) where the vertical deflection of the trace is a measure of velocity and the horizontal direction is time. As the distance from the leading edge increases the amplitude of the oscillations increases and their regularity disappears until the characteristic random velocity fluctuations associated with turbulence are seen on the lower trace. (b) shows the effect of a pressure gradient over the surface of the flat plate. Starting with the third trace the pressure is decreased as shown by the scale at the left. This damps out the oscillations, but they reappear again when the pressure is increased.

(d) A precursor to turbulence is a regular velocity fluctuation. These fluctuations can start spontaneously or be induced by sound vibrations of the appropriate frequency. This frequency is $2U^2/10^5\nu$, where U is the flow speed in metres per second and $\nu$ is the kinematic viscosity. For example, for a hull moving at 4 knots (2 m/sec) this frequency is 80 Hertz, about the same as the third E up on the piano keyboard. (One Hertz (Hz) is one cycle of vibration per second.) For a sail with apparent wind speed of 14 knots (7 m/sec) the frequency is 33 Hz. Such vibrations can easily be set up in the rigging and conveyed to sail or hull. For instance wind in the rigging or spars produces familiar tones (Aeolian tones) whose frequency is given by 0.19 U/h, where U is the wind speed and h the diameter of the rigging wire. Thus rigging of 10 mm diameter could produce a tone of 80 Hz in a wind of about 8.5 knots. (The extent to which these vibrations would be conveyed to the hull also depends on the length, tension and density of the rigging wire.)

These regular velocity fluctuations are damped by falling pressure but amplified by rising pressure. Since the pressure first falls and then rises in flow over a curved surface, we expect turbulence to start more readily just aft of the point of maximum thickness in the case of a fin keel or rudder.

## §2.8 Turbulent Boundary Layer

If one tries to measure the variation of velocity with distance from the surface in a region where the flow is turbulent, one does not get, from a single set of measurements, a smooth curve analogous to that of fig. 2.5 for laminar flow. Instead one gets results like those of fig. 2.13 which shows velocity plotted against distance from the surface with 17 separate measurements made at the same point in the flow. It is seen that individual measurements can be quite different, but if one combines a sufficient number of them a well defined statistical average emerges. This statistical average is further shown quantitatively for our specific situation of 2 m/sec air flow and 7 m/sec water flow in fig. 2.14. Two important differences between laminar and turbulent boundary layer flow are evident from (a) where the velocities for both types of flow under the same conditions are plotted on the same scale. First, the turbulent boundary layer is nearly five times as thick as the laminar one and

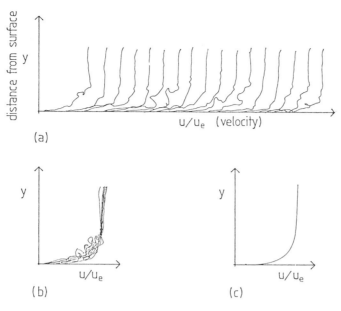

*Fig. 2.13* The velocity distribution in the turbulent boundary layer. The horizontal axis is velocity and the vertical axis is distance from the surface: (a) shows a set of measurements made at different times but in the same place in the flow, (b) shows the same set superimposed, and (c) is the well defined statistical average curve obtained from a very large number of measurements.

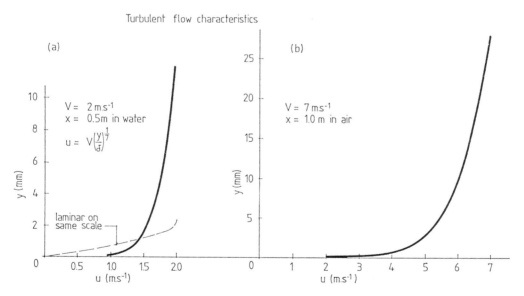

*Fig. 2.14* Variation of velocity (plotted horizontally) with distance from the surface for a turbulent boundary layer. Conditions are the same as in fig. 2.5 except that here the boundary layer is considered to be fully turbulent. In (a) the laminar velocities are plotted on the same scale for comparison. The important differences are that the turbulent boundary layer is nearly five times as thick, and that the increase in velocity is much faster away from the surface in the turbulent case. This means that the shearing forces within the liquid are greater with consequently greater drag. It must be remembered that both these plots are statistical averages of many measurements; individual measurements will give widely varying results like those of fig. 2.13.

second, the rate of increase of velocity as we move away from the surface is much greater in the turbulent case. This means that near the surface the shear forces within the fluid are greater and hence a greater drag is expected. Also, because the turbulent boundary layer is so much thicker a greater mass of fluid is affected by these shearing forces. Furthermore, in turbulent flow the fluid has a component of motion perpendicular to the surface which is not present in the laminar case. The energy for this extra motion must come from the moving boat which results in it slowing down: that is, the drag is greater.

Fig. 2.10 tells us what the drag will be and from this it is clear that laminar flow over as much of the hull or sail surface as possible should be maintained in order to minimise the surface friction drag. Unfortunately, in sailing laminar flow is the exception rather than the rule. Although one might imagine it to be the contrary, laminar flow is a rather unstable condition compared with turbulent flow. Any turbulence in the incident flow will quickly destroy the laminar condition so that only when we are sailing in very smooth water with a steady but not too strong wind is it likely that there will be regions of laminar flow on the sails and hull. When crashing along in gusty conditions only the turbulent curve in fig. 2.10 is really relevant. Thus in practice most of the surface friction drag has its origins in the region of turbulent boundary layer flow.

It is therefore important to understand fully the meaning of fig. 2.10 which gives the total surface friction drag force on objects of length L moving through a fluid at various speeds and for various surface roughnesses. The skin friction coefficient $C_F$ is related to the actual drag force D by the expression

$$D = \tfrac{1}{2}\varrho V^2 S C_F$$

where $\varrho$ = density of fluid, V = velocity and S = area. For pure water $\varrho = 1000$ kg/m$^3$ and for air it is 1.2 kg/m$^3$. The horizontal scale is Reynolds' number, $R_n = VL/v$. L is the length of the plate and $v$ is the kinematic viscosity. For water $v = 10^{-6}$ m$^2$/sec, and for air $v = 1.5 \times 10^{-5}$ m$^2$/sec.

To show how the information in fig. 2.10 may be used, let us calculate the surface friction drag of a centreboard of area 2 m$^2$, length in the direction of travel (chord length) 1.5 m, and moving at a speed of 4 m/sec (8 knots). First we must calculate the Reynolds' number, which is

$$\frac{4 \times 1.5}{10^{-6}} = 6 \times 10^6.$$

Note that this is beyond the transition point from laminar to turbulent flow on the graph. This means that transition to turbulent flow occurs somewhere on the centreboard and we should modify the calculation to take this into account. It turns out that for a speed of 4 m/sec the correct result is obtained by taking only that area of the centreboard which lies aft of a point about 5 cm in front of the transition point and assuming turbulent flow over this area only. In view of the variability of the transition point, and our remarks about the instability of laminar flow, it is not unrealistic to assume that the flow is turbulent over the whole surface of the centreboard for the purposes of calculation. Move up the graph to the line marked 'hydrodynamically smooth' and we see that the drag coefficient is about 0.0033. The actual drag is then $1/2 \times 1000 \times 4^2 \times 2 \times 0.0033 = 52.8$ newtons. (The newton N is the unit of force in the International System of units. The weight of 1 kg is approximately the same as a force of 10 N. If you feel ill at ease with SI units, 1 newton is about a quarter of a pound, actually 0.2248 lb.) The quantity k/L in fig. 2.10 is the ratio of the size of the roughness to the length of the plate. If our centreboard were sprinkled with sand of grain diameter 0.10 mm, k/L would be $10^{-4}$ m/1.5 m $= 7 \times 10^{-5}$. So for $R_n = 6 \times 10^6$ the skin friction coefficient would go from 0.0033 for a smooth surface to about 0.0042, a 27% increase in drag. The critical roughness is the point on fig. 2.10 where the dotted lines move away from the 'hydrodynamically smooth' line. For instance if the ratio of the length of a hull or centreboard to its roughness is k/L = $10^{-5}$ corresponding to dotted curve (v), we move along this curve till we reach the hydrodynamically smooth line and find a corresponding Reynolds' number of $10^7$. This means that at all speeds up to that corresponding to this Reynolds' number our surface is hydrodynamically smooth. At greater speeds the resistance is greater than that of a smooth surface.

A study of this graph reveals that the relationship between the critical roughness and Reynolds' number is $R_n = 100 L/k$. Since $R_n = (VL)/v$ this is the same as saying that the critical roughness is given by k = $100v/V$.

Graphs of this quantity are shown in fig. 2.15 for air and water, where it can be seen that for a hull at 4 knots there must be no bumps greater than 0.05 mm (0.002 in.). The sail, on the other hand, is smooth in a 10 knot breeze if there are no bumps greater than 0.3 mm. If these bumps are doubled in size the surface friction drag is increased by about 10% in both the water and air cases.

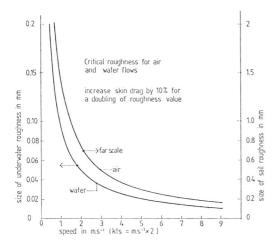

Fig. 2.15 Surface friction drag or 'skin drag' is a minimum when the surface is hydrodynamically smooth. So how smooth is smooth to fluid flow? This question is answered by the above graph. If you are ghosting along at only 1 m/s (2 knots), hull roughnesses up to 0.1 mm will be hydrodynamically smooth, but if your speed increases a smoother hull is required to keep drag at a minimum. This may be contrary to your intuition, but then many aspects of fluid flow are. The diagram also shows that most sails are not smooth in wind speeds above 10 knots since they usually have seam protuberances greater than 0.3 mm.

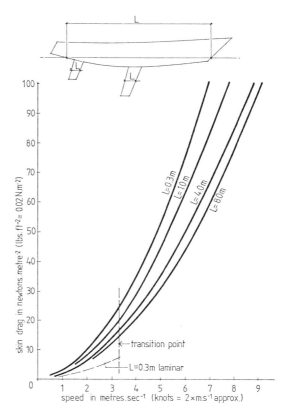

Fig. 2.16 This graph is intended to give an idea of the actual magnitude of the surface friction drag in some typical situations. The horizontal axis is speed through the water, and the vertical axis is the actual drag in newtons per square metre which may be converted to pounds per square foot by multiplying by 0.02. For instance the surface friction drag of the canoe body alone at a speed of 6 knots will be 12 newtons per square metre assuming a waterline length of 8 m. The total surface friction drag will therefore be 12 times the wetted area in square metres.

Fig. 2.16 shows the actual surface friction drag on an underwater body of length L at various speeds; each curve corresponds to a different value of L. The dotted curve is for L = 0.3 m only and implies that laminar flow might be possible over the entire surface of a small centreboard or rudder in smooth water conditions at speeds up to about 6 knots. However in practice it is unlikely that suitable conditions will prevail very often. The curves are for hydro-dynamically smooth surfaces only.

As a guide to how rough various surfaces are, here are some approximate values:

| Surface | Roughness k |
|---|---|
| smooth marine paint | 0.05 mm |
| galvanised steel | 0.15 mm |
| sandpapered bare wood | 0.5 mm |
| sail seams and stitching | 0.3–1.5 mm |

As we saw for laminar flow, the drag force in turbulent flow is also determined entirely by the shape of the velocity gradient curve (fig. 2.14) and the boundary layer thickness. When the calculation is

carried through in the same way as in the laminar case we find that:

$$\begin{bmatrix} \text{turbulent} \\ \text{boundary} \\ \text{layer drag} \\ \text{force} \end{bmatrix} \text{ is proportional to } \begin{bmatrix} U^{9/5} \times \\ \text{wetted area} \end{bmatrix}$$

Thus the surface friction drag increases almost as the square of the speed. If the speed is doubled the drag increases by nearly $2^2$ or 4 times, which is shown in fig. 2.16.

## §2.9   Normal Pressure Drag

At the beginning of this chapter it was pointed out that the phenomenon of drag could be divided into two broad categories. The first, surface friction drag, has now been discussed; the second, normal pressure drag, will occupy the remainder of the chapter. (The drag associated with lift will be treated in chapter 4.)

If one were able to determine the total drag force on a small area of hull, say 1 cm$^2$, it might have a magnitude and direction represented by the arrow in fig. 2.17a. Remember there are always two numbers associated with a force, one for its strength and the other for its direction. Any force may be thought of as being produced by two components at right angles to each other. If two people push a dinghy in directions at right angles to each other it will move off obliquely to both. Thus the components of a force always produce the same overall effect as the original single force. The component of the total drag force which is parallel to the hull is of course the surface friction drag already discussed. The component perpendicular to the hull or sail is the normal pressure drag. ('Normal' here means perpendicular.)

Any object immersed in water or air has pressure forces on it. These come from the fluid and are always perpendicular to the surface. A diver who goes 10 m below the surface experiences twice the pressure that he does in the air above, yet he doesn't feel anything.

This is because the pressure forces at every point on his body are nearly equal in magnitude but always perpendicular to the surface of his body: for every such force on one part of his body there is an equal and opposite one on the other side so that they all cancel out resulting in no nett force.

The situation is also the same for an object moving in a non-viscous fluid. This is an idealisation which is of course not true, as all fluids under normal conditions have viscosity, but we discuss this case first because it helps in understanding normal pressure drag. For a simple shape like a cylinder, mathematicians find it quite easy to calculate the fluid flow if there is no viscosity (fig. 2.18a). From such a flow pattern one can deduce the pressure distribution: since the flow pattern is symmetrical, so must be the pressure distribution. Diametrically opposite pressures are the same so the nett force on the cylinder is zero. Thus in this case of no viscosity there is certainly no surface friction drag, but also no normal pressure drag either. This state of affairs is known as d'Alembert's Paradox. Like all paradoxes, it is only a paradox if one doesn't know enough. Its resolution is simple: no real fluids are completely non-viscous. What is interesting, however, is the fact that when viscosity disappears not only does the surface friction drag, which depends directly on viscosity, disappear, but so also does the normal pressure drag. Those electric forces between

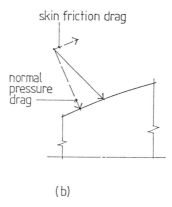

(a)                                        (b)

*Fig. 2.17* If one could isolate a small section of hull and measure the drag on it, the result might look like (a), where the length and direction of the arrow represent the magnitude and direction respectively of the drag force at that point. The total drag force is of course the vector sum of all such contributions over the whole hull surface. It is convenient to break down these contributions to the force into two components, one parallel to the hull surface and the other perpendicular or normal to it. These are shown dotted in (b). The arrow parallel to the surface is just the surface friction drag and the one perpendicular is called the normal pressure drag. This analysis is convenient because if we know the force on each small area composing a surface, the total force on that surface may be determined by simply adding arithmetically the separate components without the complication of 'adding' two forces which act in different directions.

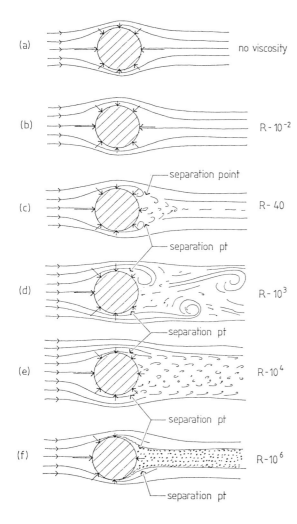

*Fig. 2.18* Flow distributions around a cylinder, such as a mast or a shroud, under different conditions. The arrows represent the direction and magnitude of the pressure forces on the cylinder resulting from the shape of the flow. The difference in arrow lengths and hence the pressure differences are greatly exaggerated in these diagrams. In (a) the drag is zero since the pressure forces exactly cancel on opposite sides of the cylinder. The fact that drag is always present in true life simply means that all fluids have viscosity, even air. Note that such pressure forces always act toward and perpendicular to the surface, as on a diver underwater. Diagrams (b) and below are realistic flows for increasing values of the Reynolds' number. They may be thought of more simply as the flows for increasing velocity; or alternatively as flows in which the relative importance of viscosity is decreasing from top to bottom. Note the asymmetry of the arrows in these cases and hence a nett drag force tending to push the cylinder to the right.

molecules do indeed have a far-reaching effect on the world we sail in.

Normal pressure drag is thus a direct result of the shape of the flow. We will see in chapter 3 how the connection between the pressure at a point and the nature of the fluid flow is made.

In fig. 2.18 the pressure differences indicated by the arrows are greatly exaggerated. To give a feel for the actual magnitudes of the pressure changes over the surface of a cylinder, these are shown quantitatively in fig. 2.19. On the vertical axis 100,000 $Nm^{-2}$ corresponds to atmospheric pressure. Looking first at the non-viscous case, it can be seen that the pressure at the leading edge of the cylinder is above atmospheric, at the side ($\theta = 90°$) it is less than atmospheric, and at the trailing edge it is again above atmospheric. Remember that these pressures always act in a direction toward the centre of the cylinder. What is interesting is that the actual pressure differences here calculated for water speeds of 2 and 4 knots are only a few per cent. Mainly because of the higher density of water, the pressure differences shown here are actually about 70 times greater than they would be on a cylinder with the same Reynolds' number in an airstream of 14 knots. Notice also that there is nothing special about a pressure which is less than atmospheric. Unfortunately there is a tendency to refer to such pressures as 'negative pressures', seeming to imply that the force acts away from the cylinder which it never does. The term is often used in engineering applications, but is really an abbreviation for negative pressure *coefficient* and should never be confused with the actual pressure force which is a simpler and more directly applicable concept. It is the actual pressure which is plotted in fig. 2.19.

Looking back at fig. 2.18 we see that the pressure near the leading edge of the cylinder has a different physical origin from that near the trailing edge. The flow over the front surface of the cylinder is characterised by the presence of a boundary layer. As already mentioned, the pressures in this region do not greatly differ for real and non-viscous fluids. For Reynolds' numbers greater than about 20, vortices are formed on the trailing side so that the pressures there are largely an outcome of this vortex formation. (If the cylinder were in the form of a moving rod penetrating the surface of the water, waves would be produced and further modify the pressure distribution in a way not shown in fig. 2.18.)

It is thus conventional to split the normal pressure drag into three parts: (i) the boundary layer normal

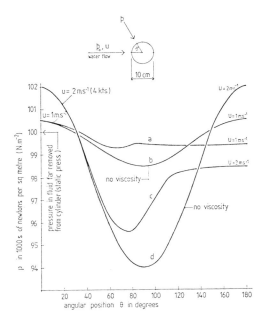

*Fig. 2.19* Actual pressure distribution around a 10 cm diameter cylinder moving in water just below the surface. The pressure, p, has a direction which is always radial toward the centre as shown in the inset. The point on the surface of the cylinder where the radial pressure acts is specified by the angle $\theta$ (theta), plotted horizontally. The graph indicates how the magnitude of p varies with position around the cylinder when the speed is $2\,\text{ms}^{-1}$ (4 knots) or $1\,\text{ms}^{-1}$ (2 knots). Curves (b) and (d) are calculated for conditions of no viscosity or so-called 'potential flow'. As can be seen the pressures are symmetrical fore and aft of 90° so that the overall pressure drag is zero. Because of the effects of viscosity, measured values of the pressure distribution differ considerably and these are shown in (a) and (c) also for 4 knots and 2 knots. Notice that there is not much difference over the leading face of the cylinder, but just before 90° where 'separation' of the flow occurs pressures are considerably higher than in the non-viscous case. From about 70° to 150° pressures are higher than in the non-viscous case, but much lower beyond 150° toward the trailing edge of the cylinder. The upshot of these unbalanced pressure forces is that there is a nett force to the right and this is the normal pressure drag force which results from the 'shape' of the flow. The static pressure in the fluid well away from the cylinder is $p_o$ and in this example is $100{,}000\,\text{Nm}^{-2}$ or one atmosphere.

pressure drag, which used to be known as form drag; (ii) vortex drag, sometimes ambiguously called induced drag; (iii) wave drag. Since the production of trailing vortices is closely associated with the production of lift, a full discussion of vortex drag will be left until chapter 4. Briefly, vortex drag is the extra

penalty paid for producing a sideways lift force perpendicular to the flow direction.

## §2.10 Boundary Layer Normal Pressure Drag

This is also known as 'form drag'. It arises from the dissipation of energy within the boundary layer, which modifies the symmetrical pressure distribution that would be obtained in a non-viscous fluid. Other terms which are often used are 'boundary layer drag' or 'profile drag'. It does not have a separate physical origin but is simply the sum of the surface friction drag and the boundary layer normal pressure drag.

## §2.11 Boundary Layer Drag

Behind the moving sail and hull there is a wake in the air and in the water. The kinetic energy of motion of the air and water eddying in this wake is abstracted from the boat and constitutes a part of the total drag. This is just the boundary layer or profile drag.

## §2.12 Wave Drag

That the formation of waves in the wake of a boat gives rise to drag is well known. That immutable law of Nature, the conservation of energy, is basically responsible. The production of waves requires energy, supplied by the moving hull. If the hull did not have to produce such large waves, as for instance when planing, more energy would be available for overcoming other forms of resistance and so it would go faster.

So we reach the somewhat obvious qualitative conclusion that the bigger the bow and stern waves the greater the wave resistance. Furthermore, the size of these waves and hence the resistance depends upon the speed and shape of the hull. A bow wave can be thought of as the pileup produced by the hull forcing its way through the water (fig. 2.20). A consequence of the law of conservation of energy is that if the water level is lowered in one region it must be raised in another to compensate. Near the stern the hull leaves a 'hole' in the water and a resultant compensating lump or stern wave.

At low speed the bow and stern waves are more or less independent and have a wavelength (distance from crest to crest) much less than the length of the boat. As the boat speeds up the height of the bow and stern waves increases and, as the stability of a water wave requires, the wavelength increases. Eventually

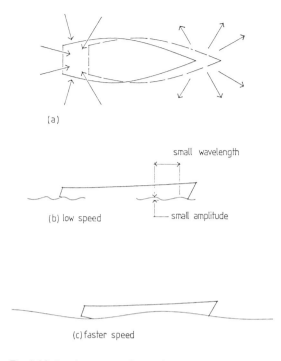

(a)

small wavelength

(b) low speed — small amplitude

(c) faster speed

*Fig. 2.20* As a boat moves forward to the position shown by the broken line the bow pushes water ahead producing a wave crest and the stern leaves a 'hole'. Because the average potential energy of the water surface is fixed, the crest at the bow must be accompanied by a trough and the trough near the stern by a crest. Thus bow waves and stern waves are produced independently of one another. At low speeds (b) the amplitude and wavelength are small. As the boat speeds up (c) the bow and stern wave amplitudes and wavelengths increase so that they coincide and reinforce one another. The production of such large-amplitude waves takes energy from the boat's motion and as the speed increases the energy requirements become rapidly greater.

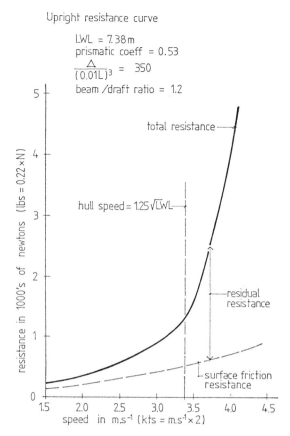

Upright resistance curve

LWL = 7.38 m
prismatic coeff = 0.53
$\frac{\triangle}{(0.01L)^3}$ = 350
beam /draft ratio = 1.2

total resistance

hull speed = $1.25\sqrt{LWL}$

residual resistance

surface friction resistance

*Fig. 2.21* The upright resistance curve of a hull whose parameters are given on the diagram. The vertical broken line is at the position of the 'hull speed' calculated from the standard formula. It is seen that it corresponds to the knee of the total resistance curve, which is the point where the resistance is starting to increase rapidly with increasing speed. The residual resistance is due to normal pressure drag, of which wave drag is the most important component in this case. Since the surface friction part of the resistance (lower line) increases only slowly with speed it is clear that wave-making is the origin of the rapidly increasing resistance beyond hull speed.

the wavelength is such that the second crest of the bow wave coincides with the stern wave (fig. 2.20c): this magnifies the size of the stern wave and hence also the wave drag. It is at this point in the speed versus drag curve of a hull (fig. 2.21) that the drag resistance starts to rise very rapidly with increasing speed: this is the much discussed 'hull speed'. It is not a speed which cannot be exceeded, but simply a speed at which resistance is increasing rapidly so that to raise the speed absorbs much more power.

The definition of hull speed can be made somewhat more precise by utilising a characteristic property of water waves, namely that they are highly dispersive, which is physicists' jargon for waves whose speed

depends upon wavelength. In the case of water waves this speed V is given by $V = \sqrt{(g\lambda)/(2\pi)}$ where g is the acceleration due to gravity (10 m/sec$^2$) and $\lambda$ (lambda) is the wavelength measured from crest to crest of the wave.

Because of the way in which the bow and stern waves are produced, the maximum wavelength for the combination wave is going to be approximately equal to the waterline length of the hull. Since the bow and stern waves are stationary with respect to the boat they must be moving across the water at the same speed as it is. For waves there is a unique relationship (above) between their speed and their wavelength, and since their wavelength is just the waterline length of the boat we get from this the well known value for hull speed:

$$V = \sqrt{(g\lambda)/(2\pi)} = \sqrt{(g \times LWL)/(2\pi)} = 1.25\sqrt{LWL}$$

|  | speed of | speed of | constants |
|---|---|---|---|
|  | water waves | boat | eliminated |

Since $g$ and $\pi$ are fixed numbers they can be replaced by 1.25 providing V is in m/sec and LWL in metres. If V is in knots and LWL in metres, then $V = 2.4 \sqrt{LWL(m)}$. When V is in knots and LWL in feet, the formula is $V = 1.34 \sqrt{LWL(ft)}$.

The generation of lift will in general alter the magnitudes of the boundary layer normal pressure drag and wave drag, and in addition give rise to a trailing vortex system. It should be emphasised that these drag concepts are not independent of one another. For instance a change in the pressure distribution caused by the presence of a boundary layer leads to a change in wave generation so that part of the boundary layer normal pressure drag may appear as a contribution to the wave drag. In general, a change in one drag component due to some flow change will be accompanied by changes in the other drag components.

## §2.13    Summary of Forms of Drag

Total drag consists of two main parts: (i) normal pressure drag, and (ii) surface friction drag. Normal pressure drag has three origins: (a) boundary layer normal pressure drag, (b) trailing vortex drag and (c) wave drag. The best way to see how these are related and upon what factors they depend is to study fig. 2.22.

Another way to get a feel for the difference between (i) and (ii), the left and right branches of fig. 2.22, is to think about, for instance, the movement of a flat plate of thin steel through water. There are two ways it can be moved, edge-on or broadside-on. Experience tells you which orientation has the most resistance. When the sheet is edge-on the shape of the water flow is unaffected by the plate so the normal pressure distribution does not change over the plate, and in any case it is only a pressure difference on the leading and trailing edges which could give rise to

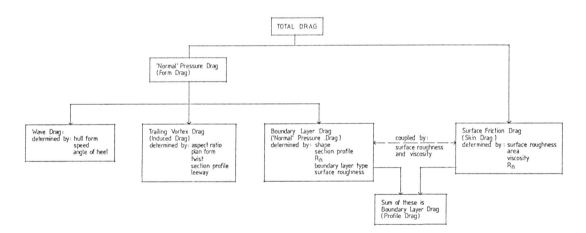

*Fig. 2.22* A summary of the various forms that fluid drag can assume. These forms are differentiated by the physical effects that produce them. For instance the three boxes on the left branch of the diagram describe types of flow which influence the normal force on the body, that perpendicular to the surface. This has to do with the shape of the flow. The right hand branch of the diagram is concerned with the component of force parallel to the surface of the body, which has its ultimate origin in inter-molecular attraction. It is important to realise that these effects are not completely independent of one another; one such coupling is shown explicitly in the diagram. Another is the coupling between vortex drag and wave drag: a change in leeway and hence vortex drag can influence wave production. The names of the drag components used here are those recommended by a panel set up for the purpose and published in the *Journal of the Royal Aeronautical Society* in 1958. The older names, which are still often used, are given in brackets.

drag. But since these edges have virtually no area the plate experiences no normal pressure drag. All the drag arises from tangential forces parallel to the plate. So this is an extreme situation in which only the surface friction forces in the right hand branch of the diagram contribute to the overall drag (fig. 2.23a).

When the plate is drawn broadside-on through the fluid the total drag is due entirely to normal pressure drag (fig. 2.23b). Although there will be some flow along the surface producing surface friction forces, these are perpendicular to the motion and do not contribute to the drag. Thus (b) corresponds to the extreme situation where only the left hand branch of fig. 2.22 is contributing to the total drag. The drag originates from the fact that the sum of the pressure forces on the left side of the plate is greater than that on the right.

When sailing directly downwind we want to maximise the drag of the sails, and the situation is similar to fig. 2.23b where the drag originates entirely from the pressure distribution over the sails.

On the other hand the hull drag when sailing directly downwind is some combination of surface friction drag, wave drag and boundary layer normal pressure drag. The various contributions to the drag cannot be measured separately. All that we can determine in a towing tank or with the full-scale craft is the total drag resistance. The contribution that the surface friction drag makes can be easily determined by calculation, however. The method of doing this was explained fully in the section on surface friction drag. All one needs to know is the wetted area of the hull, its waterline length and speed; this has been done for the hull shown in fig. 2.21. The solid line is the measured total resistance as a function of speed and the dotted line the calculated surface friction resistance. The difference between the two is known as the residual resistance. For the case shown here the hull was upright and moving without leeway, so the residual resistance is due to the sum of the wave-making and boundary layer normal pressure drags. The relative amounts of these two contributions cannot easily be separated but it is interesting to see how the relative importance of surface friction drag varies with speed.

At very low speeds most of the drag is due to surface friction effects and will therefore be reduced by minimising the wetted area of the hull. The hull section with the minimum wetted area for a given displacement is semicircular in shape and this is commonly used in catamaran hulls.

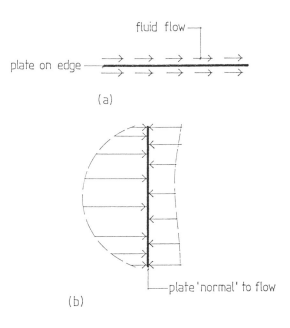

*Fig. 2.23* Two extreme drag situations arise when one draws a thin plate through a fluid edge-on (a) or broadside-on (b). Only surface friction forces contribute to the total drag in (a) whereas only normal pressure forces contribute to the drag in (b). Note that in this diagram the arrows represent the direction and magnitude of the forces (though not to scale). The nett is the algebraic sum (the sum in which the direction of the forces is taken into account) of these forces and is toward the right. The large arrows indicate the direction of motion of the fluid with respect to the plate.

In Part II, fig. 7.4 compares the wetted areas of semicircular rectangular and Vee sections. The diagram is drawn so that all three sections have the same displacement and the minimum possible wetted area for that displacement. Nevertheless the wetted areas for the rectangular and Vee sections are about 13% greater than for the semicircular one. No other shape can have a smaller wetted area for the same displacement than the semicircular one. For sailing in light airs where surface friction drag accounts for most of the total drag, a small wetted area which is hydrodynamically smooth is very desirable. Reference to figs. 2.10 or 2.15 show that the requirements for hydrodynamic smoothness are not so stringent at low speeds, which is just where surface friction drag is most important. This looks at first like a gift from Nature, but it is tempered with the fact, shown in fig. 2.22, that the boundary layer normal pressure drag is also affected by surface roughness. This comes about because of its effect on the turbulent boundary layer and the loss of kinetic energy in the flow, resulting in separation of the flow and wake production.

## §2.14 Effect of Roughness on the Total Drag

In section 2.8 the effect that surface roughness has on the surface friction drag was fully discussed. The fact that roughness also affects the normal pressure component of the drag has already been alluded to. It is exemplified in an unexpected way when one measures the drag of a cylinder (a circular section mast for instance) as a function of flow speed (fig. 2.24). At first drag increases smoothly with the square of the velocity, as expected if the drag coefficient is constant. Then, at a speed which depends upon the surface roughness, the drag actually falls off with increasing velocity. It seems at last that Nature is giving us something for nothing because there really are no strings attached other than the limitation evident in the diagram, namely that for a roughness of 0.075 mm, for instance, the drag is less than a perfectly smooth cylinder over the range 12 to 22 m/sec, and below those speeds the drag is no greater. For wind speeds greater than 22 m/sec (44 knots) the drag of the rough cylinder is much greater.

The important point is that by varying the roughness the dip in the drag curve can be moved around almost at will. Note that the relative advantage obtained by a large amount of roughness is not as great as with an intermediate amount. Fig. 2.24 is valid only for a mast diameter of 15 cm and a circular cross-section. However it is not difficult to calculate similar curves for cylinders of other diameters. Both axes of the graph simply have to be scaled in inverse proportion to the cylinder diameter. Fig. 2.24 is for a cylinder diameter of 15 cm, so for a cylinder diameter of d cm multiply the horizontal scale by 15/d and the vertical scale also by 15/d. This will give a new force and a new velocity appropriate to the chosen value of k/d. For example, if the mast diameter is reduced to 7.5 cm both scales are multiplied by 2. Thus the point at which the curve for $k/d = 4 \times 10^{-3}$ separates from the smooth curve now occurs at 14 m/sec and the drag force is 8 newtons. Since 14 m/sec corresponds to 28 knots it is seen that no practical advantage can accrue from roughening the mast of a small boat! On the other hand for a large mast of 30 cm diameter this same point on the graph corresponds to a wind speed of only 7 knots.

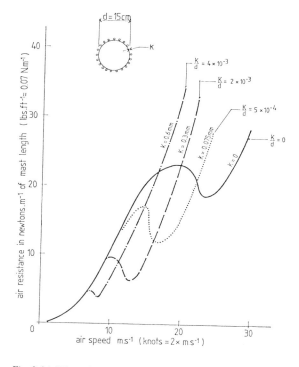

*Fig. 2.24* When the total drag of a cylinder is measured as a function of increasing speed a surprising result is found. The drag does not increase steadily with the square of the speed as it would if the drag coefficient were constant, but in a certain speed region actually decreases with increasing speed. The curves shown give the drag of a 1 m length of a circular section cylinder (e.g. a mast) as a function of the wind speed, the cylinder diameter being 15 cm (6 in.). The solid curve is for a perfectly smooth surface and the other curves are for varying degrees of roughness as indicated by the magnitude of k. As the roughness is increased the dip in the drag moves to lower air speeds. For instance the dashed curve for which the roughness size k is 0.3 mm gives a lower drag for all wind speeds between about 10 and 20 ms$^{-1}$ (20–40 knots) than does a perfectly smooth surface. Such considerations have been used in mast design, and it is possible that it results also in improved flow over the mainsail.

One must beware of applying results like this too literally to the sailing situation, however. In this case a problem arises because the dips in the curve are very sensitive to the amount of turbulence in the incident flow. As we shall see, the effect has its origin in the production of a turbulent boundary layer, so existing turbulence in the stream will push the dip to lower speeds. In fact this change in the position of the dip was at one time used to determine the amount of turbulence in wind tunnels.

Although this curious effect does not have a great deal of application in the sport of sailing, it is absolutely basic to the sport of golf. The dimples in a golf ball provide just the right amount of roughness so that at the velocity of a good drive the air resistance is much less than that of a smooth ball. Since this is a book which attempts to tell the 'whys' as well as the 'whats', I cannot leave the subject of dimpled golf balls and rough masts without an explanation of this curious behaviour of fluid resistance.

The first clue about the reason for this behaviour is obtained from fig. 2.18. Also, from fig. 2.19 it is clear that as one moves around the cylinder from the leading edge the pressure first falls to a minimum at about the 90° point and then rises steadily. Normally air will flow from a region of higher pressure to one of lower pressure. Thus air flows around the rear half of the cylinder against the pressure gradient only by virtue of its momentum or inertia. It is exactly equivalent to driving a car fast along the flat and letting it coast uphill. It will carry on up to that point at which its kinetic energy is all used up and then start to roll back. Exactly the same thing happens to the air flowing around the cylinder. The amount of kinetic energy it has determines how far toward the rear face of the cylinder it can flow against the pressure gradient before stopping and 'separating' from the surface. The distance between the two symmetrically placed separation points determines the width of the wake (figs. 2.18 and 2.25). The width is a measure of the amount of drag, so the longer we can delay separation the better. This requires that there be plenty of energy in the flow close to the surface. A glance at fig. 2.14a should convince you that the flow velocity near the surface is greater in the case of a turbulent boundary layer than a laminar one. So for the former there is more energy near the surface and thus separation occurs later and the wake is narrower. Since a turbulent boundary layer can be induced earlier by roughness, this accounts for the dips in the resistance seen in fig. 2.24. The reason a turbulent boundary layer has a greater velocity near the surface is because the transverse motion of the fluid particles allows for more mixing with the fast moving outer layers, thus transferring momentum to the surface.

## §2.15   Ways of Reducing Drag

When sailing downwind it is clearly desirable to reduce hull drag to a minimum, and in considering what might conceivably be done it is best to look at each drag component separately.

Surface friction drag has already been discussed in this context in section 2.7 so only the conclusions will be summarised here:

(a) Surface friction drag is directly proportional to the total wetted area of hull, keel, rudder and any other appendages.

(b) Surface friction drag depends upon hull smoothness (see figs. 2.10 and 2.15).

(c) Surface friction drag is very important at low speeds but its contribution to the total drag falls off as the speed is increased (see fig. 2.21).

(d) Surface friction drag depends upon the viscosity of the water in which the boat is sailing. For smooth surfaces it does *not* depend on the material of the surface.

(e) Certain chemicals when dissolved in water reduce its viscosity but their use in racing is prohibited by the IYRU rule 63, and in any case would involve phenomenal cost and tragic pollution of the environment in which we have chosen to spend our spare time.

At higher speeds boundary layer normal pressure drag and wave drag become more important. The formation of a wake greatly reduces the pressures on the aft end of a moving body so that the pressures on the forward end are no longer balanced and drag results. The points at which separation occurs are crucial in determining the width of the wake and hence the drag. Separation is brought about by too rapid curvature in the flow and by too rapid increase of the pressure in the flow. A sphere has greater curvature than a 'streamline' shape (fig. 2.25) and also more rapid increase in pressure toward the trailing edge, so separation occurs sooner than for the streamlined body with a consequent wider wake and greater boundary layer normal pressure drag. Since separation is due to the slowing down of flow near the surface caused by viscous drag, the surface smoothness is important in delaying it and hence has an effect on the boundary layer normal pressure drag. This is the coupling between surface friction drag and boundary layer normal pressure drag shown in fig. 2.22.

A streamlined body has lower drag at higher speeds but it can have relatively more at lower speeds where surface friction is important, because of its generally greater surface area.

Although the reduction of wave drag is highly desirable, hulls with the least wave drag tend to have

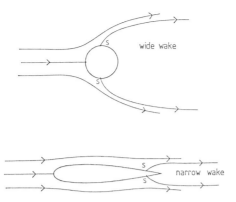

Fig. 2.25 Separation of the flow, which occurs at points marked S, is important in determining the width of the wake and hence the drag. Because of its smaller curvature and less adverse pressure gradient the streamlined body tends to have separation points much further aft, and consequently a smaller normal pressure drag. Because of its greater wetted area, however, the surface friction drag could be considerably larger.

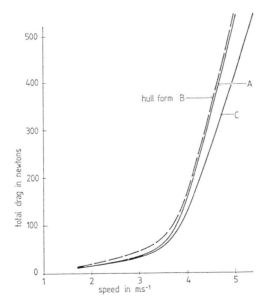

Fig. 2.26 Here the total drag of the three different hull forms shown in fig. 2.27 is plotted against speed. The hulls are moving without leeway and are upright, so the total drag includes surface friction drag, wave drag and boundary layer normal pressure drag, but not vortex drag. The three hulls all have the same waterline length but the different shapes characterised in Table 1. (Data from the Delft series models, 1978.)

shapes inconsistent with other design considerations. It is well established that for a hull which neither planes nor surfs, the wave drag is reduced by making it long and narrow. Wave drag scales like $V/\sqrt{L}$ (see chapter 5), so if a hull is made 10% longer it will have the same drag at approximately 5% higher speeds. Other design considerations and cost limit the length of a hull so it is more interesting to ask what hull features influence the resistance of hulls all with the same waterline length. Of course this cannot be answered fully because there are so many parameters, but to give a feel for the sorts of differences expected, fig. 2.26 shows total resistance curves for the three different hull forms shown in fig. 2.27 and the table below. Each hull has a waterline length of 10 m. Only upright resistance is given as the concern in this chapter is with downwind sailing. (The effects of heel and righting moment are discussed in chapter 5.) Thus one concludes from this particular comparison that the boat with the least displacement and smallest wetted area but intermediate beam has the least overall resistance. If the comparison had been made between three hull types with the same displacement instead of waterline length, the conclusions as to the effect of beam, for instance, might well have been different. Such are the problems of decision making in yacht design, but it is just such complications that make it such a maddening art and sailing such a fascinating sport.

Table 1

| Hull shape | A | B | C |
|---|---|---|---|
| Displacement (kg) | 4942 | 4881 | 4051 |
| Wetted area (m$^2$) | 32.75 | 35.9 | 31.3 |
| Beam/depth ratio | 2.90 | 4.58 | 3.41 |

Another parameter which is important in determining overall hull resistance is the prismatic coefficient. This is defined fully in chapter 5 but can be thought of simply as a measure of the fullness of the ends of the hull. A large prismatic coefficient means full ends and a small one fine ends. Broadly speaking, high prismatic coefficients of around 0.6 are better at high speeds. There is sometimes a small advantage off the wind in light airs for a low prismatic coefficient of around 0.45. For all-round performance it is likely that a prismatic coefficient of around 0.53 is best.

Speed downwind can be improved if the boat can be induced to plane or surf, of course. Surfing is caused not so much by a reduction in drag as by a

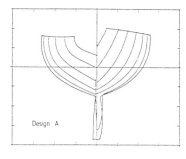

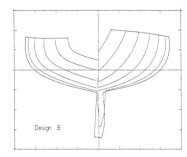

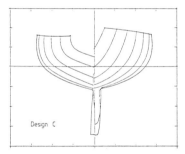

*Fig. 2.27* These three hull forms, all of 10 m waterline length, are compared for downwind performance in fig. 2.26. The displacement, wetted area and beam-to-depth ratio for each are given in Table 1.

occupied by the shaded volume of the boat and it gives rise to a buoyancy force acting upwards. Imagine a boat being lowered into the water by a crane: before it touches the water the crane takes the full weight. As it is lowered into the water and the volume displaced increases, the buoyancy force also increases. Since the buoyancy acts in the opposite direction from the weight, it subtracts from it and the crane holds an ever decreasing effective weight as the boat is lowered. At the point where the boat is floating there is no weight on the crane, so in fig. 2.28a the magnitude of the buoyancy force must be exactly equal to the weight W of the boat.

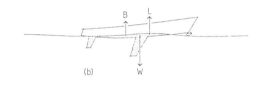

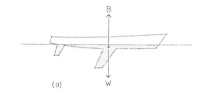

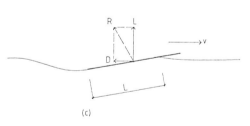

utilisation of the water motion within a wave. It is thus a matter of the dynamics of sailing in waves and as such will be treated in chapter 6.

The basic difference between a yacht sailing in the displacement mode and one that is planing is that in the latter case the weight of the hull is offset not just by buoyancy but also partly by dynamic lift forces. Fig. 2.28 shows the two situations. A boat which is stationary or moving without planing displaces a weight of water equal to its own weight. This, of course, is the famous law that Archimedes was so happy about. In (a) the mass of water displaced is that

*Fig. 2.28* For a stationary boat or one sailing in the displacement mode the buoyancy force is exactly equal to its weight. In (a) the length and direction of the arrow labelled B represent the magnitude and direction of the buoyancy force produced by displaced water. W is the weight force produced by gravity acting on the mass of the boat. When a hull is planing both buoyancy and dynamic lift are present. For stability the sum of these must be equal and opposite to the weight and act through the centre of gravity. In (a) and (b) only the vertical components of the total force acting are shown. In (c) the total dynamic force is shown where it can be seen to consist of lift and drag components. This is the vortex drag not present for the displacement hull moving downwind without leeway.

When a hull is planing we have at our disposal two upward forces, buoyancy and dynamic lift, the sum of which must be equal to its weight if the boat is in equilibrium: i.e. B + L = W. Dynamic lift has its origins in the way the water flow is shaped under the hull. One can think of the hull as acting like a stubby airplane wing of very low aspect ratio. Due to the nature of the flow (see chapter 3) pressures under an airplane wing are slightly greater than atmospheric and those above are slightly less, so although these pressures are all towards the surface there is a nett upward force producing lift. The planing hull is only half a wing, however: there is no upper surface in the water so dynamic lift comes only from increased pressure on the underside. Because of this and because of the low aspect ratio a planing hull is a somewhat inefficient lifting surface.

Nevertheless if dynamic lift is present, less buoyancy and hence less displaced water is necessary to support the weight of the boat. Less displacement means smaller bow and stern waves and hence less wave-making resistance. But there is another side to the story. A necessary companion of dynamic lift is vortex drag. By inducing our boat to plane we have reduced wave drag but introduced vortex drag. During planing the wetted area of the hull is reduced and also surface friction drag. Because the flow tends to diverge at the sides it is very difficult to know what the effective length L really is, so even surface friction calculations become unreliable in the planing mode (fig. 2.28c). Coupled with this is the effect of spray: there is evidence from measurements on flat plate planing surfaces that forward-moving spray actually reduces the drag slightly. Measurements on flat plates show that the drag-to-lift ratio is a minimum when the plate is angled at about 7° to the horizontal (fig. 2.29a). For a displacement hull the drag continuously increases with increasing speed, so the point where the curve starts to fall away is arbitrarily defined as the start of planing (fig. 2.29b). In practice it is not a well defined point; the onset of planing is not a sudden transition but a gradual taking over of buoyancy by dynamic lift. In practice it can be recognised by a change in the boat's trim, by the presence of a flat sheet of water issuing from under the transom and by a partial disappearance of the bow wave. Another characteristic is that small rudder movements become much more effective, due to its greater effectiveness caused by higher speed and the fact that with less of the fore part of the hull in the water resistance to turning is reduced.

Many planing hulls are Vee-shaped. Since the lift

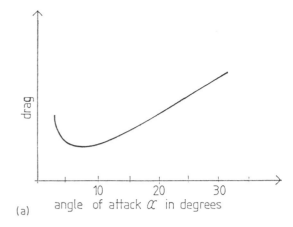

(a)

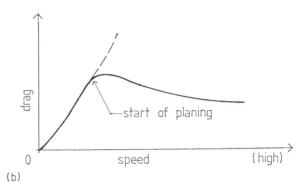

(b)

*Fig. 2.29* Both graphs shown here are meant to give a qualitative indication only of how a hull behaves when planing. The curves have been taken from measurements on flat plates which probably bear only a passing resemblance to a real hull. In (a) the drag is plotted against the angle of attack, the speed at the same time being adjusted to keep the lift constant. As the angle of attack is increased from zero the drag at first falls due to a reduction in friction drag and then rises due to an increase in vortex drag. The optimum angle of attack is found at around 7° which is probably about right for a real hull. In (b) a similar qualitative plot taken from flat plate measurements at a fixed angle of 5° shows an initial rise of resistance roughly proportional to the square of the speed, followed by a falloff to a constant value fixed by the vortex drag.

on each surface is perpendicular to it the component in the vertical direction opposing the weight is reduced approximately by the cosine of the deadrise angle (fig. 2.30a). The decrease in lift compared to no deadrise is plotted in fig. 2.30b.

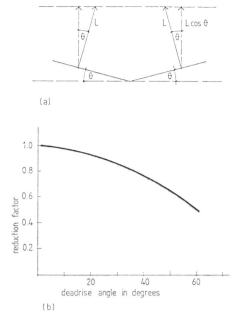

(a)

(b)

*Fig. 2.30* Since lift forces on a surface act perpendicular to that surface it is clear from (a) that the component of lift in the vertical direction is reduced by the cosine of the deadrise angle. Although this is only approximately correct in practice, a plot of the cosine reduction factor is presented in (b) to give an idea of the magnitude of the effect.

## §2.16 The Equilibrium of Downwind Sailing

You are sailing directly downwind in heavy air but smooth water. The spinnaker is tugging at sheets and halyard, the mast is bending strangely, the kicking strap is bar-taut, the creaking sounds of a boat working are heard, so if I then tell you that the sum total of all the forces acting on your boat is zero you will probably be convinced that I am indeed a mad scientist.

Most people, even some physics students, think that if an object is moving there must be a nett force acting on it to keep it in a state of movement. This was the considered view even of scientists in medieval times, and it took the genius of Isaac Newton to put them straight. Such is the power of intuition resulting from everyday experience that the truth is still hard to accept. Everyone who drives a car knows that he must constantly produce a force between his tyres and the road otherwise the car will slow down and stop. This is the origin of the false idea, an error of

omission. The car's engine does provide a forward force but it is opposed by friction in the bearings, between tyres and road, and by aerodynamic drag. What Newton said was that if the sum of *all* these forces on an object added up to zero (taking direction into account), then the object would be moving (or standing still) without *acceleration*. So the key concept is that if there is a resultant nett force on any object then that object will be *accelerated*; it will move with an ever increasing speed.

To get back to downwind sailing, imagine your boat is headed downwind flying spinnaker and main but is not moving because you have the stern attached to a dock with a strong line. The apparent wind speed over the deck is also the true wind speed, and clearly all the drag produced by the sails appears as tension in the line. These two forces are equal and opposite, so there is no nett force on the boat and hence no change of speed or acceleration. Since the initial speed was zero it then remains at zero.

Now we suddenly cut the stern line and study what happens in the next few seconds. The instant the line is cut the boat is being accelerated because now the only force acting on it is that produced by the sails. As soon as movement through the water starts, new forces appear. These are the various fluid drag forces described in this chapter, and as we have seen they increase roughly as the square of the speed. Thus as the boat accelerates downwind the drag force increases the faster it goes. On the other hand the sail force is determined by the apparent wind speed and therefore falls off the faster the boat goes. As long as the sail force exceeds the drag there is a nett forward force on the boat and it will continue to accelerate. Eventually a speed is reached where the drag force becomes exactly equal and opposite to the sail force, there is thus no nett force on the boat and she is no longer accelerated. She then continues to move at constant (unaccelerated) speed until the balance of forces is disturbed.

In order to make this sequence of events clearer the actual measured resistance without heel or leeway of the Admiral's Cup yacht *Standfast* has been plotted in fig. 2.31. The sail force falls off with boat speed since it depends upon the square of the apparent wind speed, which is that felt on board the moving boat. When sailing directly downwind it is just the true wind speed minus the boat speed. At the instant the stern line is cut the boat speed V is zero and thereafter the sail force decreases as the .hull resistance increases. The resultant force accelerating the boat is the difference between these two forces

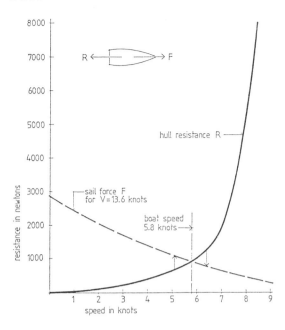

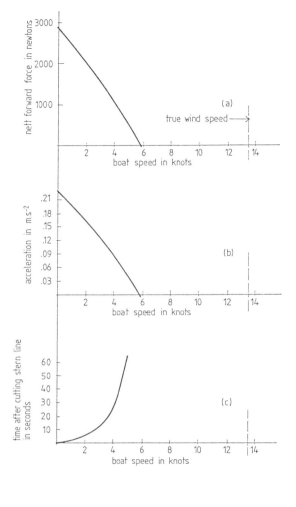

*Fig. 2.31* The solid line is the measured resistance without leeway of the Admiral's Cup yacht *Standfast*. The measurements were made at the Delft University of Technology in the mid-seventies. The dashed line is the sail force in a true wind of 13.6 knots when headed directly downwind. As the boat speeds up the apparent wind speed becomes less and so the sail force decreases. The sail force F is pushing the boat in the desired direction but the resistance R is opposing it, so the net force acting on the boat and causing it to accelerate is F − R. This is simply the difference in height between the dashed curve and the solid curve. As the boat speeds up this difference becomes less and less until at 5.8 knots there is no net force acting on the boat, acceleration ceases and she will go no faster than 5.8 knots. If she does gain speed by planing slightly and then going off the plane, hull resistance becomes greater than sail force and the boat is pushed back to 5.8 knots.

and for a speed of about 5.1 knots is shown by a vertical arrow in fig. 2.31. At this speed there is a forward force on the boat which is therefore still accelerating. Eventually, when 5.8 knots is reached acceleration ceases and this becomes the steady downwind speed. If a 20 knot wind gust came for a minute or so and then slackened back to 13.6 knots the boat would find herself going faster than 5.8 knots: hull drag would be greater than sail force so she would be decelerated by a force such as that given by the downward pointing arrow in fig. 2.31. The whole process of acceleration up to the equilibrium speed is shown in greater detail in fig. 2.32. Graphs

*Fig. 2.32* These graphs have been calculated from fig. 2.31 which was taken from actual measurements of a fairly large yacht of 12,500 kg displacement sailing directly downwind with a total sail area of 235.9 m² (spinnaker and main). The diagrams refer to the 'thought experiment' described in the text in which the boat, fully rigged in a 13.6 knot wind, is held by a stern line. At the instant of cutting the stern line the boat speed is zero and (a) gives the nett force on the boat at that instant, (b) gives the instantaneous acceleration, and (c) the time after cutting as a function of speed. The apparent wind speed at any point can be obtained by subtracting the boat speed from the true wind speed of 13.6 knots. Graph (c) shows that half of the ultimate downwind sailing speed of 5.8 knots is reached after about 11 seconds, but it takes another 60 seconds to reach 90% of the ultimate speed.

(a) and (b) are similar in shape since the forward force and acceleration are directly proportional to one another; (c) is interesting because it shows the time taken for this fairly large boat to speed up downwind.

This notion that the speed of a boat is associated with the equality of two forces is an aspect of the 'symmetry of sailing' discussed in chapter 1. One can think of a yacht as a sort of analogue computer that automatically adjusts its speed so that the water forces and the wind forces are always exactly equal and opposite. If one measures the water resistance in a test tank, then the sail forces at various equilibrium speeds are immediately known. This principle was in fact used in the first scientific investigation of the sailing yacht by K. S. M. Davidson in 1936 and gave rise to the so-called *Gimcrack* sail coefficients, which have been widely used for estimates of the windward ability of yachts whose hull resistance has been measured in a towing tank.

Thus we have seen that downwind sailing is a manifestation of the symmetry of sailing in its simplest form. In the next chapter we see how this idea of symmetry is also present in the much more complex situation of windward sailing.

# Chapter Three

# UPWIND SAILING

## §3.1 How the Albatross Goes to Windward

Everybody is familiar with the poem from which the quotation at the beginning of this chapter was taken. It is surely one of the most vivid works in the English language, particularly appealing to yachtsmen and lovers of the sea because it is a story of the age of sail and human dependence on the elements. Sailors struggling to beat to windward in an inefficient square-rigger easily believed in the supernatural powers of the incredible Wandering Albatross, which could effortlessly follow them to windward with never the slightest movement of its wings, by simply flying in great tilted circles about their ship. It was not until 1883 that the albatross's secret was laid bare. How could it fly to windward without flapping its wings as any other bird would? The first person to understand this was the Englishman John William Strutt who later became Baron Rayleigh. In his lifetime he wrote over 450 scientific papers and several books on subjects ranging from electromagnetism to the bright colours of certain animals. He was the product of an era in which it was thought that all the phenomena of Nature were explainable in terms of the laws of 'classical physics'.

Understanding the flight of the albatross was a problem which he solved in terms of the fundamental laws of physics years before a practical airplane was built.

Friction slows the wind at sea level and just above, with the result that the wind speed increases from zero to its full value in the first 30 or so metres above the waves. It is from this gradient in velocity that the albatross extracts the energy for its flight. The bird glides downward, converting potential energy to

*Fig. 3.1* The wandering albatross, or *Toroa* as it was known to the Polynesian navigators of the South Pacific. This bird, so important in the mythology of sailors of all races, never flaps its wings in flight and yet is able to go to windward and gain altitude without expenditure of its own energy. The same basic principle by which this is possible is also the means by which a sailing boat is able to extract energy from its environment and make speed good to windward.

kinetic energy and picking up air speed as well as ground speed. Just above the waves it wheels into the wind, acquiring initial lifting acceleration from an abrupt increase in wing incidence or angle of attack. As the bird rises it encounters ever higher wind speeds and, with its air speed thereby maintained, is able to rise again without any expenditure of its own energy to the altitude from which it first descended. In a similar way but with a slight change of path, it can proceed to windward again without expending its own energy. (Further details of the physics of this process are given in §8.1.)

In chapter 1 we saw that it is the *relative* speed of air and water that enables boats to sail to windward. Now we see that you don't need two media; just a velocity gradient within one will do.

## §3.2  Forces on a Yacht going to Windward

Try to move an object of arbitrary shape through a fluid (air or water) and you meet with 'resistance': we notice it every day when we stir our coffee or row a dinghy. Unless you are thinking consciously about fluid flow when stirring coffee you can be forgiven for not noticing that the direction of the 'resistance' force on the spoon is not always exactly opposite to the direction of motion.

One of the objectives of scientific investigation is to look more closely at seemingly simple phenomena such as this. What is found is indeed that the force acting on an object of any old shape moving through a fluid is rarely directly opposite to the direction of motion but at some angle to it (fig. 3.2). The situation is complicated because the direction of the force depends on the shape and orientation of the object. In order to simplify the situation the force F is divided up into two components having fixed directions.

Though any fixed directions can be used, as shown in fig. 3.3, the most simplification results when they are parallel to the undisturbed direction of fluid flow and at right angles to it.

Quantities which have both a direction and a magnitude associated with them are called **vectors**, for example forces and velocities. If one pulls a mainsheet there is a tension in the rope which is the magnitude of the force, and a direction, which in this case is the direction of the rope. A vector is therefore a quantity which has two numbers associated with it; one characterises its magnitude and the other its direction. It is thus not surprising that vectors do not obey the ordinary laws of arithmetic. A pirate pinpointing the position of his buried treasure might have written in his will, 'From the foot of the old oak tree take 25 paces due west and then 35 paces due north. At this point, buried 3 feet down you will find . . .' One cannot add 25 to 35 and say the treasure is 60 paces from the tree because it isn't: it is 43 paces. Nor does the fact that we have gone partly west and partly north mean that the treasure lies northwest of the tree: that would only be true if we had gone equal distances in these two directions. In fact the bearing of the treasure from the tree works out at 35°32′. It is not appropriate here to discuss the mathematics of this, except to mention that it can always be done by drawing a scale diagram just as one would on a chart for calculating the effect of a current on a true course.

It is intuitively obvious that if one pushes a model railway car running on straight tracks at an angle of 45°, it will move forward. Now this applied force is exactly the same as a slightly lesser force pushing directly behind the car, plus another simultaneously acting but at right angles to the track. Clearly the component at right angles to the track cannot

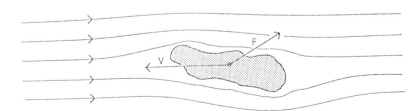

*Fig. 3.2*  If an object of any shape is moving through a fluid (water or air) with a speed and direction represented by the arrow V, then from the point of view of the object fluid is rushing past it in the opposite direction. This produces a force F on the object which is generally not exactly opposite to the direction of motion. It is found that the direction and magnitude of F depends on the density of the fluid, on the speed of the object, on its size and shape, and on its orientation.

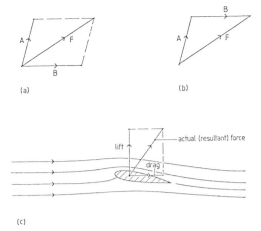

(a)

(b)

(c)

*Fig. 3.3* A quantity, like force or velocity, which has a magnitude and a direction is called a **vector**. Vectors do not obey the ordinary laws of arithmetic but must be added geometrically. Thus in (a) the two forces A and B add up to a resultant force F. The diagram could alternatively be drawn as in (b) which means exactly the same as (a). A and B are called the components of F. It is generally more useful to put the components at right angles to one another; in this way each component is independent of the other. (c) The force on an object moving in a fluid is always broken down into a component parallel to the undisturbed direction of flow, called **drag**, and one perpendicular to the flow direction called **lift**.

contribute to any motion, except perhaps to topple the car over. Acceleration along the track can only come from the component whose direction is parallel to it. Calculation of this acceleration is thus made simpler by dividing the single original force up into two appropriately oriented components. Putting them at right angles to one another also ensures that their effects are independent of each other. The component along the track has no tendency to topple the car over, whereas the other component has no effect on acceleration along the track.

This breaking down of a force into two components is always used with the force that results when an object moves through a fluid; fig. 3.3c shows their conventional names and directions. One component of the actual force is taken parallel to the direction of the undisturbed fluid flow and is usually called **drag**, although occasionally the word **resistance** is used with the same meaning. The component at right angles to the direction of flow is called **lift**. This terminology has been derived from aeronautics, where it is eminently sensible as it implies movement against gravity. In our application lift is usually horizontal, so

it would be more logical to call it **side force** or **cross-wind force**. It is nevertheless difficult to justify the use of two or three words where a graphic monosyllable will do. As 'lift' is a word that is rapidly making its way into the yachting vernacular I will therefore use it in preference to 'side force' even though in most cases a horizontal direction will be implied.

We are now in a position to look at the forces acting on a yacht going to windward. (An explanation of how these forces are produced must wait until section 3.3, however.) For a boat in smooth water on any point of sailing there are only two prime forces acting; one is due to the action of wind flow on sails, rigging and hull, and the other to the action of water on the hull, keel and rudder (fig. 3.4a). As we have seen in chapter 1 the symmetry of sailing demands that when the boat is sailing at a steady speed the total wind force and the total water force are exactly equal and opposite. Fig. 1.3 shows how this situation builds up as a boat accelerates away from rest on a windward course. To understand the details of why a boat makes leeway, why it heels, what limits its speed through the water and what governs its speed made good to windward, we must look at appropriate components of these prime forces.

It is not difficult to see that if the force on the sails $F_s$ and the force on the hull $F_h$ are equal and opposite as shown in fig. 3.4a, then the components of these forces, if taken along the same directions, must also be equal and opposite. This is shown in (b) where $F_R$ is the component of $F_s$ along the course sailed; for this reason it is called the **driving force** of the sails. This is exactly opposed by the hull drag force $D_h$ measured along the direction of undisturbed water flow, which is simply the direction of the course sailed. At right angles to these are the two independent force components $F_H$ and $L_h$ which are also equal and opposite. They are at right angles to the boat's direction of motion and so do not contribute to its speed but only to its tendency to heel.

As well as forces, the boat is subject to **torques** or **moments** (fig. 3.5). A torque is a twisting force. If the forces F are parallel and separated by a perpendicular distance d, then a measure of the strength of the torque is F multiplied by d as shown in (a). If the boat is sailing with a constant angle of heel the clockwise torque produced by $F_H$ and $L_h$ must be exactly opposed by a counterclockwise torque produced by W and B (b). This shows the situation for a small light boat which is sailed upright. W is the total weight of the boat plus crew and B is the buoyancy force

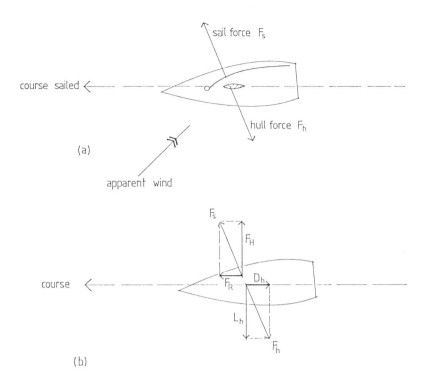

*Fig. 3.4* (a) Forces on a boat. When it is moving steadily the total sail force $F_s$ is exactly equal and opposite to the total hull force $F_h$. (b) If these prime forces are decomposed into components parallel to the course sailed and perpendicular to it we have two further pairs of forces which separately must be equal and opposite. That is, the sail driving force $F_R$ must be equal and opposite to the hull drag force $D_h$, and the heeling force $F_H$ must be equal and opposite to the hull lift force $L_h$.

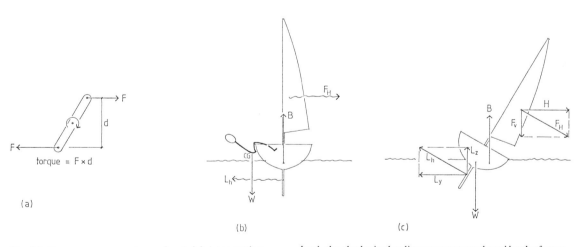

*Fig. 3.5* Torques or moments on a boat. (a) A torque is a twisting force which results when two forces are not in line. If two equal and opposite forces are parallel but displaced, then the torque is measured by the product of the strength of one of the forces multiplied by the perpendicular distance between them. (b) For a boat to remain at a steady angle of heel, the clockwise heeling moment produced by the forces $F_H$ and $L_h$ must be exactly cancelled by the counterclockwise righting moment produced by the forces B and W. (c) A keelboat must heel for even the slightest heeling force $F_H$, for only then are the forces B and W out of line enabling them to produce a righting moment.

41

resulting from the displaced water. Obviously W and B are equal and opposite, otherwise the boat would either rise up out of the water or sink farther into it.

The situation for a larger monohull which must heel when going to windward is a little more complicated and is shown in fig. 3.5c. The **righting moment** or counterclockwise torque depends on how out of line the forces W and B are, which in turn depends on the angle of heel.

Thus for a given heeling torque produced by $F_H$ and $L_h$ the boat heels to just that point where the separation between W and B is just sufficient to provide the necessary righting torque.

It is also worth noticing that when a boat heels the aerodynamic and hydrodynamic forces are considered to remain perpendicular to the surfaces producing them (fig. 3.5c). This means that the heeling force $F_H$ points slightly downward and can therefore be broken up into vertical and horizontal components. In exactly the same way the keel side force points slightly upwards and also has vertical and horizontal components. Since $F_H$ and $L_h$ are equal and opposite to one another (and therefore also parallel), their vertical components $F_v$ and $L_z$ must also be equal and opposite. It is sometimes thought that the downward force component $F_v$ causes the boat to sink further into the water, but it is always accompanied by the equal upward force $L_z$. There is thus no overall sinkage although there could well be a torque produced by these forces which could change the fore-and-aft trim.

## §3.3 Three Ways of Understanding Lift

Although there are only two prime forces acting on a boat moving through the water, we showed in the last section how it is useful to decompose these into components such as lift, drag, driving force etc. Drag is the component of the total force which is along the direction of motion and was discussed in detail in chapter 2. We have a strong intuitive notion about drag and because of this we feel that we 'understand' it and it requires no further explanation. In fact, as the quotation at the beginning of chapter 2 implies, a complete understanding of drag has not yet been arrived at.

The situation is completely reversed in the case of lift. Mathematicians can calculate and predict lift precisely in many situations, and yet most people find it not at all easy to understand why there can be such a large component of force *at right angles* to the direction of flow. The reason is that it is not part of

our package of intuitions or the prior way we have allowed our brains to become programmed by education. For this reason lift, which is so important in sailing, will be explained here in three different ways. They are all scientifically correct but will appeal to different people depending on their background of experiences and reading.

### §3.3.1 Flow Line Method

This is perhaps the standard method of explaining lift in books on sailing. One begins by drawing a diagram of the flow lines around a sail. Such flow lines can, of course, be measured and even visualised with the aid of smoke, so providing one does not ask why the flow lines go where they do but is willing to accept it the way it is, then there should be no difficulty in understanding lift from what follows.

Fig. 3.6 shows the flow lines (strictly called streamlines) around a sail without a mast, as measured by means of an electrolytic tank (see §8.2). This is a method of measurement which determines the *potential flow* or *ideal flow* around an object. It gives the correct flow distribution in situations where separation does not occur. Since this is what is desired with sails, and can be accomplished in many real sailing situations, the distribution of flow shown in fig. 3.6 can then be regarded as quantitatively correct: not invented by the artist but actually measured.

Before we can understand such a diagram it is necessary to know how to obtain quantitative information from a set of streamlines. The track of a small particle of dust in a fluid flow would trace out a 'streamline'. If the flow is steady, not turbulent, then the set of streamlines is fixed in space and time. Thus although they represent movement, the shape of the movement is not changing with time. If turbulence is present one would need a set of pictures or a moving picture to depict correctly the events. In a real flow situation turbulence is almost always present but usually only in the boundary layer. For this reason and for another that will be gone into later, flow lines are meant to depict flow around objects only in the region outside the boundary layer.

As well as the shape of the flow there is one other useful piece of information in a correctly drawn flow diagram, concerning the velocity of the flow. Fig. 3.7 represents a pipe with a constriction through which a fluid is flowing without turbulence. It is clear that the streamlines must come together at the constriction in the pipe. It is also clear that the amount of fluid

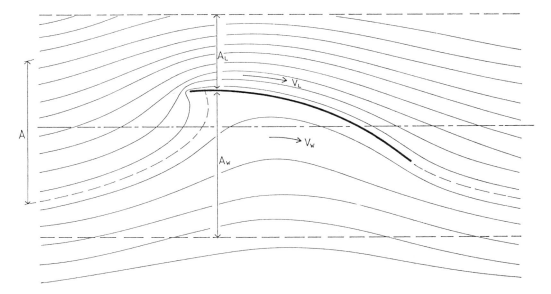

*Fig. 3.6* The flow around a sail under conditions where no separation is occurring. All the streamlines that flow through the area A ahead of the sail also flow through the smaller area $A_L$. Since the same amount of fluid must flow through both areas, the speed of flow through $A_L$ must be greater than through A. As explained in the text an increase in speed is associated with a decrease in pressure. This decrease in pressure on the leeward side of the sail coupled with an increase on the windward side gives rise to a net force on the sail having a component at right angles to the wind direction.

passing per second through the cross-section A of the pipe must be the same as that passing through cross-sections B and C. If this were not so fluid would be building up somewhere inside the pipe, which is ridiculous. All we are saying is that all the fluid which goes in past A must eventually pass through B and come out through C. If the cross-sectional area of A is

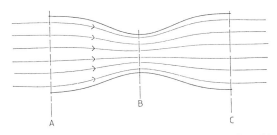

*Fig. 3.7* When fluid flows without turbulence in a pipe with a constriction the lines of flow or streamlines will also be constricted. Since the same mass of fluid flows per second through each of the areas A, B and C it follows that the speed of flow through B must be greater than that through A and C. Thus wherever flow lines are constricted the speed of flow is correspondingly greater. This fact is used in one method of understanding the origin of lift.

the same as that of C, the speed of flow in these two positions will be the same since the same amount of fluid has to cross the same area in the same time. But area B is smaller, so if all the fluid is to pass this area in the same time the flow rate must be higher. We infer from this that in any region where the streamlines come together the speed of flow increases, and where they move apart the speed of flow decreases.

To some extent this result is counterintuitive and may worry some people. This is because one often conjures up in one's mind an incorrect analogy, e.g. a situation where a large number of people are walking along a broad hallway at the end of which there is a narrow doorway. One knows immediately what would happen: people would bank up around the doorway and the overall speed of flow would be reduced. This is because people behave differently from fluid particles. Individuals are less concerned with maintaining constant the overall speed of flow of the crowd than they are with their own safety and convenience in going through the door. So instead of going faster, as they should to maintain the overall flow speed, they actually slow down. Fluid particles are concerned only with keeping constant the total volume of fluid flowing past a fixed place per second and so must speed up when going through a constriction.

Returning now to fig. 3.6 we notice that at the left of the diagram the streamlines are equally spaced, indicating that there is no variation in flow rate from top to bottom along the left hand edge. For practical reasons we limit the extent of the flow to a region with the sail placed at the centre. All the streamlines which pass to leeward of the sail go through the area $A_L$ and all those passing to windward of the sail go through $A_W$. About eight streamlines flow through the small area $A_L$. If we follow them back to the region in front of the sail we see that they all pass through an area A which is much larger than $A_L$. In other words the flow around the leeward side of a sail is such as to constrict the streamlines, just as in fig. 3.7. On the other hand, on the windward side of the sail the streamlines open out from a small area to a larger one $A_W$. This means that flow over the leeward side is faster than in front and over the windward side slower than in front. Since the same streamlines go through both A and $A_L$ the ratio of A to $A_L$ is a measure of the compression of the streamlines and hence the ratio of the wind speed over the leeward side of the sail to that in the region of undisturbed flow upwind. Since $A/A_L$ is about 2.5 then the wind speed over the leeward side is about 2.5 times that in the free flow ahead of the boat. On the windward side the speed turns out to be about 2.5 times less than it is in the undisturbed region ahead of the sail. Thus because the amount of air directed to the windward side of the sail is less than that directed around the leeward side, the flow speed on the windward side need not be so fast as that to leeward in order that the air flows smoothly and does not compress like the crowd of people trying to get through a small doorway.

All that we have established so far is that given the pattern of flow that can be observed around a sail we can predict that the speed of flow over the leeward side is much greater than that over the windward side. To make the connection between speed of flow and side force or lift, we must introduce another law of nature. It is perhaps the most fundamental of all natural laws, called the Law of Conservation of Energy.

What we mean by conservation of energy is that, given an isolated system, the total amount of energy of whatever form contained therein is fixed. The energy can change from one form to another, or be partly in one form and partly in another. For instance chemical energy can be turned into heat or mechanical energy, or electrical energy or heat or light. But unless there is communication with the outside world the total amount of energy in a closed system remains fixed. In incompressible fluid flow, the case in sailing, we are concerned only with two forms of mechanical energy, kinetic and potential. Kinetic energy is the energy an object has by virtue of its mass and speed of motion. This is most vividly exhibited by the destruction of two automobiles after a head-on crash. Potential energy, on the other hand, has less to do with the object itself and more to do with its situation. A car parked on a steep hill has potential energy: merely releasing the brake causes it to move. A closed-up jack-in-a-box has potential energy, as does a can of diesel fuel. If the system is purely mechanical and there is no exchange of energy to other forms, then conservation of energy requires that the sum of the kinetic and potential energies remains constant. This is approximately the case for fluid flow and many other mechanical systems. It is only approximately true because in all systems friction is present giving a change of energy to heat. Often friction is a very small component, however, as for the lift produced by fluid flow; but not, of course, the drag.

Think now of a particle of fluid in a region of flow. If it finds itself in a region of high pressure and there exists nearby a region of low pressure, then the particle will be compelled to flow to the point of lower pressure. It thus possesses potential energy just like the car parked on a hill. As a result of moving from the high pressure to the low pressure area the particle of fluid has speeded up. Thus we can associate high pressure with large potential energy and low pressure with large kinetic energy.

We are now in a position to write the fluid form of the conservation of energy, as follows:

### Kinetic Energy + Pressure = a Constant

Kinetic energy is proportional to the square of the velocity (the formula $\frac{1}{2}mv^2$ for kinetic energy may be familiar). So our statement of the conservation of energy in a fluid can also be written: the square of the velocity plus pressure is constant.

This means that if you could ride along on a particle of dust in the fluid flow and measure your speed and the fluid pressure around you, the number you obtained by squaring your speed and adding it to the pressure would always be the same even though the speed and pressure values alone would generally be changing. Putting it another way, the pressure

*Fig. 3.8* Daniel Bernoulli (1700–82) was the youngest member of a family of four Swiss scientists and mathematicians. He was at first a professor of anatomy and botany and afterwards of experimental and speculative philosophy. His most important work is his *Hydrodynamica* (1738) which deals with many theoretical and applied topics in hydrodynamics. Apart from his 'law' which is used here in explaining lift, one of his more interesting ideas was the propelling of a vessel by the reaction of water ejected from the stern. It took more than 200 years, however, before the jet boat became a reality. This man, who lived so far from the sea, has had a major influence on our understanding of how sailing craft work. (*Universitäts – Bibliothek, Basel*)

if you are it is because you have seen a diagram in which *pressure coefficients* are plotted rather than actual pressures. There are excellent pragmatic reasons for doing this (see §8.3), but when one is trying to understand precisely how a sail produces lift it is much better to think in terms of real pressures.

Pressure is force divided by area. If the force is kept fixed and the area reduced, the pressure increases. That is why it is easy to push a pin through a sheet of paper whereas the same force applied to a ballpoint pen would not have the same result because the contact surface area is so much greater. Since force is a vector quantity, it has a direction as well as a magnitude. But we are talking about pressures in the midst of a fluid flow: how do we know in which direction these pressures act? The curious answer is that it depends how we measure them. Think of an aneroid barometer (fig. 3.10). A hermetically sealed box has a flexible end connected by a lever mechanism to a pointer. If the atmospheric pressure is greater than the pressure which was sealed in the box the pointer is pulled to one side; if it is less, to the

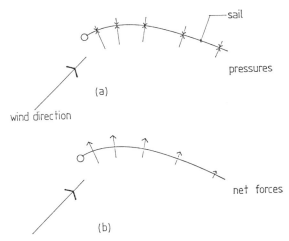

*Fig. 3.9* The distribution of pressure for the flow shown in fig. 3.6. (a) The arrows depict the real physical pressure which, in a fluid, is always toward and perpendicular to a surface. Although the pressure on the leeward side is less than that on the windward side the arrows are not drawn to scale. This is because the pressure difference is of the order of only one ten-thousandth of the atmospheric pressure available. (b) This diagram shows the difference between the opposing forces in (a). The length of these arrows represents pressures of only a few newtons per square metre. Note that they are always perpendicular to the sail and show the variation of the *net* pressure from luff to leech. The vector sum of all these forces gives the total sail force which has a component perpendicular to the wind direction. This is the **lift**.

difference between two points in the field of flow is proportional to the difference of the squares of the velocities between these two points.

This result usually goes under the name of Bernoulli's Law, formulated in 1738. If we know the pattern of the streamlines around a sail or keel we can, by means of this, determine the distribution of pressure. What we find for a sail is something like that shown in fig. 3.9. The length and direction of the arrows give the magnitude and direction of the pressure, respectively. As expected, the arrows are longer on the windward side indicating that the pressure there is greater than to leeward. But you may be surprised about the direction of these arrows;

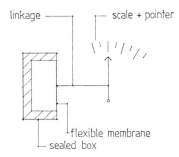

linkage ——|  |— scale + pointer

flexible membrane
sealed box

(a) membrane position on ISA day
at mean sea level

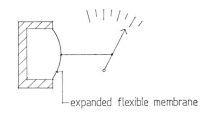

expanded flexible membrane

(b) situation for low ambient pressure

*Fig. 3.10* The raw principle of an aneroid barometer. The air pressure in the sealed box is fixed so as the external pressure changes the flexible part of the box moves in and out. If the barometer is turned over there is no change in the reading, showing that air pressure always acts perpendicular to a surface no matter how that surface is oriented.

other. Would the instrument give a different result if it were rotated through 90°? Not a bit! In one case the pressure is acting horizontally and in the other case vertically.

If you remember that the pressure on an object immersed in a fluid results from bombardment by millions of molecules in random motion, then it is not difficult to understand that the pressure in a fluid is always *perpendicular* to and *towards* the surface of an immersed object. Thus a small cube well below the surface of the water will feel almost equal pressures on each face, but the direction of the forces associated with this pressure will always be perpendicular to a face of the cube and directed toward it, as shown in fig. 3.11.

A sail is immersed in a fluid, the air, so the forces on it due to the pressure distribution in the flow are everywhere perpendicular to its surface and directed toward it. This is the physical fact and it explains why fig. 3.9a is drawn the way it is. The arrows represent the direction and magnitude of forces. Since the windward and leeward forces are opposed to each other the net force on the sail is the difference between them, shown in (b).

Although fig. 3.9a gives a correct qualitative picture of the real physical situation it has not been drawn to scale. The arrows on both sides of the sail would appear to be the same length, were it to scale, because the pressures on the leeward side are less than those on the windward side by only a few parts in ten thousand!

Ordinary atmospheric pressure at sea level is about 100,000 newtons/m$^2$, equivalent to 10,000 kg spread over an area of 1 m$^2$. If the leeward side of a sail produced a complete vacuum then a 10 m$^2$ sail area would experience a force equal to a weight of 100 tonnes, which of course does not happen because the force due to the pressure on the leeward side of the sail is less by typically only a few hundredths of a per cent. The resultant force that the sail feels amounts in this case to a mere 10 kg.

Because of these very small differences engineers usually use a quantity called the pressure coefficient (see §8.3). Since this involves the difference between the pressure at a point in the flow and the atmospheric pressure remote from the sail, it gives a

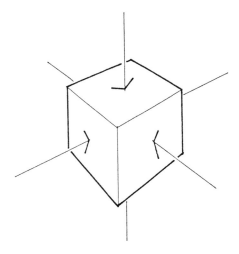

*Fig. 3.11* A small box immersed deep in water has approximately equal forces acting on each face. Although these forces represent the pressure in the water their direction is determined entirely by the orientation of the immersed object. By the same reasoning, the prime forces on a sail are perpendicular to the surface and directed toward the surface on both sides.

negative number for the pressure coefficient on the leeward side of the sail and a positive one for the windward side. This has given rise to the erroneous expression 'negative pressure': it is quite meaningless as only pressure coefficients can be negative. Sometimes people speak of 'suction' on the leeward side of a sail, but this is also an awkward concept.

To summarise the flow line method of understanding lift:

(a) We assume the character of the flow as shown in fig. 3.6. This should be based on measurement.

(b) The velocity distribution can be obtained directly from the flow pattern since the speed increases where the flow lines constrict and vice versa.

(c) Application of Bernoulli's theorem then gives the pressure distribution from the velocity distribution.

(d) The resultant pressure distribution shows that there will be a net force perpendicular to the wind direction: this is the lift.

## §3.3.2   The 'Momentum Change' Approach to Understanding Lift

This approach is probably the one that will appeal most to you as a sailor. The concept is easy to understand and although it leaves a few questions unanswered it is very helpful in sail trimming and improving boat speed.

We use here the concept of the sail as an *air deflector*. It is a rather strange kind of deflector because the air is deflected not just from the windward side but from the leeward side as well. This strange behaviour can really only be properly understood by studying the mathematical approach in §3.3.3, but for the moment it is merely necessary to accept the fact that a sail can be thought of as a device for deflecting the wind. Some confirmation of this notion can be obtained by looking at the measured flow shown in fig. 3.6. The main direction of flow just ahead of the luff of the sail is changed through about 50° by the time the air flows off the leech.

The objective in sailing is to extract as much energy as possible from the air and water flow. If energy is to be taken from the wind the wind speed will be reduced. To see just how this comes about look first at the air deflected from the mainsail of a stationary boat held by mooring lines to a dock (fig. 3.12a). A true wind of 12 knots is deflected by the sail. If we ignore frictional drag between wind and sail its speed after deflection will still be 12 knots.

Just as an aside, ignoring frictional drag requires some comment. The whole of chapter 2 was concerned with the important question of drag and yet here we choose to ignore it! The art of knowing what can be safely ignored in scientific analysis is often difficult, but vitally important: if every analysis were completely rigorous the wood would be obscured by the forest. It happens that in discussions of lift or side force the effects of drag are not important.

In fig. 3.12b the boat is moving forward at 5 knots. In the absence of any true wind the apparent wind felt on board would be 5 knots from dead ahead. If one now adds to this a true wind of say 12 knots at 62° to the yacht's heading, we find that the apparent wind is 15 knots at 45°. (This important vector addition procedure to determine apparent wind will be further explained in §3.4.)

After the wind leaves the sail it will still have an apparent speed of 15 knots. This is because apparent wind is measured with respect to the boat so the situation is as in (a). However an observer in an anchored boat would measure a true wind speed of 12 knots just ahead of the sail and a speed of 10.3 knots downwind of the leech. This is because after the sail has turned the apparent 15 knot wind we must subtract (vectorially) the boat's speed to get the true wind speed. The diagram shows that this speed is 10.3 knots if one assumes the true wind is turned through 40° by the action of the sail.

This reduction of true wind speed by 1.7 knots represents a loss of energy by the wind which has been transferred to the yacht. Since the wind that has left the mainsail is changed in direction, boats behind are not only headed but find themselves in a region of lower wind speed.

We must now examine the physics of what is happening in a little more detail in order that this 'momentum change' concept of sail force can become useful to us as sailors.

'Momentum' is the word physicists use to mean the quantity of motion. The magnitude of the momentum is simply the product of mass and velocity. A massive object going slowly could have the same momentum as a light object going fast. Although mass has no direction associated with it, velocity certainly does; it is a vector quantity so momentum is then a vector. One of the most fundamental laws of classical mechanics concerns the relationship between force and change of momentum, first described by Sir Isaac Newton. The proper understanding of mechanics that Newton

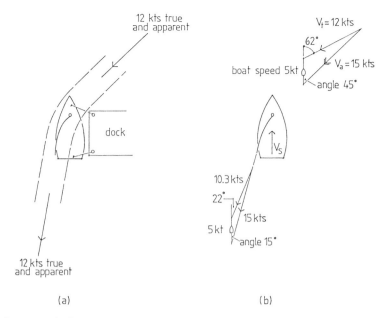

(a)                                             (b)

*Fig. 3.12* (a) The air flow around a fixed sail is equivalent to a deflection of the air stream. The true flow is shown in fig. 3.6. What is shown here is an 'equivalent flow' or deflection which gives rise to the same total sail force. If the boat is moored the speed of the air flow after it leaves the sail is unchanged; only its direction is changed. This amounts to a change in momentum of the air which gives rise to a force on the sail. (b) If the boat is allowed to move forward energy is extracted from the air flow. This is seen by the fact that the true wind speed has dropped from 12 knots to 10.3 knots.

accomplished represents perhaps the greatest single leap forward in human understanding of Nature.

Known nowadays as Newton's Second Law, it states that: **the rate of change of momentum of an object is equal to the net external force applied to it.** Rate of change refers to a change with time. Since the mass of an object is usually fixed, rate of change of momentum refers to a change in its velocity or direction or both. In other words we are speaking of acceleration or deceleration. Since momentum is a vector quantity it is just as well to remember that a change in momentum can just as easily be produced by a change in direction as by a change in velocity.

This is the situation in fig. 3.12a. Only the direction of the air flow is changing. The corresponding change in momentum over the time it takes for air to flow around the sail allows us to calculate a rate of change of momentum and hence a force on the sail.

So far so good, but force is a vector and requires to be specified not only by magnitude but also by direction. How do we determine the direction? Take the simplest situation first, namely a dead run (fig. 3.13a). Air hits the sail and from the point of view of an observer on the boat, stops completely. The air's momentum has been reduced to zero producing a force on the sail which can only be in the same direction as the wind.

Assume now the situation of fig. 3.13b. The apparent wind is turned but its speed remains unchanged. The symmetry of this situation suggests that the resultant force on the sail makes equal angles with the direction of the incident wind and with the wind leaving the sail, as shown in (c): this of course is the total sail force. The lift is the component perpendicular to the incident wind direction.

This view of the origin of the total force on a sail can be quite useful even under the stress of sailing, as opposed to sitting back comfortably beside the fire thinking theoretical thoughts. One usually has a good idea of the incident wind direction from a masthead telltale or one on the shrouds. Leech telltales show the direction of the wind leaving the sail, so the direction of the total sail force can be easily estimated using fig. 3.13c.

Consider the problem of sail trim when sailing on a reach: we would like to maximise driving force and keep heeling force as small as possible. Fig. 3.14 shows in a somewhat exaggerated form the effect of

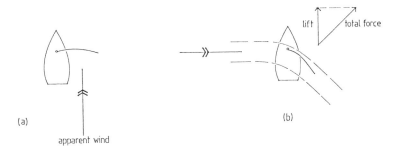

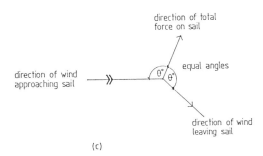

*Fig. 3.13* The 'momentum change' approach to understanding sail force connects the change in momentum of a parcel of air deflected by the sail with the force necessary for that deflection according to Newton's Second Law of Motion. What interests us in sailing is the *direction* of the total force. The simplest situation is (a) which shows a dead run. The apparent wind speed is reduced to zero on hitting the sail. The only direction involved is that of the incident wind so this must also be the direction of the force.

In (b) the beam wind is deflected through about 40°. To a first approximation the speed of the apparent wind after deflection is the same as that before so the direction of the total force is symmetrically placed with respect to the incoming and outgoing wind directions. This is shown in (c) where the direction of the total sail force makes equal angles with the direction of the apparent wind approaching the sail and that leaving.

increasing the angle of incidence of wind on sail. At zero angle of incidence the sail luffs and no lift or driving force is produced (a). (Notice that in the special case of a beam reach the lift and the driving force are one and the same.) At the correct angle of incidence (b) the driving force $F_R$ can be made somewhat greater than the heeling force $F_H$. If the sail is sheeted in still more the wind is turned more, giving an increase in the momentum change and a correspondingly greater value for the total force $F_T$. But this force is now turned more to leeward so that its driving force component is actually less than before, whereas the heeling force has become very large. Sailing like this is a common error of beginners and is reinforced by a feeling that the boat is being driven harder because it is heeling more. Sheeting the sail in still harder will result in its stalling and reducing all forces drastically.

This 'momentum change' theory of driving force

requires that we accept the idea that air has mass, which sometimes sounds strange. We can imagine that our sail is being bombarded by air and that the recoil is pushing us forward. To make this a little more quantitative, imagine that you have a 10 m² sail in a wind of 10 knots. Since 10 knots is equal to 5 m/sec, we can picture our sail as being hit each second by a wedge-shaped lump of air of cross-sectional area the same as the sail and length 5 m. The mass of this lump of air is obtained by multiplying its volume by the density of air and works out to be 60 kg: each second, 60 kg of air is being deflected by the sail. If its speed is 10 knots and the angle of deflection 40° then the total sail force works out to be about 205 newtons, a force equal to a weight of about 20.5 kg or 45 lb (fig. 3.15). §8.4 gives the formulae used for doing this kind of calculation.

Although this momentum change theory may be appealing, and is as far as it goes scientifically correct,

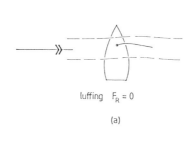

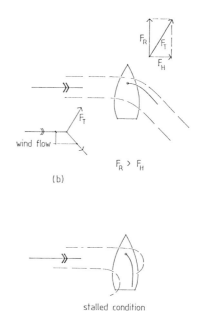

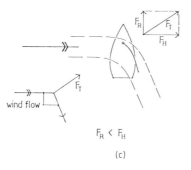

Fig. 3.14 The usefulness of the momentum change approach to understanding lift is demonstrated here for a reach. In (a) the sail is luffing and of course there is no driving force $F_R$. In (b) the incident wind direction (double arrow) and the leaving direction (thin arrow) are such as to define the direction of the total sail force $F_T$. This has a driving force component $F_R$ which is greater than the heeling force $F_H$. In (c) the sail is sheeted in too hard and although $F_T$ is somewhat greater $F_R$ is less than in (b), and worse still, the heeling force is much greater. Further sheeting in of the sail as in (b) produces a stall and greatly reduces all forces on the sail.

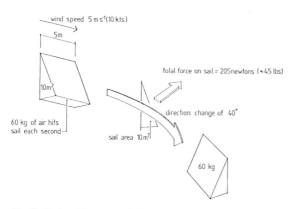

Fig. 3.15 A cubic metre of air weighs 1.2 kg. In a 10 knot breeze the air is moving at 5 m/sec. This means that for a 10 m² sail a volume of air 5 m long and of cross-section 10 m² is deflected by the sail every second. The mass of this wedge of air is 60 kg. If it is deflected through an angle of 40° the resulting force on the sail is 205 newtons or about 45 lb.

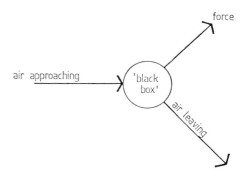

Fig. 3.16 The momentum change approach to understanding lift is a macroscopic theory. That is, it can tell nothing about how a sail deflects the wind but simply that if one knows the approaching and leaving wind directions and speeds then one can determine the resultant magnitude and direction of the total sail force. To understand how the 'black box' works one must read the mathematical approach to understanding lift in §3.3.3.

it has nevertheless many shortcomings. It is what I would call a 'macroscopic' theory; it does not explain details of the air flow around a sail but considers only the status of the air approaching the sail and the air leaving. Somewhere in between is a 'black box' which mysteriously changes the wind direction and produces a force as depicted in fig. 3.16. Although fig. 3.14 correctly portrays what happens in those situations, the diagrams are not predictions of the momentum change theory. Obviously the stall situation in fig. 3.14d has got to do with the mechanism of the 'black box'. Also, as shown in §8.4 the theory predicts that the driving force is actually maximised on a reach by sheeting in hard. This is because we are not taking into account what is happening at a more detailed 'microscopic' level, which is the price we pay for simplifying what is really a very complex problem.

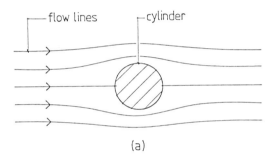

(a)

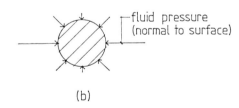

(b)

*Fig. 3.17* (a) Potential or ideal fluid flow around a cylinder. The path of these streamlines may be calculated from mathematical theory developed in the 18th century by d'Alembert. The symmetry of the flow lines predicts a symmetrical pressure distribution as in (b). Fluid pressures are always perpendicular to and toward a solid surface. In this case for every point on the surface of the cylinder there is a diametrically opposite point with an equal and opposite pressure. There is therefore no net force on the cylinder and hence no drag. This was known as d'Alembert's Paradox. The prediction of the flow as modified by the inclusion of viscous forces is a problem not yet fully solved mathematically.

The momentum change theory may be summarised as follows:

(a) the sail can be looked upon as a device for deflecting a mass of air.
(b) the recoil resulting from this deflection is the total sail force, and has a driving force component and a heeling force component.
(c) the direction of this total force makes equal angles with the direction of the incident wind and that of the wind leaving the leech.
(d) the theory is a macroscopic one which cannot explain the details of what is happening to the air flow close to the sail.

### §3.3.3 The Mathematical Approach to Understanding Lift

Although the heading to this section seems to imply that it will be full of mathematics, such is not the case. What follows is a kind of 'descriptive mathematics' which hopefully will make clear the scientific principles of lift without obscuring the essential arguments with mathematical detail.

In the 18th century mathematics had developed to the point where fluid flow could be understood for situations where viscosity was unimportant. Such 'ideal' fluid flow is called potential flow because the equations involved are the same as those describing the potential in an electric field (see also §8.2). This theory gives rise to the kind of flow shown in fig. 3.17, of a cross-section near the centre of a very long cylinder.

Potential theory shows that the flow is completely symmetrical so the distribution of pressure around the cylinder must also be symmetrical. Fig. 3.17b shows these pressure forces perpendicular to the surface and pointing toward it as they should. At top and bottom where the streamlines are crowded together the flow is faster and so the pressure is lower. Bernoulli's rule also tells us that the pressure to the left and right is somewhat greater. As is pointed out in §3.3.1 and §8.3, the length of the arrows which are normally taken to be proportional to the magnitude of the pressure is here grossly exaggerated. Nevertheless it is clear that this potential flow theory predicts that the forces produced by the flow are always equal on opposite sides of the cylinder so there is no net force on the cylinder at all. This means there is no drag and no lift.

This result was first obtained by the French mathematician Jean le Rond d'Alembert (1717–83)

51

who made many important contributions to the knowledge of calculus, mechanics and fluid flow. Because common experience tells us that drag is always associated with flow around an obstacle, this mathematical result became known as d'Alembert's Paradox.

In this chapter we are concerned with lift rather than drag. Except for vortex drag, lift and drag can to a good approximation be regarded as separate phenomena so that a theoretical approach which predicts no drag may in fact be perfectly satisfactory in explaining lift. Such is indeed the case, as the following shows.

As most people have observed, the flight of a ball can be somewhat modified by the application of spin. If a ball is caused to spin without moving laterally it will drag around with it a certain amount of air. When the ball is made to move laterally and is also spinning it will therefore have a flow pattern around it consisting of rotating air added to the characteristic potential flow shown in fig. 3.17a. Since the translational speed is normally fairly large and the effect on the air of rotation falls off rapidly with distance from the ball, no significant amount of air actually circulates through 360° but the flow is nevertheless considerably distorted by the ball's rotation.

The distorted streamlines can be predicted mathematically by simply adding in the effect of rotation at the appropriate point in the calculation, shown schematically in fig. 3.18. The ideal flow

around a ball moving to the left is shown in (a) while the circulation associated with a ball which is rotating but not translating is shown in (b). The sum of both effects appears in (c).

There are several things to notice about this result: first, the symmetry of the flow in (a) shows that there is no net force on the ball; second, the stagnation streamline which is marked 'S' separates the flow over the upper half from that over the lower half; third, when the circulation of air due to rotation is added in, the stagnation points move downward and toward each other. Since they still divide the flow between top and bottom it is clear that now more air flows above the ball than below for this particular case. More air means more speed and hence lower pressure. Although the diagram still exhibits right/ left symmetry so that there is no drag, there is now a lower pressure on the ball at the top than at the bottom giving a net upward force. Its trajectory will then be from right to left and curved upward. What we are talking about here is a net force at right angles to the undisturbed flow direction, which by our earlier definition is lift. So in the simple case of a spinning ball lift is easily understood and completely calculable. Credit is usually given to Professor Magnus of Berlin for first studying this effect in 1853, consequently it is often known as the Magnus effect.

Before showing how these basic principles can be used to explain the lift of a sail, I would like to diverge a little and describe a ship which was actually built and used tall rotating cylinders as sails. It was the

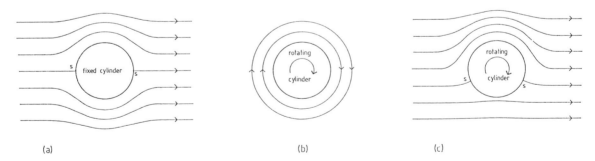

(a)                                        (b)                                        (c)

*Fig. 3.18* (a) shows the ideal potential flow around a fixed cylinder. The stagnation streamline labelled s separates the flow into two regions, one going above the cylinder and the other below. If we add to this flow pattern a circulatory flow as in (b) the result will be the flow shown in (c). Notice that the stagnation points have moved down on the cylinder so that there is a much greater amount of fluid now going above than below. Bernoulli's rule then tells us that the pressure toward the under side of the cylinder will be greater than

that toward the upper side so there will be a net upward force. This is at right angles to the free stream direction and is therefore a lift force. There is still right-left symmetry in the diagram so this potential theory predicts no drag. A spinning ball exhibits this behaviour, but for it to produce circulation as in (b) there must be viscous drag between the ball and the air. In this case lift cannot exist without drag, and yet we can accurately calculate lift with a theory which ignores viscosity!

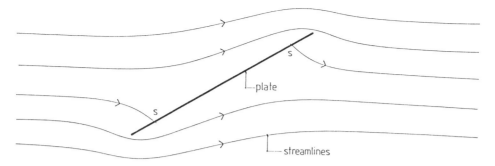

*Fig. 3.19* The potential flow around a flat plate in the absence of circulation. Such a flow which produces no lift or drag on the plate cannot occur in a real fluid medium.

brainchild of a German engineer, Anton Flettner, in the early 1920s. The auxiliary three-masted topsail schooner *Buckau* had its normal rig removed and replaced by two vertical rotating cylinders 18.5 m high and 2.8 m in diameter. They were driven at 750 revolutions per minute by 11 kW electric motors. A beam wind could thus produce a strong forward force because of the Magnus effect. Flettner's measurements showed that his rotors produced more than ten times as much lift as a sail of the same projected area. His ship could sail as close to the wind as with a normal rig but had the advantage that shortening sail was unnecessary, except to reduce windage in extreme conditions, since for a fixed rotor speed the efficiency dropped if the wind speed increased. Further details and some of Flettner's experimental results are given in §8.6 for those who are attracted by alternative methods of propulsion. Although the idea was sound it did not catch on in an age when fossil fuels were cheap and appeared limitless.

We have just seen how lift can be produced by superimposing circulation on the flow around a ball or cylinder. With that in mind let us now look at what potential flow theory predicts for the flow around a flat plate set at a few degrees angle of incidence to the oncoming stream, shown in fig. 3.19. As with all such diagrams, the stagnation streamlines marked s separate the flow into two regions going each side of the obstacle.

Now think carefully about where a streamline must go which is below but very very close to the stagnation line s. It will follow the plate close to its under surface and then must somehow get round to a point just below where the stagnation line leaves the

plate. If the fluid had no mass and no viscosity this would be possible, but real fluids aren't like that. For the fluid to flow around the sharp trailing edge and down to s very large forces would be necessary to change its direction of flow so sharply. Even if there were no viscosity this would be the case because of the mass of the fluid; with viscous drag present also the forces required are even greater. As a real fluid cannot sustain such a flow it breaks away from the surface near the sharp trailing edge.

Fig. 3.20 shows the sequence of events for the flow around a keel section when a boat starts to move forward at a constant leeway angle. (The fact that a boat accelerating from rest under sail results in a changing leeway angle should not be allowed to distract you from the train of argument here.) The movement begins at time t = 0 in (a): initially the streamlines are exactly as potential theory would have them. A short time later (b), when the water has moved about a tenth of a chord length, it is already having trouble rounding the trailing edge and is separating from the surface forming a vortex. After the keel has moved forward a distance about equal to its own chord, the vortex formed has been left astern and the flow over the keel is now smooth and attached as shown in (c). The vortex which is left behind is known as the starting vortex.

The crucial point here is that because of the inability of the water to flow around the trailing edge a vortex has formed. It is not known for sure whether this would also occur in a non-viscous fluid, but since such fluids don't exist the point is academic. Now before proceeding with the final step of the argument explaining lift, something has to be said about the strange behaviour of vortices.

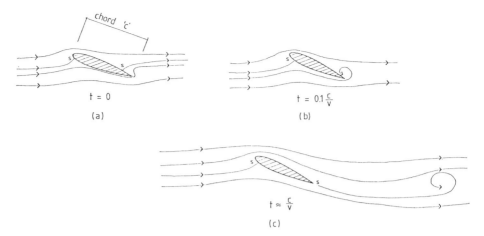

*Fig. 3.20* How the streamlines change with time immediately after flow commences; the example is a symmetrical keel section at a fixed angle of leeway. (a) shows the flow at time t = 0 just as the boat commences to make way steadily. The flow resembles the ideal momentarily, but because of the water's mass and viscosity it cannot flow close around the sharp trailing edge and up to the upper stagnation point. After the water has moved about a tenth of a keel chord length (b), flow near the trailing edge has separated and a vortex has been formed. This starting vortex remains fixed with respect to the water and is soon left behind as the keel moves forward with the stagnation point now shifted down to the trailing edge. According to a theorem of Helmholtz, a vortex created in this way must have associated with it another vortex of equal and opposite strength; this is the bound vortex or circulation around the keel section. As with the spinning ball in fig. 3.18, movement of the stagnation points is associated with circulation and the production of lift. The keel section, because of its sharp trailing edge, has moved the stagnation point and produced circulation and hence lift without having to physically rotate like the ball.

Our understanding of vortex behaviour comes mainly from the work of Hermann Ludwig Ferdinand von Helmholtz (1821–94). A truly great man of science, he was both a philosopher and a scientist and held posts as a professor of physics and of physiology. Of his four laws of vortex behaviour, the one that concerns us here states that: **circulation or vortex motion in a fluid can neither be created nor destroyed**. Such a counter-intuitive idea requires a little explanation. First we have to quantify a vortex, by defining a 'vortex strength'. Although this has a precise mathematical definition, it is simply a number which depends upon the speed and direction of rotation, and is negative if the rotation is clockwise and positive if counter-clockwise. When sailing in clear air and smooth water the apparent flow approaching the boat will normally contain no vortex strength. That is there are no eddies, just translational flow. If we now start sailing in this idyllic situation we know that the first thing that happens is that a starting vortex is created in the water. According to the theorem just stated the total amount of vorticity cannot change: since there was none to start with the total vortex strength must still be zero. The only way this is possible is that there must exist an equal and opposite vorticity or circulation (the two terms are synonymous) about the keel.

This does not show up as a bodily movement of water around the keel because it is superposed on the normal flow. This circulation when added to the potential flow of fig. 3.20a is precisely that needed to modify the flow near the trailing edge to produce the situation in (c) where the stagnation point has moved to the sharp trailing edge and flow over this point of the keel is free of discontinuities.

Now at last the true origin of the lift should be appearing out of the mist. What the keel does to the flow lines is exactly what the spinning ball does, except that the keel manages it without the need for rotation by virtue of its special shape. The key to the generation of lift is the presence of circulation in the flow and if the circulation is known lift can be calculated. In aeronautical parlance the circulation around the wing is known as the 'bound vortex' as opposed to the starting vortex which is left behind on the airfield. Everything I have said so far in this section applies of course equally well to sails also.

A point which has recurred several times in this section is the validity of ignoring drag when lift is being calculated. This practice goes back to 1904 when Ludwig Prandtl at the University of Göttingen first conceived the idea of the boundary layer. As we saw in chapter 2, the boundary layer is quite thin by comparison with the chords and cambers of sails and keels. It is within this that most of the effects of drag occur, largely a result of shearing forces within the fluid which arise because of speed differences between adjacent layers. The mathematical theory of viscous fluid flow is very much more complicated than that of the relatively simple ideal potential flow. Fortunately Nature has been kind to us and confined the difficult bits to a very thin layer. Outside the boundary layer the flow of a real fluid is indistinguishable from that of an ideal one and the equations of potential flow can be safely used.

The boundary layer concept thus greatly simplified the theoretical description of fluid flow and in the early part of the 20th century provided a strong impetus to the rapid development of aerodynamic theory. It should now be clear why the electric analogue method of determining two-dimensional fluid flow is valid for problems like understanding sail interaction in chapter 4.

This section began by discussing the lift of a spinning ball. Because of the simple geometrical shape it is not difficult to believe that given the appropriate mathematical techniques it is easy to predict the shape of the streamlines shown in fig. 3.18 for instance. You may have wondered, however, how it is possible to calculate the flow around a keel section with a much more complicated shape than a circle. This can be done by means of a very clever technique known as a Zhukovsky transformation, which uses a procedure known as 'mapping'. Imagine we have a region z in which every point is specified by its coordinates x and y (fig. 3.22). Imagine further that there is a corresponding region z' which has the properties that a short line segment in z becomes a longer and somewhat rotated line segment in z'. The amount of lengthening and rotation depends on the coordinates in z. In other words the way in which a line segment in z is mapped onto z' depends on where in z the original line segment was located. The way in which the transformation varies with position is specified by a mathematical expression called a transformation function. In this way a shape in z can be transformed into an infinite variety of shapes in z' by different choices of the transformation function.

An appropriate transformation of fig. 3.18c is shown in fig. 3.23. The circle has been transformed into a symmetric section like that of a keel and at the

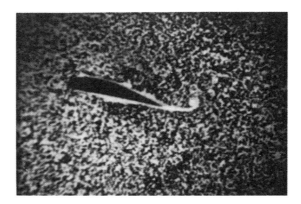

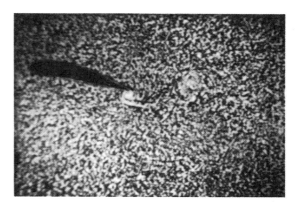

Fig. 3.21 Photographs from a movie made by Ludwig Prandtl in the early 20th century. The airfoil is moving in water containing fine aluminium powder which renders the flow visible. In the upper picture the airfoil has started from rest and is still moving. The starting vortex which remains fixed in the fluid is clearly visible. The bound vortex is carried along with the airfoil and when it stops, in the lower picture, it is shed disclosing an equal but opposite vorticity to that of the starting vortex. This is a beautiful demonstration of the reality of circulation in the flow around a foil, which is basic to our understanding of lift.

same time the streamlines are also transformed. Since we know how to calculate the flow around a cylinder we can, by the process of conformal transformation, determine the flow around a shape of more interest to us.

Thus to summarise the mathematical theory of lift:

(a) A fundamental feature of the flow around a lifting section is that the downstream stagnation line comes smoothly off the sharp trailing edge.

(b) The price for the production of this type of flow, known as the Kutta condition, is the production of a starting vortex in the fluid.

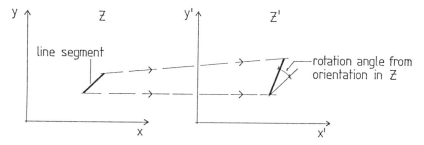

Fig. 3.22 A method of calculating fluid flows involves a mathematical procedure known as conformal mapping. A line segment in z is transformed in length and rotated. The amount of lengthening and rotation depends on position so that a shape in z can be transformed into a different shape in z'.

(c) A fundamental theorem of fluid flow tells us that the existence of one vortex must be associated with the presence somewhere in the flow of an equal and opposite vorticity. This is the bound vortex or circulation around the lifting section.

(d) Just as a spinning ball produces circulation and has lift, so a lifting section has lift by virtue of its circulation although it does not have to physically rotate to produce it.

## §3.4 The Geometry of Sailing to Windward

An aspect of sailing which always surprises the newcomer to the sport is the apparent sudden change in the weather on changing from a downwind to an upwind course. Running downwind it is warm and sunny, the boat is dry and on an even keel and the wind is light. The spinnaker is dropped, the leeward mark is rounded and suddenly the wind strength has quadrupled, spray is flying, the boat is heeling heavily and pounding into the oncoming seas. This transformation has been brought about mainly by a radical change in the apparent wind.

It is obvious that if one sails dead downwind at 8 knots in a 15 knot wind, the apparent wind over the deck will be only 7 knots. It is not, however, so obvious that after turning and now moving through the water at 6 knots with the wind direction indicator showing 33° the wind over the deck will now be 19.7 knots. Perhaps you will already know this, for instruments showing apparent wind direction also measure apparent wind speed. But what we would really like to know is: what angle are we making to the true wind direction, and most important of all, what speed are we making good to windward. How these are determined is the subject of the present section.

Perhaps in this discussion, more than in any other, it is true to say that a picture is worth a thousand words. The picture is fig. 3.24: let me first of all define the symbols used. The arrow marked $V_t$ specifies the

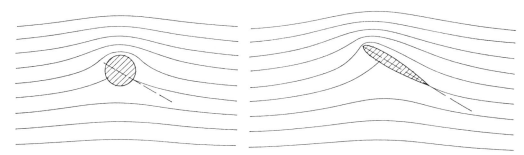

Fig. 3.23 This particular conformal mapping is known as a Zhukovsky transformation. In it the easily calculable flow around a cylinder is transformed into that of a symmetrical section like that of a keel. Such procedures are useful mainly to provide insight into how lift is produced. In practice the details of fluid flow are best determined experimentally or by numerical methods on a large computer.

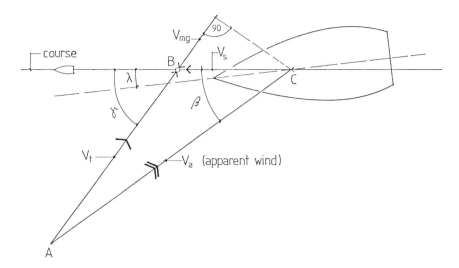

*Fig. 3.24* Variously known as the course triangle or velocity triangle, this figure gives the geometrical relationships between the apparent wind $V_a$, the true wind $V_t$, the apparent wind angle $\beta$ (beta), the true wind angle $\gamma$ (gamma), the boat speed through the water $V_s$, the leeway angle $\lambda$ (lambda) and the speed made good to windward $V_{mg}$. The true wind, added vectorially to the wind produced by the boat's motion, creates the apparent wind as 'felt' by the sails. When going to windward the apparent wind is always stronger than the true wind, as shown by the length of the arrows. The speed made good to windward $V_{mg}$ is the component of the boat speed in the direction opposite to that of the true wind. On-board instruments normally measure $V_a$, $V_s$ and $\beta - \lambda$, so if $\lambda$ is known a scale diagram such as this can be used to determine $V_{mg}$, $V_t$ and the true wind angle $\gamma$. Alternatively, more sophisticated instrumentation or a portable computer may be used.

magnitude and direction of the true wind, that measured by an observer fixed with respect to the sea. $V_s$ is the speed of the boat with respect to the water and is therefore measured directly from the knotmeter providing the leeway is not extremely large. $V_a$ is the apparent wind speed and can be measured directly on board. Remember, however, that the measured values of the magnitude and direction of $V_a$ are in a plane perpendicular to the mast. If you are sufficiently interested to want to know how to get the values in a horizontal plane, you will find the necessary formulae in §8.6.

The three quantities $V_s$, $V_t$ and $V_a$ form what is usually called a vector triangle. If there were no true wind, a boat moving under power at speed $V_s$ would experience an apparent wind in the opposite direction to $V_s$. If the true wind is not zero, then the apparent wind is the vector sum of $V_t$ plus a vector equal but opposite in direction to $V_s$. Such a vector addition is carried out geometrically in fig. 3.24: start at point A and go to B, to represent the true wind. To this we add the apparent wind speed caused by the boat speed in still air which is the negative of $V_s$, so we continue from B back toward C. Going from A to B

to C is the same thing as going directly from A to C, so the apparent wind is the combined result of the true wind and the boat's motion through the water. Obviously if the vectors $V_a$ and $V_s$ are known the true wind speed and direction can be calculated.

The other quantities shown are the angle $\beta$ (beta), which is the angle between the direction of the apparent wind and the course sailed, the angle $\gamma$ (gamma) which is the angle between the true wind and the course sailed, and finally the important quantity $V_{mg}$ which is the speed made good to windward. This latter quantity is the component of the boat speed which is directly opposite to the true wind. Given any three of the six quantities $V_a$, $V_t$, $V_s$, $V_{mg}$, $\beta$ and $\gamma$ it is possible to determine the other three from the formulae given in §8.6, from a scale vector diagram such as fig. 3.24 or by computer calculation as explained later.

Six quantities have just been spoken of, but you may have noticed that fig. 3.24 involves seven. This seventh quantity is the leeway angle $\lambda$ (lambda) and is a real spanner in the works. The problem arises because when one measures apparent wind direction, what is actually measured is the direction with respect to the boat's heading and not with respect to the course sailed (made good). In other words we do not measure $\beta$ but $\beta - \lambda$. So unless we know how much leeway we are making we cannot calculate the all-important $V_{mg}$ correctly. (A discussion of how leeway may be determined will be left to §3.5.3.)

On all points of sailing except a dead run leeway *must* be present. If the total sail force has a non-zero component perpendicular to the course sailed, or equivalently, if there is any heeling force whatsoever, there must be an equal and opposite lift force produced by the underwater body. If there isn't the boat just moves off uncontrollably in the direction of the wind like a hot air balloon. The only way a symmetrical body can produce a lift force is by setting itself at an angle of incidence to the oncoming flow when the lift generated is directly proportional to the angle of incidence, which in this case is the leeway angle. This requirement of the symmetry of sailing which tells us of the equality of wind and water forces

thus also demands that leeway exists on all points of sailing except a dead run.

If we ignore for the moment the problem of leeway, our interest in fig. 3.24 is in using the readings of common on-board instruments to determine our speed made good to windward and also the angle between the course sailed and the true wind direction. From instruments giving us $V_a$, $V_s$ and $\beta$ only (ignoring leeway), by far the best way to determine $V_{mg}$ is to invest in a set of instruments which measure $V_a$, $V_s$ and $\beta\text{-}\lambda$, and which are equipped with a microprocessor to calculate and display $V_{mg}$. Some of these instruments will calculate and display the true wind speed and direction as well.

If you have instruments that date from an earlier era and which perform no calculations on the measured data it is still possible to determine $V_{mg}$ though not on a continuous basis. Various graphical methods have been suggested, but by far the best way is by means of a programmable calculator or portable

*Fig. 3.25* One of the many types of portable microcomputer suitable as an aid to sailing efficiency. This one is based on two CMOS 8 bit microprocessors (type 6301) and has 32K each of RAM and ROM memory. For most purposes one as sophisticated as this is not necessary, however. In normal shipboard use the batteries will last two or three weeks. Such computers have the advantage that the memory is retained even when switched off and they are programmed in BASIC, a language which is used in all personal

computers and is easy for the non-specialist to become proficient in. There is a great deal of satisfaction to be had from navigating by hand bearing compass and sextant rather than Satnav, especially when the calculational drudgery and likelihood of error are removed by use of a computer. '$V_{mg}$' on the liquid crystal display refers to a program described in the text for calculating speed made good to windward from on-board instrument readings.

microcomputer. If you are computer illiterate then you owe it to yourself to rectify the situation. The remarkable thing about computers is that if you don't know how to use one it looks incredibly complicated and mystifying, but when you do know how you realise that it is absolutely trivial.

A computer can be fun on a yacht. Apart from $V_{mg}$ calculations it can be very useful for coastal and offshore navigation. Most small battery operated computers are programmed in the computer language called Basic. This language is universally understood so a program for calculating speed made good to windward $V_{mg}$, true wind speed $V_t$ and true angle between course sailed and true wind direction $\gamma$, from an input consisting of apparent wind direction $\beta$, apparent wind speed $V_a$ and the boat's speed through the water $V_s$, is given below. Note that although leeway is not included here, this results in only a small error in quantities which are themselves subject to considerable variation and error of measurement. Despite the omission of leeway $V_{mg}$ calculations will still give correct relative values which are the quantities of interest to the yachtsman.

```
10  REM Vmg CALC
20  PRINT "TYPE BETA, Va, Vs"
30  INPUT "DATA";B,V1, V2
40  LET B1#=B★3.141592654/180
50  LET T1=V1★SIN(B1#)
60  LET T2=V1★COS(B1#)−V2
70  IF T2=0 THEN 80 ELSE 95
80  G = 3.14159/2
90  GO TO 140
95  LET G=ATN(T1/T2)
100 IF T2<0 THEN 110 ELSE 140
110 LET G=G+3.141592654
120 GO TO 140
140 V3=V2★COS(G)
160 G1=G★180/3.14159
162 IF SIN(G)=0 THEN 164 ELSE 168
164 V4 = V2 + V1
166 GO TO 170
168 V4=V1★SIN(B1#)/SIN(G)
170 V$="Vt = ##.#knots"
175 PRINT USING V$;V4
180 S$="Vmg = ###.# knots"
185 PRINT USING S$;V3
190 G$="GAMMA = ### deg."
195 PRINT USING G$;G1
200 GO TO 30
210 END
```

Type this into the computer exactly as it is written here with a carriage return at the end of each line. If you have a British keyboard without the symbol # just use the pound sign instead; the program will treat it in the same way. When the program is made to run, it will ask you to 'TYPE BETA, $V_a$, $V_s$'. BETA is the apparent wind angle in degrees, $V_a$ the apparent wind speed in knots and $V_s$ the boat speed in knots. The data must be typed in the correct order with commas separating each. For example: suppose the instruments give us a wind angle of 32°, wind speed 18 knots and boat speed 6 knots, then we type in 32, 18, 6 and the computer will spew out:

$$V_t = 13.3 \text{ knots}$$
$$V_{mg} = 4.2 \text{ knots}$$
$$\text{GAMMA} = 46 \text{ deg.}$$

This program is valid on all points of sailing. However if gamma exceeds 90° $V_{mg}$ will be printed out with a negative sign in front, indicating that speed is being made good *downwind* rather than upwind. Of course there are instrument packages which will compute these quantities directly from measurements, but nevertheless a separate computer is a handy thing to have on a boat, especially for navigation, and there is still the vexing question of how to handle leeway (see §3.5.3).

It is interesting to speculate on how much the speed made good is increased or decreased by small changes in $\beta$, $V_t$ or $V_s$. However, it must be borne in mind that virtually all the variable quantities that govern the motion of a vessel are interdependent so that the 'scientific approach' of studying the effect of changing one variable on the performance is in practice virtually impossible to carry out. It should further be remembered that the crew has direct control over only two variables: the apparent wind angle via the helm, and the angle of incidence of wind on sails via the sheets. Of course a number of minor adjustments are possible, mainly concerned with sail shape. Nothing can be done about the underwater shape unless a trim tab is fitted to the keel.

Bearing these limitations in mind let's look at some specific cases to get a better feel for the velocity triangle. Suppose we are sailing in 14 knots of apparent wind, with apparent wind angle $\beta$ 30° and boat speed 5 knots: speed made good to windward is 3.6 knots and true wind speed is 10 knots. If the true wind suddenly increases by 10% to 11 knots, what is the effect on the speed made good? If nothing else is changed by this gust (same $\beta$, same leeway) then we find that $V_{mg}$ is now 3.9 knots, an increase of 8%.

You might well feel that this argument is based on unrealistic premises; for example the angle of heel will certainly change in the gust. It has been found, however, that in light wind conditions hull and sail characteristics are unaffected by wave-making and heeling. Now assume that the true wind remains fixed but by sail adjustment we are able to maintain the same boat speed while reducing the apparent wind angle by 3% from 30° to 29°. This results in a $V_{mg}$ of 3.7 knots, an increase of close to 3% also.

Thus, as a rough rule of thumb one can say that a 1% increase in the true wind will result in a 1% increase in speed made good to windward, and a 1% decrease in apparent wind angle while keeping a constant boat speed will also give about the same increase in speed made good to windward.

## §3.4.1   Sail and Hull Polar Diagrams

Optimising a boat's speed to windward is one of the more fascinating aspects of sailing. It is in many ways an intellectual challenge, with many variables involved. Although one has direct control of only a few of them, others such as leeway, sail drag angle or heel can be indirectly affected. The purpose of the remainder of this chapter is to study the factors which affect speed made good to windward so that when trying to get your own boat 'in the groove' you will have some logical basis for deciding what to adjust.

Since it is a combination of water forces and wind forces that drive a yacht to windward, it is therefore necessary to understand the workings of her two powerhouses, the underwater body and the sails. On the wind, forward thrust has its origins in lift, and the ever-present drag, though unwanted in windward sailing, has an important influence. We will therefore look first at the relationship between lift and drag for sails and hulls. It is clear that the magnitude and direction of the total force on a sail will depend on the *angle of attack*, here defined as the angle between the direction of the undisturbed wind and the chord of the sail. The chord is the horizontal straight line drawn between the luff and the leech of a sail, or between the leading and trailing edges of a keel or rudder. Thus the three quantities that we want to relate are lift, drag and angle of attack. This may be done by plotting lift against drag and labelling points on the curve with the corresponding angle of attack. A typical curve for a mainsail alone of area about 8 m² measured on a full size boat is shown in fig. 3.26. Such a curve is generally known as a polar diagram, though here only a quarter (90°) of the full 360°

heading range is given, and it is plotted in a conventional graph form.

You will notice that the actual lift and drag are not plotted in the diagram but in their stead the quantities $C_L$ and $C_D$, the *lift coefficient* and *drag coefficient* respectively. Up till now I have tacked away from introducing 'coefficients', since we experience real forces on our boats not coefficients! But a time inevitably arrives when their convenience counterbalances their apparent complications. Look for a moment back at §2.1 and at §7.1. There it is shown that drag consists of some function of Reynolds' and Froude's numbers multiplied by fluid density, area and the square of the velocity. In the succinct notation of mathematics this is written:

$$D = f(R_n, F_n)\varrho V^2 S.$$

D is the drag force. $f(R_n, F_n)$ means 'some function of Reynolds' and Froude's numbers'. $\varrho$ is the density of the fluid, S is the surface area and V is the velocity. The part of the drag which depends on $R_n$ and $F_n$ cannot in general be predicted but must be determined from measurement, so one replaces $f(R_n, F_n)$ with a coefficient. In this case it is the drag coefficient $C_D$ and we write:

$$D = C_D(1/2)V^2 S\varrho$$

As the last three quantities can be measured, if the total drag is measured also this fixes the drag coefficient. The usefulness of such coefficients depends on the fact that they vary only slowly with the velocity V. Having made a drag measurement at one velocity it is then possible to predict the drag at a different velocity because of the simple dependence on $V^2$. A word of warning is due: fig. 2.10, for instance, shows that the drag coefficient is not by any means completely independent of speed, but for many estimates the assumption is justified. As to lift coefficients, the situation is a little brighter: they can be used with reasonable confidence over a wide range of velocity providing one does not exceed the stall angle. You may have wondered why the factor 1/2 was introduced along with the coefficient. No really good reason except that it is conventional and $1/2\varrho V^2$ represents the kinetic energy of the air. (It was in fact left out in some of the older literature thereby halving $C_D$.)

Now back to fig. 3.26. Along the curve are the angles of incidence or attack in degrees. Each angle corresponds to a certain lift and drag, e.g. an angle of 60° which might be used when running has a lift coefficient of 1.08 and a drag coefficient of 1.3. Our

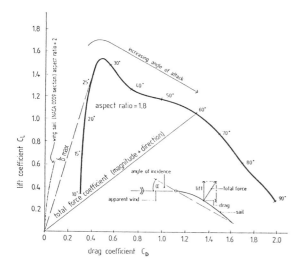

Fig. 3.26 A 'polar diagram' for the mainsail alone of a 3.7 m centreboard boat. The measurement was made in actual winds on a full scale boat by determining the forces in a tethering line to a fixed mooring. $C_L$ and $C_D$ are the measured lift and drag coefficients which are related to the actual lift and drag forces as explained in the text. The numbers along the curve refer to the angle of incidence or attack of the apparent wind on the sail. Although the sail had some twist, the angle of attack was measured between the direction of the apparent wind and the sail. The dashed line tangent to the curve from the origin gives the angle of attack for which the lift/drag ratio is a maximum, in this case about 26°. The dotted curve is that of a symmetric section wingsail of similar aspect ratio. No hull or rigging drag is included in this curve which is why the sail measurement made under realistic conditions is pushed so far to the right along the drag axis. The arrow marked total force coefficient determines the magnitude and direction of the force on the sail when the angle of incidence is 60°. In this case the direction of the force is at about 40° to the apparent wind direction.

ancient forebear Pythagoras told us how to get the total force coefficient $C_T$ from these two: $C_T^2 = C_L^2 + C_D^2$. What one measures is the total force and the angle at which it acts, so this process is reversed in constructing the graph.

The dashed line on the graph is a tangent to the curve from the origin. The point where it touches the curve corresponds to the maximum value of the ratio of lift to drag, which occurs in this case at an angle of attack of about 26°. The lift coefficient at that point is about 1.37 and the drag 0.4 so the maximum value for the ratio L/D for this particular sail is 1.37/0.4 = 3.4.

It is instructive to compare this soft sail polar diagram with that of a wing sail of about the same

aspect ratio, shown by the dotted curve in fig. 3.26. Beyond 12° the wing sail stalls and the lift coefficient falls off rapidly. One is unlikely to build a wing sail with such a low aspect ratio, but this figure was used purely for comparison with the soft sail.

The most notable difference between the two sails is the very much lower drag of the wing sail. This leads to a maximum lift/drag ratio of 16.5, nearly five times that of the cloth sail. As we shall see, however, this does not necessarily guarantee better windward performance. Although not shown, the total force coefficient at large angles of attack is much less than for the cloth sail so that the performance downwind will be inferior for the wing sail. In some respects this comparison is a little unfair because the wing sail curve does not include the parasitic drag of the hull, whereas the soft sail curve includes the extra drag produced by the hull, the rigging and the circular section non-rotating mast. This is called 'parasitic drag' because these parts of the boat produce only drag and no lift. It offsets the whole curve to the right along the drag axis.

Parasitic drag is something that generally receives scant treatment in books about sailing, yet it is of vital importance for windward work as we will see in the next few pages. The hull is the biggest source, especially when heavily heeled, but mast and rigging can also make a large contribution particularly in strong winds when a boat is well reefed down. Then the ratio of parasitic drag area to lifting surface area is greatly increased. The ratio of heeling force to forward drive increases and the boat must respond by making more leeway. It has been estimated (H. M. Barkla, 'The behaviour of the sailing yacht') that on average the proportion of drag contributed by the hull probably exceeds 20% and may approach 50%, when sailing closehauled.

Fig. 3.26 shows that the total force coefficient tends to be greater at larger angles of incidence. But remember that this force can be broken down into two components, as shown in fig. 3.4, the heeling force which we always want to minimise and the driving force which we always want to maximise. Thus the polar diagram only has significance when it is combined with the yacht's direction with respect to the apparent wind.

This latter point is made clearer by fig. 3.27 which is worth a little careful study. Three points of sailing are shown and for simplicity it is assumed that the sail shape is kept the same for all. It is further assumed that the apparent wind strength is the same for each, which of course means that the true wind is rather

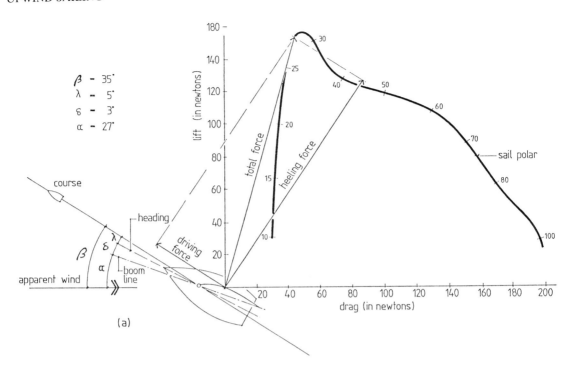

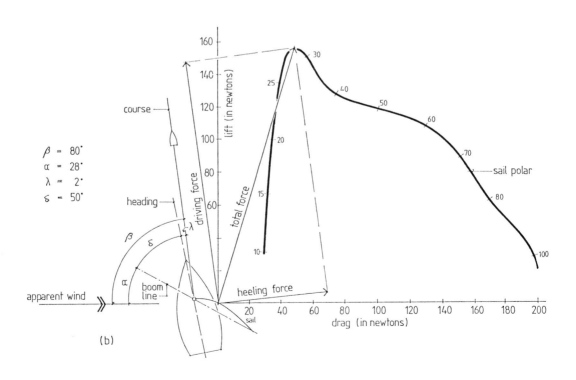

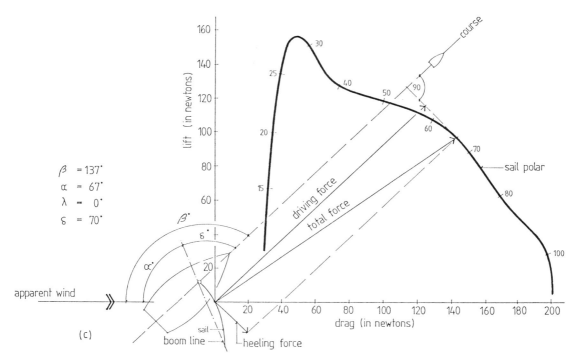

*Fig. 3.27* Using the sail polar diagram of fig. 3.26 to determine the conditions for maximum driving force on various points of sailing. (a) shows a boat on the wind where the course angle $\beta$ is taken as 35° with a leeway $\lambda$ of 5°. The angle of attack corresponding to the maximum driving force for this course is obtained by drawing the dashed line perpendicular to the course and just touching the curve. This point of contact corresponds to an angle of attack $\alpha$ of 27°. From the figure it is clear that $\beta = \alpha + \lambda + \delta$, where $\delta$ is the boom angle. Knowing the first three enables the correct sheeting position for maximum drive to be established. In

(b) the course angle is 80° and the leeway 2°. The maximum driving force is determined as in (a) and occurs at an angle of attack of 28°. This condition will be met for a boom angle of 50°. Note the requirement that the sail be at about the same angle of incidence as when on the wind. Now, however, the driving force is much greater. (c) corresponds to a broad reach with a course angle of 137°. If the leeway is taken as zero the required boom angle works out to be 70°. Notice that the driving force now has most of its source in drag forces rather than lift forces.

different in each case. The lift and drag values shown have been determined from fig. 3.26 for a sail area of 8 m² and a wind speed of 4.56 m/sec or about 9 knots. In fig. 3.27a the course angle $\beta$ is 35° which is typical for this type of boat when making its best speed to windward. If this fact is accepted the armchair sailor can lean back, look at his polar diagram and tell the skipper that his boat speed will be optimised if he adjusts his sail for an angle of attack of 27°. This is obtained by drawing the dashed line perpendicular to the course and tangent to the curve. It gives the maximum possible driving force, which is the component of the total force in the direction the boat is going. It is clear, for instance, that if the angle of attack were 25° or 30° the total force arrow would end at those points and the driving force component

would be less even though the total force in the 30° case is marginally greater.

In the diagram a number of angles are shown designated by their usual Greek letters. The course angle $\beta$ (beta) is already familiar. $\lambda$ (lambda) is the leeway angle measured between the boat's heading and the actual course it is sailing. For the purposes of fig. 3.27a this has been taken as 5°, but its value in reality depends not only on the sail forces but also the underwater characteristics of the boat. As the symmetry of sailing emphasises, these two are always uncompromisingly connected (there will be more about the determination of leeway later). $\alpha$ (alpha) is the angle of incidence of the wind on the sail and is measured between the direction of the undisturbed apparent wind and the boom. $\delta$ (delta) is the sheeting angle and is simply the angle between the boom and the boat's centreline. The skipper has direct control over this and on a fixed course changing the sheeting angle effectively changes the angle of incidence and hence the point on the curve that is being used. A glance at the diagram shows that the course angle $\beta$ is equal to the sum of the other three angles. Thus, with $\beta = 35°$, $\lambda = 5°$ and $\alpha = 27°$ determined from the polar curve, our armchair sailor can be more explicit and tell the skipper that his sheeting angle must be 3°.

The component of the total force which is at right angles to the course is the heeling force. It is not surprising to see that when sailing to windward this is three times as great as the forward driving force.

Move now to fig. 3.27b, for a reach with a course angle of 80°. It is assumed that the leeway is reduced to 2° which is consistent with this point of sailing. Again the sail operating point is given by the dashed line perpendicular to the course and just touching the curve. It is interesting that the optimum operating point has changed little from the windward case. The required angle of incidence is now about 28°. This plus the leeway makes 30°, add to this the sheeting angle and the total must equal the course angle which here is 80°. The sheeting angle is therefore 80° − 30° = 50°. This is the reason for the advice always given to beginners that when paying off to a reach adjust the sheets for roughly the same angle of attack as when beating. Notice the much greater driving force and the fact that, rather than being less than the heeling force, it is about twice as great. For many boats this is their fastest point of sailing.

Fig. 3.27c shows the situation when on a broad reach. The optimum operating point is again determined from the dashed tangent perpendicular to the course and we find that we require an angle of attack of 67°. On this point of sailing leeway is very small and we will assume that it is close enough to zero. The course angle is now 137° so the sheeting angle becomes 70°. As expected the heeling force is now almost negligible and most of the total sail force goes into driving the boat forward.

It is interesting to note that in diagram (c) most of the total force has a fundamentally different origin from that in (a) and (b). In (c) the total force consists of about 140 newtons (divide by 5 to get pounds roughly) of drag and about 90 newtons of lift. By contrast, in (a) and (b) the lift component of the total force is about 150 newtons whereas the drag part is only about 45 newtons. Thus when sailing on the wind and on a close reach the sail is acting mainly as a lift producing device whereas on a broad reach or

dead run it is its drag producing aspect which is more important. For a single sail to be a good all-round performer it must be capable of producing high lift with little drag, and with a mere change of angle of attack of producing a large drag. For angles of incidence greater than 30° it is operating in the stalled condition: separation of the flow on the leeward side is occurring because the air flow has insufficient energy to overcome the adverse pressure gradient. This destroys lift but increases drag due to the production of a turbulent wake.

What has been shown so far is for a single sail in light winds, which should have illuminated the broad principles of sail trimming. The amount of quantitative information about sail forces available from a single polar diagram like this is, however, really very limited. It has already been mentioned that the coefficients $C_L$ and $C_D$ depend on Reynolds' number or velocity, so one cannot scale up or down in wind speed or sail area with too much confidence. In fact the true situation is worse than this. Although it was not stated, the polar diagram in fig. 3.26 is for zero angle of heel, which is fine for a small boat which is normally sailed without heel, balanced by the crew. For a larger fixed-keel boat any amount of wind must produce some heel if only hull form and permanent ballast are producing the righting moment. So the question arises: does the sail polar diagram remain the same when the boat is heeled? Unfortunately, no. (These effects will be discussed in more detail in chapter 4 on sails.) As wind strength increases the shape of the sail changes because of stretch, which alters the lift and drag coefficients. Also, on most boats the crew will purposely change the sail shape on different points of sailing. Add to this the fact that wind strength and direction vary with height and it is clear that a single polar diagram even for a boat that sports only one sail is inadequate to fully predict all the wind forces that may occur.

I started by saying that this state of affairs was unfortunate. But for sailing and the art of yacht design it is what makes it so fascinating. The application of physics can tell us much, but the well tried pragmatic approach of slow evolution abetted by scientific input is still the most practical method of improving yacht performance.

So far only wind forces have been discussed, but as we know, for a boat sailing at a constant speed there must exist equal and opposite water forces. As with sails the best way to characterise these is also with a polar diagram in which lift is plotted against drag for various angles of incidence. The nautical term for angle of incidence is *leeway angle*. A further difference is that the fluid density which is involved in calculating lift and drag from their coefficients is nearly a thousand times greater for water than for air. This means that the same forces can be produced by smaller areas and lower velocities.

To keep matters straightforward we will consider the underwater forces on a small boat which can be sailed without heel. Such boats normally have shallow hulls of light displacement with deep, efficient centreboards and rudders. Such hull forms are now also common in many larger boats so the considerations that follow have quite wide application.

The broad characteristics of the underwater body of a light displacement boat may be understood by making the following simplifications:

(a) All the lift or side force is produced by the keel and rudder.
(b) The hull contributes no lift, but most of the drag which is independent of leeway angle.
(c) A significant portion of the total drag has its origin in the vortex drag of the keel and rudder (explained in chapter 4). It is essentially the penalty paid for generating lift and is therefore dependent on the leeway angle.

If these assumptions are justifiable it enables one to determine the general nature of the underwater characteristics. Such methods cannot of course be used to make quantitative comparisons of designs, but the exercise helps to clarify our understanding of the factors involved in producing the underwater lift and drag forces.

Although assumptions (a), (b) and (c) seem reasonable, something a little more concrete than reasonableness is required. This was supplied in 1961 through a classic paper by T. Tanner in the *Transactions* of the Royal Institution of Naval Architects. Full-scale tank tests were made on an International 10 Square Metre canoe. Briefly, these measurements showed the following:

(1) The lift of the hull alone without centreboard was very small, being about 2½% of the lift with centreboard.
(2) The lift could be well predicted by the application of standard aerodynamic theory to the centreboard.
(3) The hull itself contributes about 90% of the drag at zero leeway and about 50% at 7°. The hull drag itself increases due to this leeway by about 20%.

We see therefore that the assumptions above are well founded, especially if the leeway is not too great. Thus for the purposes of arriving at an illustrative underwater polar diagram the hull drag without leeway was calculated from the sum of two terms (details are in §8.8). The first term represents the surface friction contribution to the total drag, which is proportional to the 9/5th power of the boat speed. The second term is proportional to a higher power of the velocity and describes the so-called residual or wave-making drag. A rough argument given in §10.3 implies that wave-making resistance is expected to depend mainly on the 4th power of the boat speed. However a light displacement boat such as the one being considered here sails with a certain amount of dynamic lift even at low speeds, offsetting somewhat the purely displacement drag. The second term has accordingly been taken as proportional to $V^3$. A graph of the resulting drag at zero leeway is shown in fig. 3.28 which shows the estimated drag as a function of boat speed for a 3.7 m boat having a wetted area of 3.38 m².

Since the polar diagram involves lift versus drag for various leeway angles, it is still necessary to calculate the effect of leeway on the underwater drag. Assuming that the drag of the hull alone does not change significantly with leeway, it is merely necessary to calculate the increase in drag of the keel

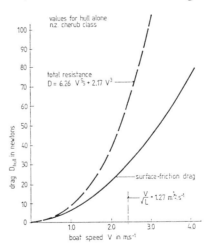

*Fig. 3.28* Hull resistance for the 3.7 m boat used to demonstrate performance prediction. The resistance without leeway is assumed to consist of two parts, surface friction drag proportional to the 9/5 power of the boat speed and a second part depending on the cube of the speed. This is typical of a light displacement hull with a tendency to plane.

due to its angle of incidence (i.e. leeway) and this may be done by standard aeronautical methods.

Most of the resistance increase due to leeway comes from vortex drag (chapter 4). In fact, it is interesting to note that for the boat being considered here, with a centreboard area of 0.47 m² the centreboard total drag becomes equal to the hull drag at a leeway angle of about 6°. This fact comes as a surprise to many sailors, who instinctively feel that the extra drag associated with leeway comes from the crabwise motion of the hull through the water. Instead, it is the vortex drag (induced drag) resulting from the lift produced by the keel. From time to time it has been fashionable to build keels capable of rotating on a vertical axis or centreboards that twist from side to side in the casing, the idea being that the hull can now point directly along the course with the necessary lift obtained by giving only the keel an angle of attack to the water flow. Little is to be gained by this complication in a light displacement boat and

incorrect adjustment could easily nullify that small gain.

Following the procedure outlined in §8.8 we can now construct a set of polar curves for the hull. Instead of a single curve for one apparent speed as was done for the sail we plot a 'family' of curves for various boat speeds $V_s$ shown in fig. 3.29. As for the sail these are plots of lift against drag. Each curve corresponds to a particular boat speed with respect to the water. Remember that 'leeway angle' is just the angle of attack of the water on the keel. According to well established aerodynamic theory the side force or lift is directly proportional to this angle.

Looking at fig. 3.29 we see that if the boat is making a speed of 2.5 ms⁻¹ (metres per second) with a leeway of 2°, then the total underwater force on it has a lift component of 240 newtons and a drag component of 87 newtons. The magnitude of the total force is given by the length of the straight arrow drawn from the origin to this operating point. It could be determined by measuring its length with a ruler and laying it off along one of the axes, or by following Pythagoras: the square of the hypotenuse is equal to the sum of the squares on the other two sides. $F_T{}^2 = 240^2 + 87^2$ which gives $F_T = 255$.

Thus the total water force on a boat is produced in exactly the same way as the wind force; it is just that the underwater characteristics look a little different from the sail characteristics. This is partly because a boat should never be sailed with the keel stalled but may often be with the sail stalled. So only those parts of the curves for the water forces are shown which correspond to small angles of attack.

Now comes the crucial point in understanding windward forces on a boat. Having sheeted his sail and turned his boat to what he judges is the best angle to the wind, the skipper fixes the operating point on the sail polar (fig. 3.26). This generates a total force which by the symmetry of sailing must be exactly opposed by the water force. It is represented in fig. 3.29 by an arrow whose length corresponds to the total wind force and whose direction is determined by the helmsman's choice of apparent wind angle. This will normally pick out in fig. 3.29 a unique boat speed and angle of leeway. Thus the sailor who controls directly the boat's heading and sheeting angle has in this sense only indirect control over boat speed and leeway angle.

You will have guessed by now that knowing the wind force and water force characteristics of a boat enables us to predict its resultant boat speed and heading. The details of this performance prediction

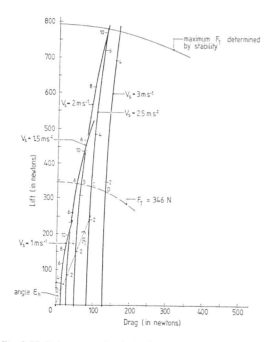

*Fig. 3.29* Polar curves for the hull. This is the water force equivalent of the wind force curve in fig. 3.26. The difference here is that several curves have been drawn for different boat speeds $V_s$. The numbers along the curves are leeway angles in degrees. Since the underwater force produced must be equal and opposite to the sail forces this will specify an arrow such as $F_T$ on the diagram, the end point of which determines the corresponding boat speed and leeway angle.

process will be covered in §3.5, but first it is necessary to look at a rather remarkable, important and often misunderstood geometrical property of all boats when sailing to windward.

## §3.4.2 The Beta Theorem

In windward sailing it is the ratio of lift to drag that is important in determining the characteristics of the boat. You may be getting the feeling that a description of windward sailing is somewhat like an exercise in geometry, so you will not be surprised to learn that lift/drag ratio can be represented by an angle, the *drag angle*. When one does this a fascinating relationship known as the 'beta theorem' or the 'course theorem' can easily be demonstrated.

Fig. 3.2 showed that the total force acting on an object in a fluid flow is generally at some angle to the undisturbed flow direction. If this force slopes aft a long way the drag component is large compared to the lift, whereas if it is near to the vertical the lift component is relatively larger than the drag. The drag angle $\varepsilon$ is therefore defined as the angle between the direction of the lift force and the total force (fig. 3.30). Clearly a large drag angle implies a large drag, or to be more precise a large drag/lift ratio. If you remember a little trigonometry you will realise that the ratio D/L is the tangent of the drag angle. The drag angle is always represented by the Greek letter $\varepsilon$ (epsilon). A subscript s is used when it pertains to the sail, and h for the hull including keel and rudder.

Since we will more often be speaking of the ratio of lift to drag rather than drag to lift we define drag angle by:

$$L/D = 1/\tan\varepsilon \quad \text{or} \quad \varepsilon = \tan^{-1}(D/L).$$

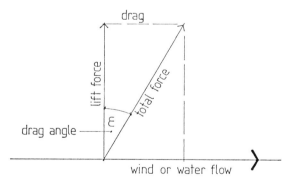

*Fig. 3.30* Drag angle is a measure of how far the total force on sail or hull points downstream and is the angle between the direction of the lift and the total force. Mathematically it is related to the lift L and drag D by $\tan\varepsilon = D/L$.

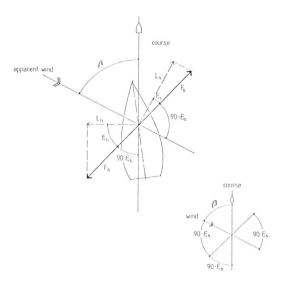

*Fig. 3.31* The geometry of the 'beta theorem'. A study of the figure shows that the course angle $\beta$ is simply the sum of the sail and hull drag angles. This deceptively simple summary of windward geometry is always true, but it must be remembered that the drag angles are those corresponding to the 'operating points' on the sail and hull polar diagrams. These points rarely correspond to the minimum drag angles of the hull and sails taken in isolation.

These forms are used rather than cotangents since calculators don't have cotans.

Fig. 3.31 shows only those sail and hull forces required for our purpose: $F_s$ is the total sail force, $F_h$ is the total hull force. They must be of equal magnitude and point in opposite directions. As there tends to be a slight optical illusion in such diagrams you should convince yourself that they are in a straight line by laying a ruler on the figure. The dashed lines are the lift forces for the hull and sails and are therefore drawn at right angles to the course and apparent wind respectively. The drag angles are then as defined in fig. 3.30. Thus the angle between the total sail force $F_s$ and the apparent wind direction is $90° - \varepsilon_s$, and the angle between the total hull force $F_h$ and the course line is $90° - \varepsilon_h$. Now look at the inset diagram in fig. 3.31 where only these angles are drawn. $90 - \varepsilon_s$ can be transferred to the left hand side since it is formed by the crossing of the same pair of lines as that on the right. Thus we have three angles on the left which add up to 180°, so we may write:

$$\beta + 90 - \varepsilon_s + 90 - \varepsilon_h = 180.$$

From this we arrive at the very important result that $\beta = \varepsilon_s + \varepsilon_h$.

This says that the angle between the apparent wind and the course actually sailed is equal to the sum of the sail and hull drag angles. The course angle $\beta$ is obviously a measure of how high the boat is pointing.

Although this statement of the beta theorem or course theorem is always true, it should not be misconstrued as it seems often to be. A false impression of a boat's pointing ability is given when it is incorrectly assumed that the drag angles in the above formula are the minimum drag angles that the sails and hull can separately produce. This is quite wrong. The drag angles in the beta theorem are those corresponding to the actual 'operating points' on the sail and hull polar diagrams when the constraint is imposed that the total sail and hull forces must be equal and opposite. As we shall see later, this generally means that the sail is operated at a lift higher than that at which the drag angle is a minimum, whereas the hull is operated at a lift lower than that corresponding to the minimum drag angle.

The beta theorem is a beautiful example of the symmetry of sailing. The course we sail to windward depends equally on two quantities, the drag angle of the sails and that of the hull. Both have equal relevance: one may never argue that the sails are more important than the hull or vice versa. If crew adjustments result in a 1° reduction of either the hull or sail drag angles at the operating points then the boat will point 1° higher. This does not necessarily mean that it will go to windward faster as boat speed is also involved. The relationship between drag angles and the adjustments the crew are able to make is a complex one, however, and the apparent simplicity of the beta theorem tends to obscure the complex interactions of physical effects which are involved when a boat sails to windward.

## §3.5 Performance Prediction

The purpose of this section is to explain how the performance of a boat can be related to sail and hull polar diagrams. To keep things simple the discussion will be confined to the polars introduced in §3.4.1 for a small boat which is normally sailed upright. The concern here will not be on what factors make a boat go fast but rather on why it goes at all. A clear understanding of the interaction of hull and sail forces is a prerequisite to any discussion on what may improve say windward ability. These aspects will be left to chapters 4 and 5.

There are several ways in which performance prediction may be carried out. A number of graphical

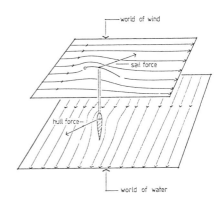

*Fig. 3.32* This schematic diagram depicts the basic principle of sailing. The movement of two fluid media with respect to one another is exploited by sailing craft which, by transferring momentum from the air to the water, are able to propel themselves through the water.

methods have been propounded. One involving the rotation of polar curves appears in C. A. Marchaj, *Sailing Theory and Practice*. Another method using sliding tracings appears in H. M. Barkla's paper 'The behaviour of the sailing yacht', and a yacht 'slide rule' was described by P. R. Crewe in 'Estimation of effect of sail performance on yacht close-hauled behaviour'. These days it is more appropriate to do it numerically with the aid of a computer. The method given here may not be the best for routine performance prediction but is, I believe, the easiest way to understand what is going on.

The controllable movement of a sailing boat requires that there be a relative movement of wind and water, shown in fig. 3.32. When a yacht is accelerating away from a tack unbalanced sail and hull forces are present, but as the boat reaches a steady speed they build up in such a way that they eventually become exactly equal and opposite. It is this condition of equality which is exploited by all performance prediction methods.

Since the crew has greater control over the sail forces than over the hull forces, it is reasonable to begin from the known conditions set up by the crew and deduce from these the important quantities of boat speed and leeway which the crew can only control indirectly.

Fig. 3.33 shows only those forces and angles which concern us here. Remember that the sail lift force is perpendicular to the apparent wind direction and the hull lift force is perpendicular to the course sailed. Essentially the crew has control over only two

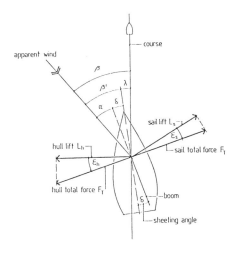

apparent wind

course

*Fig. 3.33* When sailing steadily the total sail force must be exactly equal and opposite to the total hull force, so both are called $F_T$. The angle between the direction of the lift and the total force is called the drag angle. $\varepsilon_s$ is the sail drag angle and $\varepsilon_h$ is the hull drag angle. $\lambda$ is the leeway angle. The apparent wind direction as measured by on-board instruments is $\beta'$. The quantity needed for calculating the speed made good to windward is $\beta$ which is equal to $\beta' + \lambda$. $\delta$ is the boom position with respect to the centreline of the boat and $\alpha$ is the angle of incidence of wind on the sail. It is clear from the figure that $\beta = \alpha + \delta + \lambda$.

quantities: the boom angle $\delta$ and the angle between the apparent wind and the boat centreline $\beta'$. A glance at the diagram shows that the angle of incidence of the sail $\alpha$, which is measured between the apparent wind direction and the boom direction, is equal to $\beta' - \delta$. Thus setting up the sail and the course $\beta'$ is equivalent to setting up the sail's angle of incidence. Using this angle of incidence and the sail polar diagram (fig. 3.26) we can read off the corresponding lift and drag coefficients and hence determine the sail total force coefficient and its drag angle.

The simplest way to continue now is to take a specific example. Let us suppose that the skipper considers that he will go to windward best if he sheets the boom in so that $\delta = 0$, i.e. the boom is amidships. By means of the tiller he adjusts his course so that the apparent wind makes an angle of 28° with the centreline of the boat. This would be measured with instruments on a larger boat, as would the apparent wind speed. Let us assume that this wind speed $V_a$ is 6 ms$^{-1}$ or about 12 knots.

*Step 1:* Determine sail angle of incidence. This is:

$$\alpha = \beta' - \delta \qquad = 28 - 0 = 28°.$$

*Step 2:* From the sail polar diagram (fig. 3.26) we see that 28° corresponds to a lift coefficient $C_L$ of 1.54 and a drag coefficient $C_D$ of 0.46. The total force coefficient could be measured by ruler or compass to determine the distance from the origin of the graph to the 28° mark, and scaled along one of the axes or by using the formula:

$$C_T^2 = C_L^2 + C_D^2. \quad \textit{Either method gives} \quad C_T = 1.6.$$

*Step 3:* Another quantity we will need is the sail drag angle, between the total force and the lift direction. It could be measured with a protractor on the sail polar diagram or, if you like, calculated from

$$\varepsilon_s = \tan^{-1}(C_D/C_L). \text{ This gives } 16.6°.$$

*Step 4:* From the sail coefficient we must now determine an actual force which will depend, of course, on the sail area and the wind speed. The appropriate formula is:

$$F_T = 1/2 A_s V_a^2 C_T \varrho.$$

The air density $\varrho$ is about 1.2 kg/m$^3$. With a sail area of 10 m$^2$, say, all the quantities on the right hand side of the formula are now known so we can multiply them all together and get a sail force of 346 newtons (78 lbs).

*Step 5:* Go to the hull polar diagram (fig. 3.29) and draw an arc of a circle of radius corresponding to 346 N (newtons) centred at the origin of the graph. This is shown as a dashed curve. Two conditions must now be met. The first is that the boat's motion through the water must be such that it generates an underwater force exactly equal to 346 N. This condition is satisfied all along the arc and in particular by the points A, B, C and D in fig. 3.29. These points are at the intersection of the dashed arc corresponding to a force of 346 N and the curves corresponding to various boat speeds. One can also determine from the points of intersection the corresponding leeway angles.

*Step 6:* We now make a table giving the values of the three quantities obtainable from the hull polar at each of the intersection points. The hull drag angle $\varepsilon_h$ is determined in exactly the same way as was the sail drag angle.

| Intersection point | $V_s$ | $\lambda$ | $\varepsilon_h$ | $\beta' = \beta - \lambda = \varepsilon_s + \varepsilon_h - \lambda$ |
|---|---|---|---|---|
| A | 1.5 | 7.9° | 9° | $16.6 + 9 - 7.9 = 17.7$ |
| B | 2 | 4.4 | 11 | $16.6 + 11 - 4.4 = 23.2$ |
| C | 2.5 | 2.8 | 15 | $16.6 + 15 - 2.8 = 28.8$ |
| D | 3 | 1.8 | 23 | $16.6 + 23 - 1.8 = 37.8$ |

The right hand column of the table is the second of the two conditions alluded to above and states a requirement which can be seen as necessary from fig. 3.33. The apparent wind direction as measured to the centreline of the boat is equal to the difference between the 'true' apparent wind direction and the leeway angle. In symbols, $\beta' = \beta - \lambda$. But we know from the beta theorem that beta is just the sum of the drag angles, $\beta = \varepsilon_s + \varepsilon_h$. Thus it is a second requirement that $\beta' = \varepsilon_s + \varepsilon_h - \lambda$. It is just the one which says that the hull total force must be in exactly the opposite direction to the sail force.

*Step 7:* Since the angle $\beta'$ is held at 28° by the skipper, the right hand column of the table shows that the only physically possible intersection point is at C. Thus the predicted boat speed is 2.5 m/sec, the angle $\beta$ is $\varepsilon_s + \varepsilon_h = 31.6°$ and the apparent wind speed $V_a$ was 6 m/sec. From these three numbers one may now determine the speed made good to windward graphically, or with a computer using the program in §3.4.1. The result is $V_{mg} = 1.6$ knots, true wind speed $V_t = 4.1$ knots, and true wind angle $\gamma = 50°$. This last quantity determines the angle (100° in this case) through which the boat tacks and also the lay line to a windward mark.

This may have seemed a complex procedure but it is worth going back over those seven steps again carefully, making sure you understand fully the reason for each. Most people will tell you that they sail to windward by 'instinct'. But it isn't instinct: it is the result of practice which is acquired knowledge. The above is simply some more acquired knowledge which you will not get from sailing but will certainly improve your ability and enjoyment.

The initial conditions of course angle and boom position were my guess as how best to sail this boat to windward. To know whether I was correct one would have to change the course or sheeting angle or both and do the whole analysis over again to see if it improved the speed made good to windward. This is in effect what the experienced sailor is doing in the process of getting his boat 'in the groove'. The above performance prediction method is not restricted to windward sailing but is applicable to any course.

At the top of fig. 3.29 there is an arc of a circle labelled 'maximum $F_T$ determined by stability'. This has been determined on the assumption that the boat is sailed by two people each of 70 kg weight and that one of them is on a trapeze. (How this is estimated is in §8.9.) When sailing to windward this limit is determined basically by the maximum heeling force sustainable.

If performance calculations are carried out for all points of sailing using the polar curves of figs. 3.26 and 3.29 a great many numbers can soon be amassed. Not all of these correspond to the maximum speed on a particular course, but if the optimum results are sifted out and converted to boat speed as a function of *true wind* speed, and course with respect to the true wind, a fairly simple pattern results (fig. 3.34). This is a true polar diagram and to distinguish it from the earlier diagrams which are called 'polar', it will be referred to as a *performance polar diagram*.

This diagram applies only to a true wind speed of 3.5 m/sec (about 7 knots). The direction of the true wind is taken to be 0° coming in from the top of the diagram. The arrow marked $V_s$ represents the vector describing the boat speed at an angle of 66° to the true wind. Its length gives the boat speed; the scale to be used is marked along the horizontal axis in metres per second. In this case, at 66° the boat speed is 2.5 m/sec or about 5 knots. Such diagrams are useful in various ways. They can be used in determining the value of $\gamma$ for best speed made good to windward. Since the boat speed on a broad reach with spinnaker is greater than that dead downwind, it is possible to gain by tacking downwind: the optimum tacking angle may be determined from the diagram (described in §3.5.4). If a number of diagrams are drawn up for different wind strengths and sail combinations they can be very useful in racing to determine the best sails for the prevailing conditions. Producing such diagrams is, however, not a trivial task. Thousands of individual measurements may be required in steady wind conditions as well as access to a computer to process the data into the form shown in fig. 3.34. Some ideas about how to do this are given in Part II.

Although the beta theorem states that the course sailed to the apparent wind is equal to the sum of the sail and hull drag angles, it may seem surprising that the best speed made good to windward does not occur at the minimum drag angles or highest lift/drag ratios. It is not difficult to see why this is so.

In fig. 3.35 A is the best operating point of the sail because it is associated with the maximum possible driving force component. Point B is where the lift/drag ratio is a maximum; where the broken line from the origin just touches the curve. If B were the operating point, it is clear that the forward driving force would be less. Maximum driving force results in maximum boat speed, but speed through the water alone does not ensure maximum speed made good to windward. Reducing the course angle slows the boat but may result in improved $V_{mg}$. Continued reduction of the

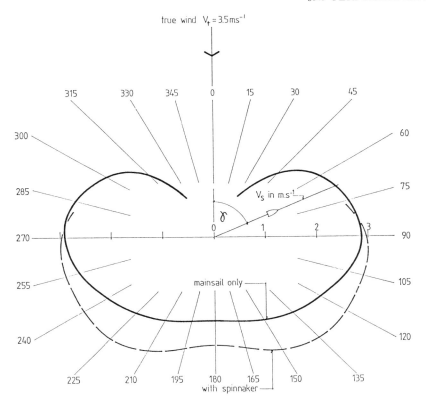

*Fig. 3.34* The performance polar diagram of the small centreboard boat described in the text. This was calculated from the sail and hull polar diagrams given in figs. 3.26 and 3.29.

course angle brings the boat to the point where the driving force is zero; the line which is tangent to the curve and perpendicular to the course goes through the origin of the graph. But this is none other than the broken line to point B. Thus, to make any progress to windward at all the sail *must* be operated at a lift coefficient higher than that corresponding to the maximum lift/drag ratio.

Look now at the hull polar curves of fig. 3.36. They use the same data as fig. 3.29 but have been plotted with an expanded drag axis for clarity. A characteristic of the hull curves noticeable in fig. 3.29 is that two curves for closely adjacent boat speeds will always cross at some point; two such curves have been plotted in fig. 3.36. Another fact evident from fig. 3.29 is that a ruler through the origin and just touching the curves does so at a leeway angle of 8° and corresponds to a drag angle of 8.9°. This corresponds to a maximum lift/drag ratio of 6.4. Thus it is always possible to plot curves for two adjacent

boat speeds such that the curve for the higher boat speed crosses the other at its point of maximum L/D. If the required hull force is 336 newtons and the boat speed is 1.5 m/sec we find that the hull operating point is at a leeway of 8°, the point of maximum L/D. This total hull force would be made necessary by operating the sail of fig. 3.26 in an apparent wind of 5.9 m/sec at an angle of attack of 29°. This is associated with a sail drag angle of 18° so that the course angle $\beta$ is 26.9°. Calculating from these facts the speed made good to windward gives 1.2 m/sec.

Since the two curves in fig. 3.36 cross at the operating point, it is clear that the hull force requirements are satisfied also for a boat speed of 1.6 m/sec and a leeway angle of 7°. It is self evident that this will give rise to a better speed to windward since not only is the boat speed greater but the leeway is less. In fact, when the calculations are done we now find that $V_{mg} = 1.3$ m/sec, an improvement of nearly 10%. The hull is now operating at 7° leeway

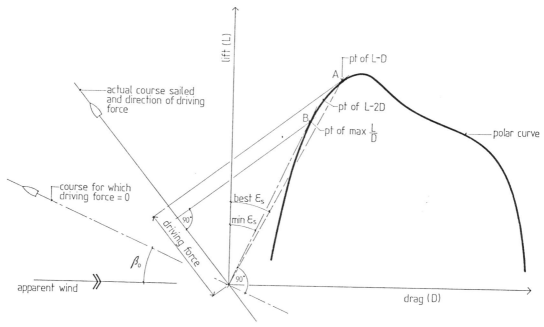

*Fig. 3.35* The dotted line in this diagram is a tangent from the origin to the polar curve. Point B, where it touches the curve, is therefore the point at which the lift/drag ratio is a maximum. However, to get maximum drive for the course sailed the sail must be operated at point A which has a lift well above that corresponding to the maximum L/D ratio. For a light displacement boat this point is located

approximately where the quantity L − D is a maximum. Heavier displacement boats go to windward better when pointed higher and have operating points about where L − 2D is a maximum. Whatever the type of boat, best $V_{mg}$ will always correspond to an angle of attack greater than that of point B.

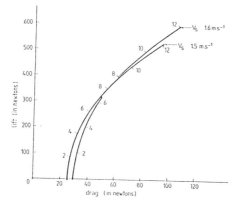

*Fig. 3.36* The maximum lift/drag ratio for these hull polars occurs at a leeway of 8°. This, however, is never the best operating point for the hull. This can be seen by plotting a second polar for a slightly greater boat speed such that it crosses the first at its 8° point. Thus the same hull lift can be produced by operating at greater boat speed and less leeway. So for best $V_{mg}$ the hull operating point is at a lift less than that for which the L/D ratio is a maximum.

corresponding to a lift coefficient which is less than that associated with the maximum L/D ratio. Although the hull characteristics cross, they correspond to different crew-imposed conditions. In the poorer case $\beta'$ is about 19° and in the other about 20°. Coupled with this is a change of about 1° in the boom angle.

Thus *for maximum speed made good to windward* the sail must be operated at a lift coefficient greater than that corresponding to the maximum L/D ratio, and the hull must be operated at a lift coefficient less than that corresponding to its maximum L/D ratio.

Since the hull lift has been assumed due entirely to the centreboard, it is interesting to ask how the $V_{mg}$ is affected if the centreboard area is halved. Fig. 3.37 shows the hull characteristics at a boat speed of 2.5 m/sec with full and half centreboard areas. If, for instance, the sail force were 615 newtons, then at a boat speed of 2.5 m/sec the leeway with full centreboard would be 5° and just double that amount with half the area. Assuming as before that the sail angle of incidence is 29°, a total force of 615 N results from a wind of 8 m/sec with a drag angle of 18.2°.

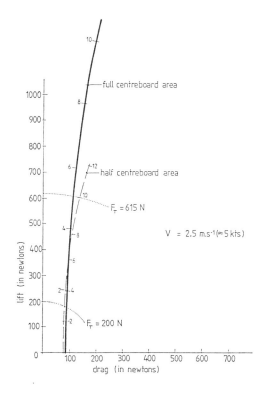

*Fig. 3.37* The effect of halving the centreboard area is to double the leeway angle. In light winds or on points of sailing where leeway is small the resulting reduction in wetted area is an advantage. However when greater hull lift is required, as when going to windward in fresher winds, the overall drag of a half-area centreboard is greater due to increased vortex drag resulting from the bigger leeway angle.

From fig. 3.37 the hull drag angles for full and half centreboard areas turn out to be 10.6° and 13.1° respectively. From this the speed made good to windward can be calculated in each case, and one finds that for the full centreboard it is 1.9 m/sec whereas for the half area it is 1.8 m/sec. In practice the difference would be greater than this because to make the analysis easy both the boat speed and the apparent wind speed were constrained to remain the same. Because of this the full centreboard case corresponds to a true wind speed less than that for the half area board.

On the other hand, if we are sailing in lighter winds on a reach where the rather high boat speed of 2.5 m/sec can still be maintained with a total sail force of say 200 N, then the hull drag is slightly less for half the centreboard area and the boat will be faster. In this

case the increase in leeway from 1.5° to 3° is of little significance. The reason the curves cross over in this way is because vortex drag becomes very important at large leeway angles. Thus there is always a balance to be struck between the increased surface friction drag produced by increasing the centreboard area and the relatively smaller vortex drag resulting from reduced leeway. Thus variation of keel area is as important as variation of sail area if a boat is to be optimised in all wind strengths and on all points of sailing.

A more mathematical way of looking at this topic is given in the book by H. F. Kay, *The Science of Yachts, Wind and Water*. There it is shown that the best sail operating point for going to windward depends on the shape of the hull drag curve. A very efficient hull has a drag which increases with the square of the boat speed; this is approximately true for hydrofoils and catamarans and most hulls at sufficiently small speeds. For such a hull the optimum sail operating point turns out to be where the quantity lift minus drag is a maximum on the polar diagram (see fig. 3.35). For hulls whose drag increases more rapidly with speed, as occurs for heavier boats where wave-making drag is more important, the best sail operating point turns out to be where the quantity lift

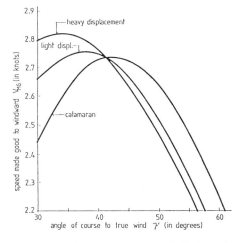

*Fig. 3.38* The optimum true course angle for best $V_{mg}$ depends upon hull type. The determining factor is the 'power law' of the resistance curve. At low speeds a catamaran hull's resistance is roughly proportional to the square of the boat speed, whereas light and heavy displacement boats have resistance curves approximately proportional to the cube and 4th power of the boat speed respectively. Heavy displacement boats must point higher than catamarans for best $V_{mg}$. In this graph the absolute values of $V_{mg}$ are not significant.

minus twice the drag (L − 2D) is a maximum, or even where L − 3D is a maximum. Such points always occur at smaller angles of attack. This is common experience: heavy boats are best sailed high on the wind with the sail close to luffing, whereas a multihull is best sailed freer and at higher sail incidence. This is shown explicitly in fig. 3.38 which is taken from a diagram in a paper by H. M. Barkla, 'The behaviour of the sailing yacht'; it is seen that the optimum course angles $\gamma$ are 42.2° for catamarans in light winds, 38° for light displacement boats and 34° for heavier types. The absolute values of $V_{mg}$ in the graph are not relevant in this context and simply reflect the way in which the calculations were made.

It should therefore be clear that there is an intimate relationship between sail and hull design. A sail which performs well on a catamaran cannot be expected to do so on a heavy displacement cruiser.

## §3.5.1 'Symmetry Breaking' Effects

In the physics of fundamental particles it has been found that the breaking of some, but not all, of the symmetries inherent in a system has come to be seen as a major property of the furniture of the world: 'There is no excellent beauty that hath not some strangeness in the proportion,' Francis Bacon philosophised long ago. Even in the mundane application of physics to sailing some of the more interesting points arise in situations where exact symmetry is broken.

Certainly it is still true that the same laws of fluid flow are applicable to the sails just as they are to the hull. It is absolutely always true that when the boat is moving steadily the total sail force must be equal to the total hull force. It is when one comes to look at the details of how to optimise these forces that we see 'symmetry breaking' effects.

Most of the differences between the air and water force producing mechanisms are rather obvious, but I will list them nevertheless:

(1) Sails are used at angles of incidence ranging from about 10° to 90°. Hulls, on the other hand, are normally at angles of incidence which are restricted to a much smaller range between 0° and about 10°.

(2) Hulls are never operated continuously under stall, whereas on broad reaches and dead runs sails are often completely stalled.

(3) The angle of attack of the wind on the sails is controlled by the crew, but he has no direct control over the angle of attack (leeway angle) of water on the hull. This is determined by hull and keel design. An uncommon exception would be boats fitted with trim tabs on the trailing edge of the keel.

(4) When course and wind strength change it is normal to change sails. Only if the boat has a centreboard is there any chance of changing keel area.

(5) Then there is the theorem proven in the last section: for best $V_{mg}$ sails must be operated at a lift *greater* than that corresponding to the maximum lift/drag ratio, and hulls must be operated at a lift *less* than that corresponding to their maximum lift/drag ratio.

## §3.5.2 Effect of Heel

That the characteristics of a boat are changed by heeling is well known to all those who sail. Quite a number of different physical effects are involved. To sort these out, the effects of heel will be separated into geometrical effects, aerodynamic changes and hydrodynamic changes.

Let's look first at the geometrical effects. Fig. 3.39 shows the forces which keep a heeled boat in equilibrium. As before all hull lift will be assumed to be due to the keel. The forces produced by the sail are assumed to lie in a plane perpendicular to the mast and those produced by the keel are similarly in a plane perpendicular to its surface. The heel angle $\varphi$ results when the righting moment due to buoyancy and ballast balances the heeling moment due to $F_T$.

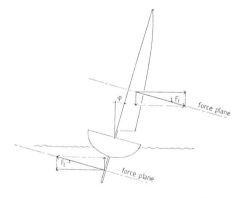

*Fig. 3.39* The total force on an airfoil is perpendicular to the plane of the foil. Thus if we rotate our view point by the amount of the heel angle the behaviour of the boat appears the same as when upright. It is only when we want to interpret this behaviour in the horizontal plane that our conclusions must be modified by a geometric factor.

This is no different from the situation when the boat is sailed upright and balanced by crew weight.

This is true providing our considerations are limited to the inclined plane. As soon as we ask, 'where is the boat going as measured in the horizontal plane?' a geometric correction must be made. For instance, in the inclined plane it is exactly true that the total sail force is equal to the total hull force. It is also true that the sum of the drag angles is equal to the course angle, but this course angle is now measured in the inclined plane. The relationship between the velocity vectors in the inclined plane and in the horizontal plane are shown in fig. 3.40. The superscript i refers to quantities in the inclined plane. Thus the beta theorem in the inclined plane is $\beta^i = \varepsilon_s + \varepsilon_h$.

From this point of view the drag angles are properties of the boat and do not depend on the angle of heel. Mathematically the course angle $\beta$ in the horizontal plane is related to $\beta^i$ in the inclined plane by the expression $\tan \beta = \tan\beta^i/\cos \varphi$ where $\varphi$ is the angle of heel.

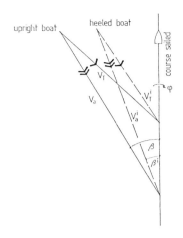

Fig. 3.40 Geometric effect of heel: the wind velocity vectors for a boat sailing upright are shown by the solid lines. Rotate the diagram about the course direction through the heel angle and the dashed lines now give the wind velocity vectors as seen by the heeled sail. Thus a boat capable of a course angle $\beta^i$ when heeled is accomplishing only the larger angle $\beta$ in the plane of the sea.

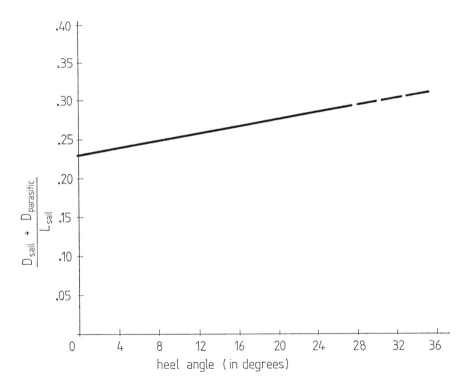

Fig. 3.41 Aerodynamic effect of heel: one important change is the increase in parasitic drag due to the hull. This has been estimated here for an International Dragon class yacht. For this boat an increase in D/L of 6% resulted in a decrease of $V_{mg}$ by 1.4%.

It is possible to interpret the geometric effect of heel on performance in two ways. The drag angles of the craft determine the inclined plane wind triangle geometry and the true geometry is determined as in fig. 3.40. Alternatively the heel angle may be considered to alter the effective drag angles. In other words, the heeled boat at an angle $\varphi$ has the same performance as an upright boat with drag angles increased by $1/\cos\varphi$. Whichever way one looks at it, there is a loss of performance due to heel which arises solely from the geometry of the situation. A possible exception to this general rule is that of a catamaran just lifting one hull and thereby suddenly reducing the wetted area.

Fig. 3.40 should convince you that a loss of performance must always occur. A boat which is able to point at an angle $\beta^i$ in the inclined plane is only managing the larger angle $\beta$ in the real world of the horizontal water plane.

As always in sailing one is led to a tradeoff. Heel can be reduced by decreasing the angle of attack on the sails, but this reduces drive. Somewhere there is a compromise point where the gain due to reduced heel more than offsets the loss due to heading up higher. Trimming the craft in this way means using a sail operating point nearer to the position of maximum lift/drag ratio.

The aerodynamic changes which occur when a rig is heeled have not been thoroughly studied, largely due to the great complexity of the problem. A number of qualitative statements can be made, however. The assumption that sail characteristics are independent of heel needs to be modified if there are changes in effective camber and aspect ratio as the wind gains a component upwards and parallel to the mast. In a seaway, a heeled yacht experiences greater fluctuations in angle of incidence when pitching.

An aerodynamic change with heel which is probably somewhat more significant is the change in parasitic drag due to the hull. If we suppose that the drag/lift ratio of a yacht is the sum of two independent parts $(D/L)$, due to the rig alone, and $(D/L)_h$ due to hull drag relative to *sail* lift, then it is possible to estimate the effect of a change in the ratio of hull profile area to effective sail area as the boat heels. This has been done in the paper by Barkla for the old International Dragon class and the results are shown in fig. 3.41. For this boat an increase in D/L of 6% resulted in a decrease of $V_{mg}$ by 1.4%. So in heeling to 20° the speed made good to windward is reduced by about 4.6%. This is a very substantial amount and is in addition to the geometric effect discussed above.

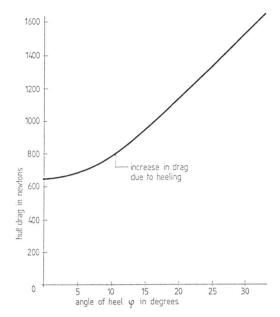

*Fig. 3.42* Hydrodynamic effects of heel: as a yacht heels a different shape is immersed in the water usually resulting in an increase in drag. This graph is taken from published values of towing tank measurements of the yacht *Standfast*.

As a hull heels a new shape is immersed in the water, so a change of hull drag is to be expected. This is shown in fig. 3.42 which resulted from towing tank measurements of a model of the Admiral's Cup yacht *Standfast* built in the 1970s. The drag increases only slightly for heel angles up to 5°; beyond 12° it increases in direct proportion to the heel angle.

Apart from the change of immersed hull shape, the keel which produces most of the lift will be moving nearer the surface resulting in an increase in its wave resistance. In the extreme case air could be drawn down to the low pressure side of the keel, which is always the side nearest the surface, giving a drastic loss of lift due to ventilation.

One final point concerning heel arises because the sail is operating in the atmospheric boundary layer which is just at the sea surface and about 100 m thick. At the bottom the wind speed is reduced by friction with the water. Heeling brings the rig more and more into the slow-moving air further reducing driving force.

### §3.5.3 Leeway Measurement

Under normal sailing conditions leeway angle is usually less than 10°. One might therefore well ask why a few degrees matter. They are important if absolute values of $V_{mg}$ are required, and they are particularly important when navigating by dead reckoning. The need to know the leeway is not so vital if only comparative $V_{mg}$ figures are required, which is usually the case when small adjustments are being made to get the boat in the groove. As leeway is not likely to change much during such adjustments relative values of $V_{mg}$ will still be valid.

If absolute values of $V_{mg}$ are required then the leeway angle must be known, as the following example makes clear. To calculate $V_{mg}$ the leeway angle must be added to the measured apparent wind angle $\beta'$. Let us assume that $\beta'$ is 25°, the apparent wind speed 15 knots and the boat speed 5 knots. If a leeway angle of 6° is assumed, $V_{mg}$ works out at 3.6 knots. On the other hand if we take a leeway of 3° we get $V_{mg} = 3.8$ knots, a difference of 5%.

More commonly one needs to know leeway when passage making. An error of 3° in leeway can change one's estimated position after a run of 10 nautical miles by half a mile, which could easily be the difference between being in safe water or ending up on rocks.

Leeway is not a constant quantity for a given boat but depends on wind strength, sail combination and point of sailing. In heavy conditions when well reefed down it can be astonishingly large while hard on the wind. This is because of the large relative increase in parasitic drag area of the hull compared to the sail area. Wave action could also be a contributing factor.

There are a number of ways in which leeway can be measured and they can be divided into what I will call direct and indirect measurements. The simplest direct measurement is to measure the angle of the wake with respect to the hull centreline. Choose a position which will probably be part way down the companionway into the cabin, so that when looking aft the stern pulpit is level with the horizon. Mark it with tape so that a sighting from your position across the marks corresponds to 5°, 10° and 15° on the port and starboard quarters. Lining the wake up with the marks will give a crude measurement of leeway.

It is also possible, though again not very accurate, to measure leeway directly with on-board instruments. A vane in the water can be used to determine leeway in exactly the same way as a wind vane determines the apparent wind direction. Another method is to use a pitot tube mounted at the bow, which has three sets of holes placed at the front and at positions of 45° to either side. Speed is measured by the difference in pressure in a tube

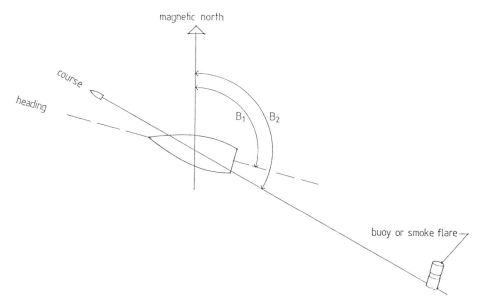

Fig. 3.43 One method of measuring leeway is to take bearings on a smoke flare placed in the water. From a position at the mast, $B_1$ is the bearing of the backstay and $B_2$ that of the flare. Several readings should be taken while sailing on a steady course and the average value of $B_2 - B_1$ is the leeway for that particular boat speed, sail combination and heel angle. If a buoy is used, correction for any stream is necessary.

connected to the centre hole compared to the average pressure in tubes connected to the port and starboard 45° holes. Leeway is proportional to the difference in pressure in the tubes connected to the port and starboard openings. Neither of these methods is easy to set up and they are plagued by large fluctuations due to wave action and motion of the boat.

A refinement of the wake method is to drop a smoke signal or just a floating marker over the stern and observe its position with a pelorus.

A more practical and accurate direct measurement method using equipment normally available on a yacht is shown in fig. 3.43. Round a convenient mark and sail off on the required point of sailing. Stand at the mast with a hand bearing compass and take the bearing of the backstay, then take the bearing of the mark. Repeated readings should give the same bearing difference, which is the required leeway providing there is no tidal flow. If correcting for current is a problem then a floating smoke signal is to be preferred.

None of these methods of determining leeway is the kind of thing that one would do routinely and record in the log. What is required is a method of leeway determination that depends only on standard instrumentation. Direct measurement with vanes does not work well, however it is possible to establish the leeway with reasonable accuracy from a measurement of the boat speed and a simultaneous measurement of the angle of heel with a home made mechanical inclinometer. The justification for this method depends on the relationship between the heeling force, the stability and the hull forces. Fig. 3.44a shows how the heeling force $F_H$ combined with the hull force $L_h$ produces an overturning moment on the yacht. It is opposed by a righting moment produced by the boat's weight acting through its centre of gravity and the buoyancy acting upwards. This is called static equilibrium and the situation is changed only slightly when the boat is moving through the water. Thus the righting moment is equal to the heeling moment. Fig. 3.44b makes the now often repeated statement that the total sail force is equal to the total hull force, from which it follows that components of these forces along the same line are also equal. The heeling force $F_H$ is the component of the total force $F_T$ along a line perpendicular to the course. The component of the total hull force which is perpendicular to the course is the hull lift $L_h$. These two act along the same line so that $F_H = L_h$. Now the hull lift force results from water flow of speed $V_s$ approaching the keel at an angle of leeway $\lambda$.

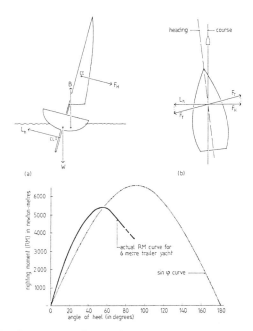

Fig. 3.44 To a good approximation leeway depends only on heel angle and boat speed for a given rig. This is because hull lift $L_h$ depends on leeway, but hull lift is exactly equal to heeling force and angle of heel is uniquely related to heeling force. So there is a simple relationship between leeway and heel.

According to aerodynamic theory the lift force is proportional to the angle of attack or leeway and to the square of the fluid flow velocity.

In addition the heeling force is related to the heeling angle. Chapter 5 gives more details of this relationship but what we need here can be quickly stated. The righting moment, which is clearly the same in magnitude as the heeling moment for a steady angle of heel, depends in some way on the heeling angle. Mathematically one says that the righting moment is a 'function of the heeling angle' which just means that there is a one-to-one relationship between the heeling angle and the righting moment. If one were to measure the righting moment one could determine the heeling angle and vice versa. This is not quite true for a boat that is moving, and certainly there would be fluctuations due to waves, but these corrections are small. So mathematically we may write: $RM = f(\varphi)$ where RM is the righting moment and $f(\varphi)$ means 'is a function of the heeling angle $\varphi$'. In terms of fig. 3.44a the righting moment is just the heeling force $F_H$ multiplied by the distance between the centre of lateral resistance CLR and the centre of effort CE. Thus the heeling force, and hence the hull lift force, $L_h$, is proportional to $f(\varphi)$.

This is not much use unless we know the form of $f(\varphi)$. In chapter 5 it is shown that $f(\varphi)$ is approximately equal to $\sin\varphi$. This is a good approximation for heavy displacement boats but poorer for light and beamy ones. A graph of $\sin\varphi$ in fig. 3.44c is compared with the actual righting moment curve of a 6 m trailer yacht. At normal angles of heel up to about 30° the two curves are in reasonable correspondence, so we have reached the point where the hull lift $L_h$ is proportional to $\sin\varphi$. We also know that the lift force is proportional to the product of the leeway and the square of the boat speed. Combining all these facts, we may write:

Righting Moment = $f(\varphi)$ = $\sin\varphi$ approximately,

$\qquad$ is proportional to heeling force $F_H$

$\qquad$ = hull lift $L_h$

$\qquad$ is proportional to $\lambda V_s^2$

Bringing all this together, we may write:

$$\lambda = K\sin\varphi/V_s^2.$$

Here K is called the *constant of proportionality* and must be determined by experiment. Leeway need be measured only once, by one of the methods described above, and if angle of heel and boat speed are measured at the same time the factor K can be determined. The same K can then be used to determine leeway from any other heel angle or boat speed. K is a characteristic of the particular rig and hull; a different sail combination or hull shape, raising the centreboard for instance, will alter it.

Although this method of determining leeway is eminently practical, one should be aware of the approximations inherent in it. First, we assume a static righting moment instead of a dynamic one. Second, we approximate this by a simple sine function. Third, we assume that the hull lift is proportional to $V_s^2$. This is correct for an aeroplane wing, but the underwater body of a boat is expected to differ somewhat because the immersed shape changes with heel angle, and the presence of the water surface has an effect. Fourth, we assume that the distance from the CE to the CLR is independent of heel angle.

Despite these limitations leeway measured by this method is good enough for most purposes, as is shown by fig. 3.45 compiled from published data for *Standfast*, 12.2 m overall and displacing 12,498 kg. In

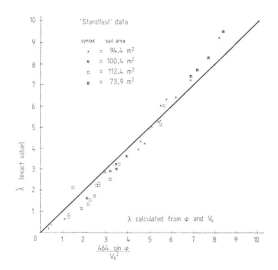

Fig. 3.45 Model tests of *Standfast* were used to determine the exact leeway, plotted vertically. From full-scale trials in which boat speed and heel angle were recorded, leeway was calculated from the simple formula shown below the graph and plotted horizontally. If the agreement were perfect all points would fall on the diagonal line. The correlation is good enough, however, for this to be considered a useful method for on-the-water determination of leeway.

this case the factor K has the value 464 when boat speed is measured in knots; this was determined from towing tank measurements on a model. The points on the graph were calculated from full-scale heeling angle and boat speed data taken while racing; values for the horizontal axis of the graph could then be determined and these were compared with the model's exact leeway values plotted vertically. The agreement is reasonable despite the fact that a range of sail areas from 73.9 m² to 112.4 m² has been used.

## §3.5.4 Tacking Upwind, Acrosswind and Downwind

The art of tacking to get to a point directly to windward is highly developed. A successful skipper's senses are highly tuned to how a boat feels and should be sailed to get upwind fastest. Most of the time spent on an Olympic course is in sailing to windward. A crew practises going about hundreds of times in a season and the operation is usually accomplished fast and flawlessly.

By comparison, how many times does the same crew gybe their spinnaker? It is more complex and difficult to be sure, but how many times does this

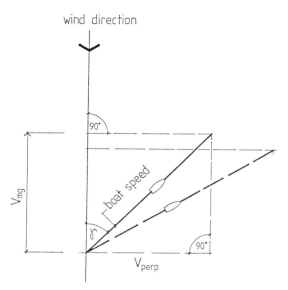

wind direction

$V_{mg}$

boat speed

90°

90°

$\gamma$

$V_{perp}$

*Fig. 3.46* Speed made good to windward is the component of the boat speed along the wind direction. Although the dashed line speed is greater, the boat's $V_{mg}$ is less because of the larger wind angle.

same crew do the job fast and flawlessly? Although top skippers know 'instinctively' how to get from A to B in the minimum time, the object of this section is to explain to the rest of us why it is important to be able to tack downwind and across the wind as well as upwind.

Let's look at upwind sailing first since this is a very familiar concept. The important quantity is the speed made good to windward, i.e. the component of the boat's speed in the direction against the wind (fig. 3.46). The length of the 'boat' vector represents the boat's speed and direction through the water. The speed made good to windward or the windward component is obtained by drawing a perpendicular line from the end of the 'boat' arrow across to the wind direction. Of course by sailing only on port tack we have a speed component perpendicular to the wind and to starboard. This displaces the boat far to the right of the mark, so by sailing for the same time on starboard tack we get back to the rhumb line and our average speed component perpendicular to the wind cancels out while our speed component against the wind remains fixed. If the angle $\gamma$ is increased (dashed line) the boat will go faster through the water, but this may result in a smaller value of $V_{mg}$. The trick is to choose $\gamma$ correctly. The skipper of a

boat without instruments uses his experience to optimise $\gamma$, but this says nothing about what he is really doing.

To understand this we must look at the yacht's polar diagram. Polar diagrams are specific for different types of boat and sail combinations, so fig. 3.47 is meant to give the trend only. For sailing at an angle $\gamma$ to the true wind the speed of the boat is given by the distance from the origin O out to the appropriate curve.

Fig. 3.48 shows that the speed made good to windward is a maximum when the heading is such that the perpendicular from the head of the boat speed vector to the wind direction just touches the polar curve rather than crossing it. That is, it is a tangent to the curve.

Now to understand better what follows let's look at the polar diagram as representing not how *fast* the boat goes in different directions but how *far* it goes in a certain length of time, say one hour. Fig. 3.49 shows the windward part of such a polar diagram. Because the diagram is symmetrical about the wind direction, the tangents to both the port and starboard tack sides which are perpendicular to the true wind are one and the same straight line ABC. By sailing for one hour along either OB or OA, the distance made good to windward is OC. If the windward mark is at C one's windward component of distance has reached the mark but the boat is at B or A! Obviously C can be reached by making an even number of equal tacks, say two. If we spend half an hour on port tack we will reach point D. Then half an hour on starboard will take us along the line DC which of course is parallel to OA, the optimum starboard tack direction.

Our speed through the water on both tacks is the same, so the distance covered in the second half hour along DC is the same as would have been covered along DB if we had stayed on port. All this may seem a complicated way of saying something trivial, but the point is that it gives us a general geometrical way to determine the optimum directions and length of time to sail on each tack. This is not quite so obvious when the mark is not directly to windward or when it is on a tight reach, as we shall see.

The geometry of what we are doing when we sail correctly to windward is therefore as follows:

(a) Draw a tangent to the port and starboard bulges of the polar diagram (this is the dotted line AB).
(b) For a mark directly to windward the scale of the diagram is expanded so that the mark is at point C (fig. 3.49).

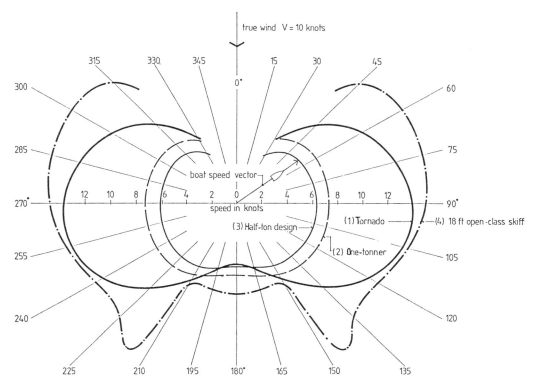

Fig. 3.47 Typical performance polar diagrams of three types of boat: (1) a Tornado catamaran, (2) a One-Tonner, (3) a Half-Tonner and (4) an 18 ft skiff with a crew of three. Boat speed at a given angle to the true wind is measured by the length of the line from the origin of the graph to the appropriate curve using the scale along the horizontal axis. Notice that the actual speed made good downwind by the 18 footer is faster than the wind! (See §3.5.5 for further discussion of this ultimate in sailing.)

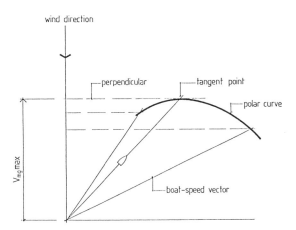

Fig. 3.48 Best $V_{mg}$ is obtained when the boat is sailed on that course where the tangent to the polar curve is also perpendicular to the true wind direction. Any other course gives a lower $V_{mg}$.

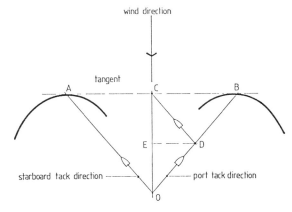

Fig. 3.49 To reach C from O in the minimum time a boat must sail along OD parallel to the port tack optimum direction and then along DC parallel to the starboard tack direction. In this trivial case it is clear that equal times must be spent on each tack. This provides a rule for determining the best course in more complicated situations.

81

(c) Assuming we start on port tack (though starboard might be safer), draw a line through the mark C parallel to the starboard tack direction OA. Then ODC is the optimum path to sail making two tacks.

Look now at fig. 3.50. The mark M is no longer directly upwind, but we can still use the same formula for finding the optimum path ODM using two tacks. No other path will get us there more quickly. Now instead of two equal tacks we spend more time on the favourable tack, as expected.

It is a geometrical property of triangles that if MD is parallel to OA, then DM always equals DB if the speed on both tacks is the same (fig. 3.50). By sailing on port for one hour we reach B, but by going about after 42.5 minutes we reach M after one hour. Of course one can make more than two tacks so long as 42.5 minutes is spent on port and 17.5 on starboard. The solution is unique: there is no faster way of getting to M. This also shows the importance of the dotted tangent line. *Whenever the desired course intersects this tangent the maximum speed made good in this direction is some combination of tacks along directions parallel to the lines from O to the tangent points.*

How often have you struggled to get a spinnaker filling on a close reach only to find that, after all the straining and danger of broaching, the boat is no faster than without it? The appropriate polar diagram now has two parts, one with spinnaker and one without (fig. 3.51). Assume the direction to the mark lies along OM. If we sail along the rhumb line OM we will reach C in one hour, but if we sail partly in the

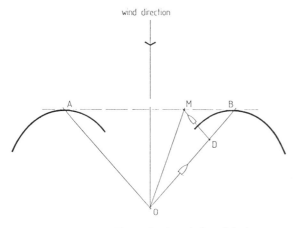

Fig. 3.50 If the mark M is not dead to windward the best course is found by drawing DM parallel to the optimum starboard tack direction OA.

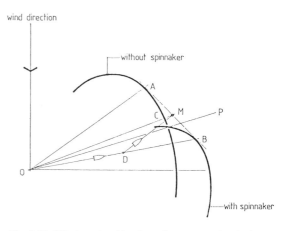

Fig. 3.51 Windward tacking is made necessary by the fact that the performance polar has 'humps' in it. On a reach the use or not of a spinnaker also produces humps in the polar. We therefore expect that the same principle of tacking will also apply to getting from O to M in the minimum time. First sail along OD with spinnaker, then sail along DM parallel to OA without spinnaker.

direction OB with spinnaker and partly in the direction OA without, we will reach point M in one hour. Thus whenever the direction we want to sail intercepts the tangent line AB it is *always* faster to sail partly with spinnaker along OB and partly without along a direction parallel to OA.

To determine the optimum path we use the same rules as before. Draw a line through M parallel to OA. Then the optimum path is ODM. Because the speed on the two legs is different DM is no longer equal to DB, but the time to go along ODM is always the same as the time to go along OB or OA. If you like geometry you can prove this for yourself.

Now you may well ask: how do I know that the course to the mark lies along a line which intercepts the tangent to two bumps on a polar diagram which, for small boats at least, is difficult to determine? You don't, but most skippers have a pretty fair idea when they are sailing along OP (fig. 3.51). This is the direction where the boat speed through the water is not increased by flying a spinnaker. In that situation it is *certainly always* better to pay off with the spinnaker and then close reach to the mark without it. Whichever is done first and how many 'tacks' are made will depend on wind shifts and other factors.

Finally, in general one should tack downwind. Fig. 3.52 shows that the situation is basically identical to tacking to windward. OA and OB are the directions which give maximum downwind component. By drawing the line DM parallel to OA we find that, for

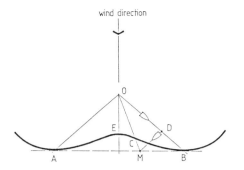

wind direction

Fig. 3.52 Tacking downwind is just a repetition of the same principles used when tacking to windward or across the wind. Again it is essential that the polar diagram have two humps: then the best course is along OD on one gybe followed by DM parallel to OA on the other.

this situation, we should sail 67% of the time in the direction OB and 33% in the direction OA.

When should one tack downwind? It is necessary to know whether the polar diagram has a concavity so that the tangent line AB falls outside the curve over some region. Fig. 3.47 shows the extent to which this occurs with three types of boat. It is certainly also true for all light centreboarders that carry spinnakers. For all boats the concavity in the downwind part of the polar diagram is greater in light winds than in heavy, so downwind tacking is more important in light conditions when gybing is easier and the time taken in the manoeuvre is relatively less important. Notice that to make tacking downwind really pay off it is necessary, in the case of a Tornado for instance, to sail about 50° each side of the downwind direction. For other boats it is less, but still rather more than one might at first think necessary or prudent.

For larger boats with instruments a polar diagram could be constructed by sailing on various points with different sail combinations in a steady wind. There is a difficulty, however: what one measures on the boat is the apparent wind direction rather than the true. This can be corrected for by using the computer code of §3.4 for instance, or very sophisticated instrumentation, but it might be easier to sail along fixed compass courses and plot the results with respect to the true wind bearing.

In fact a variant of this method can be used while actually racing if the downwind leg is reasonably long. Assuming the mark is dead downwind, sail directly toward it and take the compass bearing and the boat speed making sure the boat has first reached equilibrium speed. Then change course to the direction you judge is best for downwind tacking. Hold that course for a minute or so, record the heading and boat speed. Now take out a pocket calculator and key in the two courses, take the difference and find the cosine of that difference. Multiply this result by the second boat speed and compare the answer with that obtained when sailing dead downwind. If it is greater, then tacking at that angle is to your advantage. Another angle might be better still of course, and only a knowledge of your boat's polar can give you that.

Even if it is not practical to consult a polar curve for your boat while racing, just the knowledge that it *can* help to tack across and down wind should inspire you to try these tactics, and with practice and experience the human computer will become 'programmed' to the point where its performance will be difficult to beat.

## §3.5.5 The Wind Turbine Boat

In this section the principle of the symmetry of sailing will be taken to its ultimate conclusion. It will be shown that it is not only possible to sail directly into the eye of the wind, but that it is possible to sail downwind faster than the wind. Measurements on a full-size wind turbine vessel have already suggested the validity of these theoretical conclusions.

All that is required for the propulsion of a sailing vessel is that there be a relative motion of air and water. The situation is completely symmetrical in that both the air-borne and water-borne elements of the propulsion system have the same function. Sailing upwind is analytically identical to sailing downwind with the roles of wind and water reversed.

Fundamentally there are three ways of extracting thrust from the relative velocity between wind and water. The first is when a sail is used as an airfoil. The air flow direction is changed by the action of the sail, resulting in a change of *momentum* which produces thrust. The side force or heeling force component of this thrust is balanced by the keel, which being a hydrofoil generates its own side force or lift by inducing a change of momentum in the water. This, of course, describes a normal yacht sailing to windward. The second way is to use the sail as a drag element. Energy is extracted from the air flow and dissipated in turbulence; no energy is transferred to the water through the keel. This is the case when a boat runs straight downwind at speeds less than the wind speed.

The third method involves extracting *energy* from

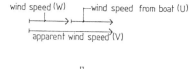

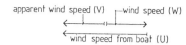

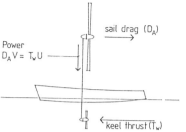

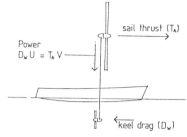

net thrust (T) = $T_w - D_A$

since    V > U ,    $T_w > D_A$

therefore    T = $T_w - D_A > 0$

net thrust (T) = $T_A - D_w$

since    V < U,    $T_A > D_w$

therefore    T = $T_A - D_w > 0$

(a)  Upwind  condition

(b)  Downwind  condition  faster than wind

*Fig. 3.53*  A wind turbine boat can sail directly into the wind. The power absorbed from the wind is the product of a large apparent wind speed and a small drag. This same power is communicated to the propeller and is now the product of a large thrust and a small apparent water speed.

The water thrust is therefore greater than the sail drag and the boat moves through the water. Downwind, the roles of wind and water are reversed and the propeller drives the windmill so that the craft moves through the air.

one of the fluids and imparting it to the other. Conventional boats do not do this, but it can be accomplished if the sail is replaced by a wind turbine and the keel by a propeller. When sailing directly downwind drag would still be produced, but instead of dissipating the energy in turbulence the energy would be extracted as shaft power. It would then be imparted as thrust to the water by means of the propeller, giving an increase in total thrust compared with that of a spinnaker. Energy extraction methods have the advantage that the vessel can sail straight upwind and also, with sufficiently low hull resistance, sail downwind faster than the speed of the wind.

Fig. 3.53 shows schematically how this is possible. When sailing upwind the relative velocity of the air over the hull is greater than the speed of the water under the hull. Since power is the product of force times velocity, power can be extracted from the air at low drag and imparted to the water at high thrust. The situation is reversed when going downwind. Energy is extracted from the water at lower drag than the thrust achieved when this energy is imparted to the air, because the relative velocity of air over the

hull is less than the speed of the water under the hull. A careful study of fig. 3.53 should convince you of this.

In general, power is extracted from that fluid which moves fastest relative to the vessel and is imparted to the slower fluid. From this view point sailing downwind faster than the wind is just the mirror image of sailing upwind. This is perhaps the ultimate expression of the symmetry of sailing.

An approximate theoretical analysis of the speed attainable when a wind turbine is mounted on a standard sailing hull is shown in fig. 3.54, to be compared with the performance of the same hull with a normal sail of the same area as the turbine and in the same true wind. The horizontal axis of the graph is the true wind angle $\gamma$. The vertical axis is boat speed, but for the case of the normal sail at $\gamma = 0°$ it corresponds to $V_{mg}$. Notice that a normal sail (dashed line) is only more effective on a beam reach, where nearly all the total sail force produced has a forward component. The theoretical wind turbine curve (solid line) does not include inevitable frictional losses in transmission between the air and

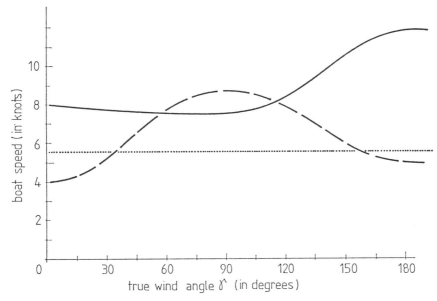

*Fig. 3.54* The dashed curve is a measured boat speed as a function of the true wind angle, using a normal sail. The solid curve is the calculated performance of the same hull with a wind turbine of the same area as the sail. The dotted line is a measurement on a full-scale wind turbine craft with a somewhat different hull shape. The theoretical (solid) curve is higher because it does not include transmission losses. Transition to the mode in which the water propeller drives the air propeller was clearly not achieved during these measurements. Note that a conventional sail still seems best on a beam reach.

water propellers. Apart from the rise in boat speed at 180° where the functions of the two propellers reverse, the wind turbine boat should have a fairly constant speed on all points of sailing.

An experimental wind turbine vessel, pictured in fig. 3.55, has her normal rig replaced by a three-bladed variable pitch air screw. The wind turbine is mounted on top of a short mast which is rotated to keep the air screw downwind of the mast with its axis roughly in line with the wind. It is connected to a conventional fixed pitch water screw 0.9 m across via a revolving shaft inside the mast.

In a steady true wind of 14.5 knots boat speed was measured on various points of sailing with a carefully calibrated knotmeter. 7.0 knots was logged and found to be independent of heading, whether directly into the wind, on a reach or downwind. A large number of measurements were made, giving an average boat speed just a little less than half the true wind speed. This result is included in fig. 3.54 as the dotted horizontal line.

Comparison with theoretical predictions can be regarded only as qualitative since a different hull

*Fig. 3.55* The successful wind turbine *Te Waka*, designed and built by Jim Bates of Whangarei, New Zealand. It has a variable pitch airscrew 8 m across mounted on a 9.4 m hull. Boat speed was found to be independent of the point of sailing and was nearly half the true wind speed.

design was involved. Lower than theoretical speeds would be expected due to transmission losses and the fact that the water screw pitch could not be adjusted to the optimum value. The constant measured speed despite wind angle probably partly reflects the expected characteristic of a wind turbine vessel as well as the fact that during the trials *Te Waka* was sailing at her 'hull speed' where a further speed increase would be associated with a very rapid increase in resistance due to wave-making. This would be one reason for the lack of expected rise in speed at 180°. Another might be the fact that the changeover to the water screw driving the air screw may need to be accompanied by a change in blade pitch.

So far, in the real world of sailing only ice yachts and the remarkable 18 ft skiffs sailed in Australia and New Zealand have, to my knowledge, succeeded in tacking downwind faster than the wind. It now seems

clear that a well designed wind turbine mounted on a planing hull or catamaran could make this a reality without the need to tack.

Curve (4) in fig. 3.47 shows the measured performance polar of an 18 ft skiff. These highly developed boats can normally sail around an Olympic course at an average speed made good along the rhumb line which is greater than the wind speed. The sharp bulge in the polar is due to the use of the spinnaker: because the boat speed is so great the apparent wind angle is never large and the spinnaker can only be set when the true wind angle is at least around 130°. In fact the peak of the spinnaker lobe shown occurs at a true wind angle of 143° while the boat speed is 14.5 knots in a 10 knot true wind. Under these conditions the apparent wind speed over the boat is 8.8 knots and the apparent wind angle is only 44°! Close-reaching spinnakers which work by producing lift rather than drag are therefore vital for

*Fig. 3.56* Possibly the most efficient sailing machines around a triangular course, these boats are capable of exceeding the wind speed on all points of sailing, and downwind can make speed good which exceeds the true

wind speed. The enormous righting moment provided by the three crew on trapezes enables the boat to be sailed upright while carrying a large rig in strong breezes. (*Photo: Sea Spray*)

this kind of sailing. Note that the speed made good downwind by tacking at 37° away from the leeward direction is about 11.6 knots in a 10 knot true wind.

## §3.6 Some Practical Tips on Windward Sailing

Perhaps the most obvious feature of a keelboat sailing to windward is that it heels. To non-sailors invited out for a day on the water, the awkward and slightly frightening lean is the most vivid impression they take back to the fixed earth. Yet for those who sail often it is taken for granted and rarely given any thought. This is a pity, because a boat can tell you a lot by its angle of heel.

We saw in §3.5.2 that any amount of heel is bad. Tell this to the skipper of a light centreboard boat and he responds by putting his crew, and himself if necessary, out on a trapeze (fig. 3.56). There seems to be a kind of rule that sailboat performance is inversely proportional to crew comfort. On a long passage crew comfort may mean crew survival, so most boats are not sailed like this and must be content with the evil of heeling. (Perhaps the catamaran enthusiasts might feel it their duty to skip the next few pages!)

Since heeling of a fixed ballast monohull is essential when going to windward, serious consideration should be given to the amount of heel just as it is to the amount of boat speed. Fig. 3.57 shows the hull polars for a typical keelboat. The inset on the same graph shows the relationship between hull lift force and heel angle for this boat. (The direct connection between these two quantities was discussed in §3.5.3.) This means that the hull lift axis of the polar curves may be replaced by heel angle as has been done on the right hand side of the graph. Now going back to §3.5, you will remember that the optimum operating point on the hull polar is 'below' the point where L/D is a maximum. This suggests that there exists an optimum heel angle for sailing to windward.

Calculations by Barkla support this conclusion and he gives the relationship shown in fig. 3.58 for the optimum heel angle and sail angle of attack as a function of true wind speed. Of course this will differ from one boat to another but the general form is likely to be correct.

The physical reason for an optimum heel angle is that although the driving force also increases when the heeling force increases, hull drag increases with heel (fig. 3.57) and so also does the leeway. There is

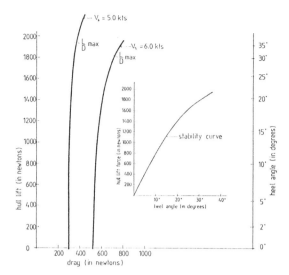

Fig. 3.57 Hull polar curves for a keelboat. Heel angle is produced by the magnitude of the heeling force. But heeling force is exactly equal to hull lift force, so we have the relationship given by the inset curve connecting hull lift and heel angle. The vertical axis of the hull polar diagram can therefore be drawn in terms of heel angle using the stability curve. Since best windward performance is obtained using a hull operating point 'below' the point of maximum L/D, we see that heel should not exceed 25 to 30° for this particular boat.

thus always a compromise which gives best performance at a well defined heel angle. Fig. 3.58 also tells us another fact: since heeling force increases with the square of the wind speed, the angle of heel will go on increasing rapidly if course and boom angle are fixed. The fact that the optimum heeling angle curve has an S shape means that above a true wind of about 10 knots air is being spilled from the sails. This is clear also from the angle of attack curve which shows a decrease in this region.

So the tip is: fit an inclinometer to your boat. Take a mental note of its readings while racing in the normal way. Take particular note on the days when you feel you are in the groove and tacking ahead of the opposition. One cannot say *a priori* what angle of heel will be best for your boat, but after sailing with an inclinometer for a while you will have built up a pretty fair idea in your mind.

The requirement of fig. 3.58 that angle of incidence should be decreased in stronger winds can be accomplished in two ways. Either one holds a fixed course and increases the boom angle, or the course angle is decreased keeping the boom angle fixed. In gusty conditions it is usually best to keep the sheeting

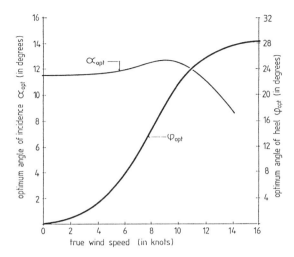

*Fig. 3.58* These calculated curves will be qualitatively true for most keelboats. The curve marked $\alpha_{opt}$ gives the optimum sail angle of attack for best speed made good to windward as a function of true wind speed. A reduced angle of attack in stronger winds means a reduction in sail forces. This reduction is reflected in the optimum heeling angle $\varphi$ which at first rises rapidly in the increasing wind strength and then flattens out as the rig is de-powered. The maximum strong wind heeling angle of 28° for best windward performance is larger than would be the case for more modern and beamier boats. This optimum angle of heel is best found by experiment.

angle fixed and use the boat's momentum to run a few metres up to windward. In steady conditions no clear-cut answer can be given. Increasing the sheeting angle gives a greater driving force component than when the course angle is decreased. As always the right compromise must be struck if other circumstances such as sea condition are not already the overriding factor.

In this age when yachts are equipped with all kinds of high-tech devices for determining their speed, direction and position on the surface of the Earth, it is somehow satisfying for most sailors to realise that a simple device, whose cost is infinitesimal, is very high-tech indeed if it is judged by the usefulness of the information it gives. I am speaking of sail telltales, or wools, streamers or tufts as they are variously called. They usually consist of a short length (about 10 cm) of light wool yarn or nylon fastened at one end to the sail. They indicate instantly and continuously the complex flow pattern of the air over the sails. Unfortunately the indication is not a flashing liquid crystal display that tells you precisely what to do next: the human must be programmed to interpret what the telltales are telling.

This is best explained by means of a diagram (fig. 3.59). In (a) three regimes of flow are shown both in terms of the range of operating points on a sail polar diagram and in terms of the appearance of the corresponding air flow. In regime (1) the sail is just beginning to luff and separation of flow occurs on the windward side. Regime (2) corresponds to the region of maximum lift where the flow is attached, that is there is no separation of the flow from the surface, on both sides of the sail. Regime (3) corresponds to a condition of stall where separation occurs over most of the leeward side.

Telltales are best placed near the luff and at the leech. Fig. 3.59b shows their response for each of the three cases. With attached flow the luff telltales stream out and lie flat against the sail. A telltale in a region of separation will twirl or oscillate irregularly with its unattached end moving out from the sail surface. Leech telltales are a very sensitive indicator that the flow has remained attached all across the surface of the sail. In regime (2) the leech telltale should stream out in the plane of the cloth: if the sail is stalled it will twirl down to leeward and it will tend to stream up to windward when the sail is about to luff. Since a sail has twist, and since the wind direction varies with height (chapter 4), it is necessary to have about three sets of telltales more or less equally spaced up the sail. Adjustments to the shape of the sail should be made as described in chapter 4 so that all sets are streaming simultaneously.

It is important not to have too many telltales, otherwise you will be overcome with the kind of despair you might feel when sitting in an airline pilot's seat for the first time. Also, since clear thinking is sometimes difficult when crashing to windward in a small boat it is a good idea to have a pre-programmed guide as to how to react when a tuft twirls. Normally, when sailing to windward the sheeting angle is fixed and the boat is steered to retain the optimum angle of attack. A useful guide is then the following: *always move the tiller toward the direction of the twirling tuft.* Convince yourself of this by checking it out with fig. 3.59. When reaching, the situation is usually different: one steers the desired course and then trims the sails to give maximum lift. In this case *the sheets must be trimmed in the direction of the twirling tuft.*

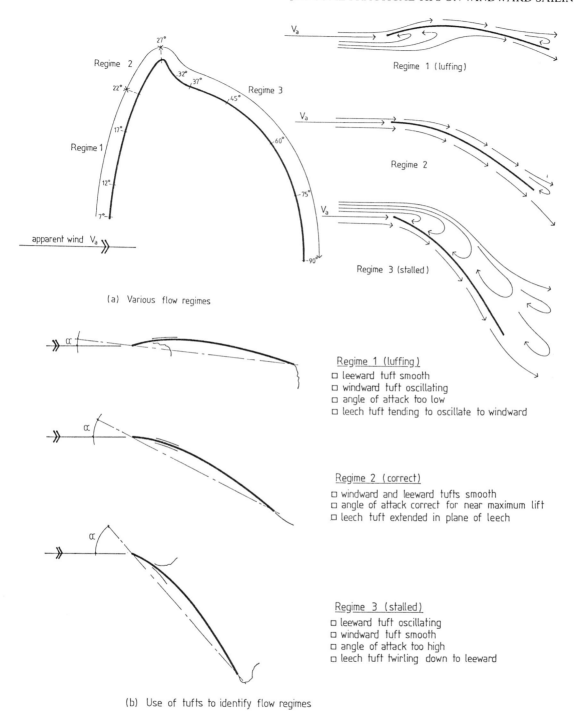

(a) Various flow regimes

Regime 1 (luffing)

Regime 2

Regime 3 (stalled)

Regime 1 (luffing)
□ leeward tuft smooth
□ windward tuft oscillating
□ angle of attack too low
□ leech tuft tending to oscillate to windward

Regime 2 (correct)
□ windward and leeward tufts smooth
□ angle of attack correct for near maximum lift
□ leech tuft extended in plane of leech

Regime 3 (stalled)
□ leeward tuft oscillating
□ windward tuft smooth
□ angle of attack too high
□ leech tuft twirling down to leeward

(b) Use of tufts to identify flow regimes

*Fig. 3.59* Telltales give information about the three flow regimes indicated in (a). Regions of separation are associated with backflow and result in the tuft twirling out from the surface of the sail. In most steady conditions attached flow can be maintained over the whole sail and the tufts will then stream steadily close to its surface.

89

So far telltales have been used to maximise the driving force. An unwanted spinoff from this is that the heeling force is also maximised. In boats which are sailed on a heel it is important to adjust the 'power' of the rig so that the optimum angle of heel is always maintained. Assume one is on a correct windward course with the sails properly sheeted but the angle of heel is too great. Heeling could be reduced to the correct value by luffing, and when this is done the windward telltale will lift and begin to twirl. To get a smooth flow *and* the correct heel angle the sail needs to be flattened. Similarly an underpowered boat can be brought up to the correct heel by bearing away slightly, but then the leeward telltale will lift. This indicates that the sails should be made fuller. Thus it is seen that tufts of yarn can give us at least as much important information about how to sail our boat as the latest microprocessor from Silicon Valley.

One final windward sailing tip concerns the apparently anomalous behaviour of sails in very light weather. The standard practice then is to use full sails for maximum possible lift. This is fine until the wind gets below about 4 knots, when flow around the heavily cambered sail finds it difficult to remain attached. This is because slow-moving air with little kinetic energy is having to move against a large adverse pressure gradient. Separation results in a radical loss of lift. If the camber is reduced by flattening the sail, the adverse pressure gradient is reduced and separation can be avoided. Lift is once again restored and the boat moves away faster in the very light with a flatter sail. Again the telltales will give you warning of this situation.

It is often written that perfecting one's windward ability simply requires practice and more practice until one has developed the 'feel'. Perhaps the cardinal point of this chapter is that there can be no feel without feedback. You need to know what changes to make, why you are making them and what effect they are having.

## §3.7 Summary

The various ways of understanding lift have already been summarised at the end of the relevant sections. The most important fact that emerges from the geometry of sailing to windward is that the course angle $\beta$, or 'pointing ability' measured between the apparent wind and the course sailed, is equal simply to the sum of the hull and sail drag angles. These angles are a measure of the lift/drag ratio but do not correspond to the maximum sail and hull L/D ratios. The requirement that sail forces and hull forces be equal and that sail driving force must be maximised at small leeway means that, when sailing to windward, sails are best operated at a point on the polar curve where the lift is greater than that corresponding to the maximum L/D ratio. Hulls, on the other hand, are best operated with a lift less than that corresponding to the maximum L/D.

For those wishing to determine the leeway made by their boat a method of measurement is outlined. Once a heeling factor, K, has been determined leeway can be obtained easily from boat speed $V_s$ and angle of heel $\varphi$, using the approximate formula: $\lambda = K \sin\varphi/V_s^2$.

Angle of heel is also important in sailing to windward. Any heel is bad and therefore crew-ballasted boats should be sailed upright. For bigger monohulls which must heel there exists an optimum heel angle for winds more than about 10 to 12 knots. It depends on the boat and can best be found by controlled experiments while racing. Once known, the boat is best sailed by inclinometer.

Tacking downwind is common practice in light boats and most effective in light winds: when to do it is the problem. The criteria are explained in §3.5.4.

Although it is now a proven fact for at least one class of boat, most sailors would be inclined to smile indulgently when told of sailing downwind faster than the wind. The principle of the symmetry of sailing predicts this possibility, which is commonplace for ice yachts and the 18 ft skiffs.

# Chapter Four

# SAILS

*. . . behold the threaden sails*
*Borne with the invisible and creeping wind*
*Draw the huge bottoms through the*
*    furrowed sea*
*Breasting the lofty surge . . .*

WILLIAM SHAKESPEARE,
*The Life of King Henry V,* Act III, Prologue

## §4.1   Three-dimensional Lift Effects

The kind of ships referred to by Shakespeare could not go to windward. Only drag forces were involved in determining the driving force of the sails and the resistance of the hull. Boats that sail this way have no need of keels, in fact they are a disadvantage. In chapter 3 we saw that fluid flow, as well as generating drag, can also generate lift. Although not properly understood until the twentieth century, its existence was slowly permeating human consciousness perhaps two hundred years earlier when leeboards and keels were first added to boats. Polynesian and Melanesian double canoes and proas probably sailed to windward long before Europeans did, but the environment had forced upon them a hull form with a large lateral resistance obviating the need for a keel.

The origin of lift which acts at right angles to the flow was fully explained in chapter 3, however entirely in terms of two-dimensional flow. This is the kind usually measured in a wind tunnel where the effects of the ends of the foils are eliminated. For yachts these *end effects* are very important indeed, both under water and especially in the air.

End effects or three-dimensional effects arise simply because there is a difference in pressure between the two sides of a sail or keel. Fluid always wants to flow from a region of high pressure to a region of low pressure. If a sail is porous air will flow straight through it from the windward to the leeward side. Good sail cloth doesn't allow this so the only other course open is for the air to flow under the boom or over the head of the sail in order to equalise the pressure difference. This gives rise to an additional form of drag, as we shall see. The farther the ends are from the main body of the sail, the less is this extra drag: a tall sail (high aspect ratio) is more efficient aerodynamically than a short one (low aspect ratio).

Let us now look at the physics of this in more detail. I will use a sail as a concrete example although the same arguments are applicable to keels and rudders (surface piercing rudders have added complications). Fig. 4.1a shows a sail set so close to a flat deck that there is a hermetic seal between it and the deck. This means that the air flow near the foot of the sail is going to be close to the two-dimensional ideal; its direction is parallel to the sail chord and deck.

The discussion of chapter 3 on lift is now directly applicable in this lower region of the sail. Because of the sail's shape and the angle of incidence of the undisturbed flow, a circulation or bound vortex is set up around it. The lift produced in this region is directly proportional to the strength of the circulation. If we imagine the sail as being divided into a series of horizontal strips each with a width about a tenth of the height of the mast, then we expect the circulation and hence the local lift to decrease with height. The main reason for this is simply that the sail chord is decreasing and so also is the area of each strip. So the picture we have of the sail is a system in which the strength of the circulation is decreasing with height. It was pointed out that the amount of vorticity in the kind of fluid flow we are interested in is fixed. When an aircraft wing begins to move a bound vortex is formed and an equal and opposite starting vortex is left behind. Thus when the circulation decreases from one region to the adjacent region of a sail the difference in vorticity must appear somewhere. It does so in the form of a shed vortex from the leech of the sail (fig. 4.1a).

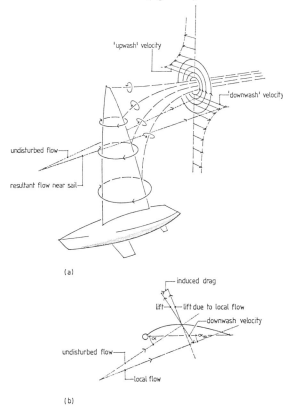

(a)

(b)

*Fig. 4.1* The flow of air around a sail results in circulation which is directly responsible for its lift. A fundamental principle of fluid flow first enunciated by the German scientist Helmholtz and the Englishman Lord Kelvin is that the total circulation in the flow must remain constant. As the circulation decreases with height up the sail the difference is made up in the form of vortices shed from the leech. These all combine downwind to form a trailing vortex whose motion modifies the local flow direction. The lift due to this local flow is no longer perpendicular to the undisturbed flow and has a component parallel to it (b). This is the induced drag or vortex drag brought about by the existence of lift in a system where the flow is no longer just two-dimensional.

Of course the process is not discontinuous as implied by the diagram but consists of an infinite number of infinitesimal vortices which all add up to a finite amount. At the top of the mast, which is an 'end', there is a component of flow from the high pressure side over to the low pressure side which is just part of the overall system of shed vortices. That the shed vortices combine downwind to form one large vortex appearing to emanate from the tip of the sail is beautifully demonstrated by vapour trails from high-flying aircraft. The pressure at the centre of a vortex is less than that of the surroundings, resulting in condensation of the water vapour in the air along the axis of the shed vortex.

The variation of the circulation and lift with height is often referred to as the *distribution of loading*. The term comes from aeronautics where the wing lift must support the weight of the 'plane. As we shall see, the loading distribution, which is a three-dimensional effect, is important in determining the overall drag.

### §4.1.1   Vortex Drag or Induced Drag

The consequences of this trailing vortex will now be looked at. The flow in a vortex is, of course, circular and the tangential speed of the flow varies inversely as the distance out from the centre. This vortex motion is now imposed on the incoming flow, modifying it in the region of the sail.

If we concentrate our attention on the velocities near the plane of the sail we see that the contribution of the trailing vortex is the tangential velocity distribution at the bottom of the vortex flow. This velocity is perpendicular to the oncoming flow and falls off with distance from the centre of the vortex, as shown schematically in fig. 4.1a. For aeronautical reasons it is referred to as *downwash velocity*.

Looking now at a cross-section part way up the sail (fig. 4.1b), the downwash velocity adds vectorially to the undisturbed flow giving the local flow direction as shown. This reduces the effective angle of incidence so that for the same lift as one would get for a two-dimensional sail of infinite aspect ratio at angle of incidence $\alpha_\infty$ one requires the greater geometrical angle of incidence $\alpha$. Thus the effect of a finite aspect ratio is to generate a trailing vortex which produces downwash which reduces the effective angle of incidence.

But there is a more serious effect. Lift is the component of the total force which is perpendicular to the direction of the undisturbed flow. Because of the downwash velocity the perpendicular to the local flow direction is angled downwind. From the point of view of the undisturbed flow the local lift has a component parallel to the flow. As we have seen, such a component is a contribution to the drag. Since this drag has its origin in lift it is often referred to as *induced drag*. The correct modern name is *vortex drag*.

Clearly if the trailing vortex were not there, then there would be no vortex drag. This is the case for an infinite wing of constant section so the data given in figs. 8.16 to 8.21 contain boundary layer drag only. The vortex drag depends on the vorticity shed from the leech, which in turn depends on the change in circulation with height up the sail, so it should be not surprising to discover that the overall vortex drag depends on the sail form and shape.

It is natural then to ask what sail shape will give the minimum vortex drag for a given lift. Perhaps surprisingly, there is no unique answer to this, though that is because it is not a good question. We must ask a slightly more fundamental one: what loading distribution will result in minimum vortex drag? This refers to the way in which the lift varies with height. Since lift is related to circulation K by the expression: lift = $\varrho$VK (see §9.2), we are concerned with the variation of circulation with height. Where the circulation changes there will be a shed vortex which contributes to the downwash. As we have seen, it is the amount of the downwash which ultimately determines the magnitude of the vortex drag, so the really basic question we should be asking is: what distribution of downwash results in the minimum vortex drag for a fixed amount of lift?

According to §3.3.2, lift is determined by the perpendicular momentum change in the flow. This is none other than the downwash. Since we are using aeronautical terms and theory first developed for flight it will perhaps be simpler to visualise the problem as in fig. 4.2. This represents an edge-on view of an aircraft wing. The arrows show the distribution of downwash velocity w. The total lift produced is proportional to the area outlined by the arrows. In (a) the downwash velocity distribution is constant across the wing. (b) represents the same total lift since the area is the same, but the downwash distribution is not constant across the span.

(c) shows how the non-uniform distribution of (b) can be thought of as a combination of a uniform distribution the same as (a) plus a variable distribution consisting of both positive and negative

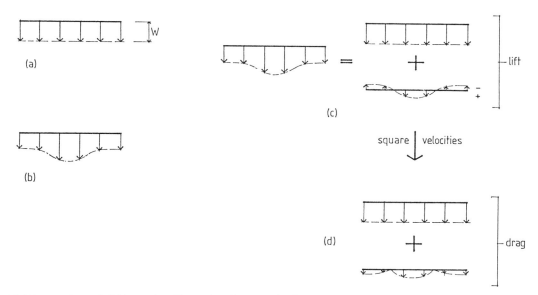

*Fig. 4.2* Edge-on views of a lifting surface like an aircraft wing. Trailing vortices from the tips induce a downward component in the flow called downwash, labelled w. The rate of change of vertical momentum produced by the downwash is responsible for the lift. (a) is a wing with constant downwash and (b) is one with the same lift but variable downwash. (c) shows that (b) can be thought of as having two parts: one the same as (a) and the other with positive and negative downwash. Vortex drag depends upon the squares of the downwash velocities and since the squares of all real numbers are positive, the drag for (b) is greater than that for (a). Thus minimum vortex drag is associated with constant downwash velocity.

downwash velocities. Just as the rate of change of the vertical momentum gives rise to lift, so the energy transfer or rate of change of kinetic energy gives rise to the vortex drag. The kinetic energy and hence the drag is proportional to the square of the downwash velocity.

This is shown in (d) where the downwash velocities have been squared. Now the variable part of the downwash, instead of cancelling at the ends as it does to keep the overall lift the same as in (a), always adds a contribution to the drag that was obtained with a uniform distribution of downwash. This is because the square of any real number, negative or positive, is always positive. Thus for a given amount of lift the vortex drag will always be greater if the downwash varies over the span.

The downwash velocity is produced by the trailing vortices. Each vortex contributes a velocity which varies inversely as the distance from its centre, so the relationship between the change in circulation across the span and the resultant downwash velocity is not an obvious one. It can, however, be determined mathematically in a straightforward way when one finds the very important result that: minimum vortex drag results from constant downwash velocity which in turn results from an elliptical load distribution. This requires a little explanation. 'Load distribution' simply refers to the amount of lift per square metre supplied at different positions between the root and the tip. 'Elliptic' means that the variation of lift or circulation along the span obeys the same equation as describes an ellipse. It is important to realise at the outset that this does *not* mean that the wing or sail has to be elliptical in plan. The circulation depends upon section shape, angle of attack and chord, any of which can be changed along the length.

The elliptical variation of loading along the span is shown in fig. 4.3. This is for a symmetric wing of total span 2s. $K_0$ is a measure of the circulation at the wing root chord, where the wing joins the fuselage. K is the circulation at a distance y from the centre. Circulation drops to zero at the wing tips as expected.

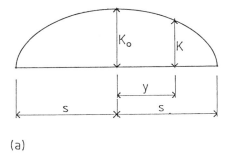

(a)

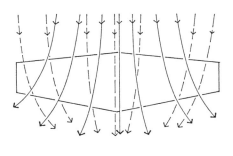

(b)

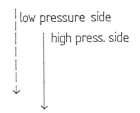

low pressure side

high press. side

Fig. 4.3 Minimum vortex drag is associated with an elliptic lift distribution. This means that the circulation K is related to that at the centre, $K_o$, by the equation $K = K_o \sqrt{1 - (y/s)^2}$. s is the semi-span of the wing and y is the position of K measured from the centre. This equation is just that of an ellipse of semi-minor axis $K_o$ and semi-major axis s. The plan view (b) shows how the flow has a span-wise component everywhere except at the centre. Solid arrows correspond to the flow over the high pressure side and dashed lines over the low pressure side.

Before returning to the practical world of sails it is worthwhile to record here some important theoretical results applicable to the ideal elliptic lift distribution. The first is that the downwash velocity, which is constant across the span, is given by: $w = K_0/4s$, where $s$ and $K_0$ are defined as in fig. 4.3. The second is the vortex drag coefficient which is given by: $C_{DV} = C_L^2/\pi(AR)$. As expected, this depends upon the square of the lift and also varies inversely as the aspect ratio (AR). A large aspect ratio therefore implies a small vortex drag for the same amount of lift; in other words a higher lift/drag ratio. Aspect ratio is defined here as span$^2$/area $= 4s^2/A$, using the symbols of fig. 4.3. The third formula gives the relationship between the lift coefficient, the angle of attack and the aspect ratio. It is: $C_L = 2\pi\alpha/(1 + 2/(AR))$. These have been used to calculate the graphs in §8.7.

Returning now to sails, can we expect to obtain an elliptic loading? Before answering this we need to see how sails differ from airplane wings. The air flow over the central section of the wing in fig. 4.3 is essentially two-dimensional. This is because this is a line of symmetry, each side of which there is a component of flow in opposite directions out along the span and over the tips to the low pressure side of the wing. Right at the centre there must be a neutral zone where no spanwise flow is present. Thus if one were to place a large, thin, flat plate on the centreline in fig. 4.3 and perpendicular to the plane of the wing, the flow would not be affected (apart from a very thin boundary layer region next to the plate) because the flow is everywhere parallel to the plate and hence undisturbed by its presence. Since there is no flow from the right hand side to the left hand side of fig. 4.3, we could physically remove one half of the wing and, providing the plate were left in place, the flow over the remaining half would remain unchanged. The flat plate effectively removes the 'end' and prevents any flow from the high pressure side to the low pressure side.

Now rotate this half diagram through 90° and we have something resembling a sail. In fact if the boom were hermetically sealed to the deck and the hull had negligible effect, then the air flow around the sail would be the same as that around a sail of twice the luff height (fig. 4.4). This means that the vortex drag is that appropriate to the sail plus its 'image' which has a higher aspect ratio than the sail alone. In reality a hermetic seal between the boom and the deck is impractical so a second trailing vortex normally forms in the region of the boom. This doubles the vortex drag, as the following argument shows.

The vortex drag coefficient is given by $C_{DV} = C_L^2/\pi(AR)$. Aspect ratio is given by span$^2$/area. So for the sail plus image of fig. 4.4, $AR = 4s^2/ls = 4s/l$. For the more normal case of fig. 4.5 the aspect ratio is $s^2/(1/2)sl = 2s/l$. Half the aspect ratio means double the vortex drag coefficient. Note that the definition of aspect ratio used here is strictly only for use with an elliptic lift distribution. It is common practice with sails to take the ratio of luff height to foot length, that is $s/l$, as a measure of aspect ratio. This is not unreasonable since the lift distribution of a sail generally differs greatly from the elliptical form, as we shall now see.

Fig. 4.6 shows the measured polar diagram of a

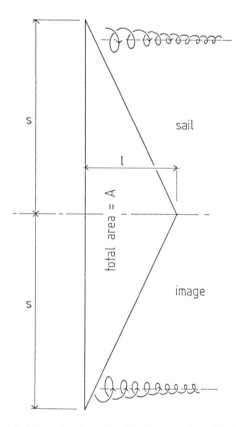

*Fig. 4.4* A large flat plate placed at the centreline of fig. 4.3b would not interfere with the flow since there is no span-wise component at the centre. Therefore the flow over a sail where no leakage between the boom and the deck is allowed is exactly the same as that over a sail plus its 'image' as shown here. In this case the vortex drag is that associated with the higher aspect ratio, $4s/l$, of sail plus image rather than sail alone.

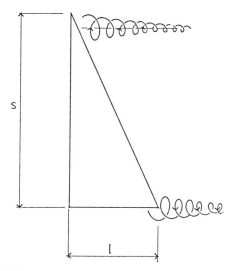

*Fig. 4.5* Since a hermetic seal between the sail and the deck is impractical, there is normally a flow of air under the boom from the windward to the leeward side. Tip vortices are thus produced at both the head and the foot of the sail. This doubles the vortex drag, as one expects from the fact that the aspect ratio is now only 2s/l. Note the comment in the text on this definition of aspect ratio.

Finn dinghy sail taken from C. A. Marchaj's *Sailing Theory and Practice*. The total drag of a sail consists of boundary layer drag plus vortex drag. The former comprises surface friction drag plus drag due to the pressure distribution at infinite aspect ratio. This 'section drag' is usually measured in a wind tunnel and is what is given in the data of §8.7. The vortex drag for an elliptic lift distribution may be calculated using the formula given above and the two added together should give the total drag. As can be seen from fig. 4.6, the measured drag is greater than the sum of these two because the assumption of an elliptic lift distribution has underestimated the vortex drag. Because the lift distribution is not elliptic the vortex drag is greater by the amount shown. A sail with more twist would have even more additional drag so that calculating the drag using an aspect ratio defined as luff height divided by foot length will give a more realistic measure of vortex drag. There is no theoretical basis for this, however. In §4.9 are given formulae for situations where the loading differs from elliptic.

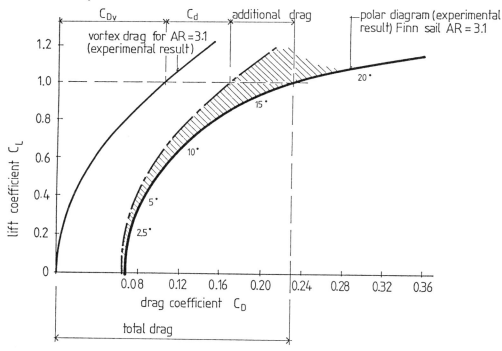

*Fig. 4.6* The curve on the right is the measured polar diagram for a reasonably efficient sail. Total drag is the sum of vortex drag and boundary layer drag. If the vortex drag is calculated as explained in the text on the assumption of elliptic loading, the total drag comes out too small. The remaining drag is additional vortex drag which arises because the sail lift distribution does not correspond to the elliptic distribution which gives minimum drag.

## §4.1.2 Sail Twist and Wind Gradient

The circulation or lift produced by a section of sail depends upon three things: angle of attack, sail shape and chord length. In general all three of these quantities vary with position up a sail. Not only that, they also vary with the strength and direction of the wind as well as with sail tension adjustments and spar bend. Since for a given point of sailing there is undoubtedly a best lift distribution, there are probably a number of ways of obtaining it. Even if we knew how to trim the sail for a given lift distribution, knowing what distribution to use is much more complicated than the equivalent situation in aircraft design.

As pointed out in §3.5 a large lift/drag ratio does not necessarily ensure good windward ability. The constraint imposed by stability and the need to maximise forward thrust also suggest that the sail departs considerably from elliptic loading. Nevertheless it is interesting to see how the total drag is affected by sail twist, which changes the angle of attack and hence the lift distribution with height. Fig. 4.7a shows a sail with very little twist. Twist is measured by comparing the angle of a straight line between luff and leech with that of the boom. It is best estimated in practice by positioning oneself directly behind and in line with the boom and comparing the line of the leech with that of the mast (the aft view diagram shows this). The graph on the right shows the percentage increase in the total drag over that resulting from an elliptic lift distribution as a function of the overall sail lift coefficient. (b) shows the same sail with greater twist. In both cases the air flow is assumed to be uniform over the sail.

Since sails are usually operated with a lift coefficient around 1.0 the sail with less twist has a smaller increase in drag. Furthermore, if it is necessary to luff and operate the sail at a lower lift coefficient the advantage of the untwisted sail is even greater. However the conclusion that a sail should always have a minimum of twist is not necessarily valid. There are some other factors which must be taken into account: the effect of the wind gradient, the constraints of stability, and the influence of the genoa on the flow over the mainsail.

Air flowing over the fixed sea surface has a gradient in velocity, i.e. its speed will vary with height. Unlike laminar flow over a smooth surface where the physical conditions are well defined, the velocity gradient over water will clearly depend on the sea state which will introduce turbulence near the surface

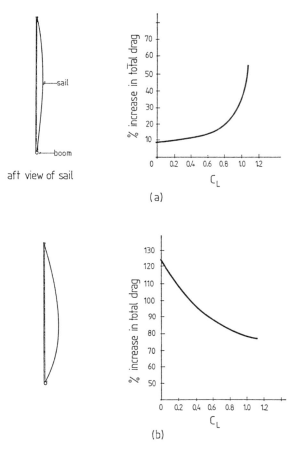

(a)

aft view of sail

(b)

*Fig. 4.7* In practice the degree of twist in a mainsail can be estimated by sighting the leech from directly behind and in line with the boom. This view of a sail with little twist is shown on the left in (a). On the right is the per cent increase in the total drag over that for an elliptic lift distribution. (b) shows the same for a sail with greater twist; note the suppressed zero in the scale. The twist causes a greater deviation from an elliptic lift distribution and hence greater drag. This is for a uniform air flow. In practice the wind turns with height in a way which tends to offset the effect of twist.

with periodic strength and direction variations. Meteorological conditions will also have an influence. Stable conditions result when warm air lies over cooler layers in contact with the sea. Since the temperature is increasing with height the more dense air is at the bottom and there is little tendency for vertical mixing: the velocity change between the upper air and the surface air is fairly gradual. Unstable conditions result when cool air moves over a warm sea. There is then an increase of density with

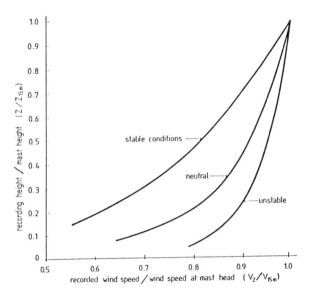

*Fig. 4.8* Friction causes the air flow to slow near the water surface. The effect depends both on sea state and meteorological conditions. For this reason measurements often give widely divergent results. This diagram shows the trend. The vertical axis is height as a fraction of the mast height which is assumed to be 15 m above the water. The horizontal scale gives the wind speed as a fraction of that at the masthead. Stable conditions correspond to normal fine weather in which the air density decreases with height, discouraging the possibility of vertical air movements and hence mixing of fast moving air with lower slow moving air.

This is made clear in fig. 4.9 where the velocity vectors for a boat moving at 5 knots are drawn. The true wind speed at masthead height is 15 knots at an angle of 60° to the yacht's centreline. This results in an apparent wind measured at the masthead of 18.1 knots at an angle of 46°. Now at the foot of the sail the true wind speed drops to about 66% of that at the masthead, that is 10 knots, while the true wind angle remains the same. An anemometer at this level would show an apparent wind speed of 13.3 knots and angle of 41°. The forward motion of the boat then produces an apparent wind which not only increases in strength with height but also increases in angle. In this case the twist in apparent wind direction is 5° so a sail twist also of 5° will tend to keep the angle of attack up the sail more nearly constant. To keep it exactly constant all the way requires the sail twist to vary in exactly the same way as the wind twist. This can be accomplished in practice by the use of luff telltales to ensure that the whole sail luffs at the same time, its twist being controlled by the boom vang.

The second reason why an elliptic lift distribution is not necessarily desirable has to do with the fact that best windward performance is obtained by maximising forward drive while not exceeding an acceptable heeling moment. A paper by C. J. Wood and S. H. Tan in the *Journal of Fluid Mechanics* uses lifting line theory to come up with a surprising result for the optimum load distribution of a sail. It was this theory which showed in the case of aircraft that the elliptic lift distribution gave the highest lift/drag

height and the upper heavier layers move downwards, mixing the fast-moving upper air with the slow-moving air near the surface and producing a rapid increase in wind speed closer to the sea.

Fig. 4.8 shows the trend in wind gradients. Attempts are often made to fit measurements to logarithmic curves determined by parameters said to have some physical significance. Unfortunately measurements vary greatly and are influenced by many factors. Fig. 4.8 assumes that the wind speed is measured atop a mast 15 m above the water surface. In stable conditions (normal fine weather) with the masthead anemometer reading 10 knots the wind speed near the foot of the sail would be in the vicinity of only 5 knots. The average wind speed over the whole sail will thus be greater in unstable conditions for the same masthead reading. The reason this is relevant to sail twist is that a change in direction of the apparent wind is associated with the velocity increase of the true wind.

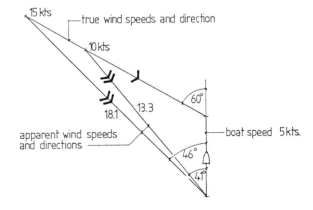

*Fig. 4.9* Friction with the water surface causes the true wind to increase in speed with height, although there is generally no change in the direction of the true wind. When added to the wind caused by the boat's motion through the water, an increase in true wind speed produces both an increase in the apparent wind speed and a change in its direction.

ratio, which for aircraft is clearly desirable. Boats, on the other hand, require a maximum driving force for a given heeling force. Using this criterion the authors were able to determine from an approximate theoretical calculation what the optimum lift distribution of a sail should be for this criterion.

The results are given in fig. 4.10, of circulation K plotted against the 'dimensionless mast height'. It depends on the true mast height, the righting moment and the wind speed. To appreciate the physical significance of dimensionless heights of 2, 4 and 6 shown in the diagram, a comparison may be made with a typical racing dinghy having a righting moment of 1500 Nm, obtained by the crew stacking out, and a mast height of 7 m. In a wind of 10 m/sec this corresponds to a dimensionless mast height of about 3. Thus the left hand graph of fig. 4.10 corresponds to a rather under-rigged boat. The centre one corresponds to a rig somewhat taller than conventional for the conditions, whereas the right graph corresponds to an extreme rig about twice as high as normal. Curve A in each case gives the calculated optimum circulation distribution with a

normal boom height. Curve B gives the optimum with no gap between the boom and the deck. Curve C is for elliptic loading, also with no gap between boom and deck. The calculations have been made for close windward sailing. Under each diagram the resultant overall driving force $F_R$ and the heeling force $F_H$ are given. For the low rig or light winds elliptic loading seems as good as the optimum loading. With the taller rig in the centre the optimum loading gives a driving force improvement associated with a big reduction in lift from the upper portion of the sail. This is commonly achieved by allowing a large amount of twist in the upper part of the sail when beating in a strong breeze.

For the very tall rig in the right hand diagram this idea is taken to the extreme. Because of the disproportionate overturning effect of a conventionally loaded sail, the magnitude of the elliptic loading distribution becomes very small and the corresponding forward thrust is severely reduced. In contrast, the optimum loadings A and B give much greater driving force. The striking result is that this is achieved with an actual *reversal* of lift at the top of the

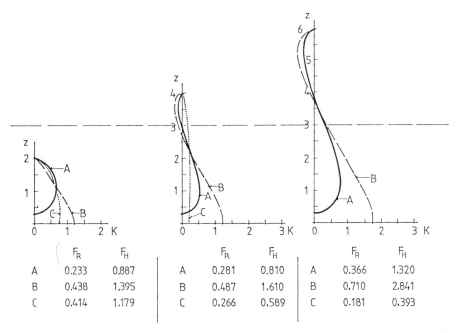

| | $F_R$ | $F_H$ |
|---|---|---|
| A | 0.233 | 0.887 |
| B | 0.438 | 1.395 |
| C | 0.414 | 1.179 |

| | $F_R$ | $F_H$ |
|---|---|---|
| A | 0.281 | 0.810 |
| B | 0.487 | 1.610 |
| C | 0.266 | 0.589 |

| | $F_R$ | $F_H$ |
|---|---|---|
| A | 0.366 | 1.320 |
| B | 0.710 | 2.841 |
| C | 0.181 | 0.393 |

*Fig. 4.10* Airplanes function best with an elliptic distribution of lift across the wing. The criteria for optimum speed to windward of a sailing boat are different. One requires to maximise the sail drive $F_R$ for a given maximum heeling force $F_H$. The same theory which predicted elliptic lift distributions for airplanes then predicts distributions A and B for yachts. The vertical axis is mast height and the horizontal one the circulation. These calculations show that improved windward efficiency can be obtained by using very tall rigs in which the upper part of the sail keeps the boat on its feet by producing reverse lift.

sail, which is not used to produce thrust. This local negative lift, by virtue of its long moment arm, has the effect of holding the craft upright, at the expense of some drag, in order to support a very much increased sail loading lower down.

What these observations imply is that very tall masts incorporating a method of forcing the upper portion of the sail to leeward would give greater windward efficiency. The structural difficulties of building such a rig and controlling it probably rule this out as a practical idea. The theory that gave rise to these predictions is an approximate one and the numbers given in fig. 4.10 must therefore be regarded as optimistic. The main point is that sails with little twist and close to elliptic lift distributions are fine for sailing in light and stable weather conditions. When increasing wind makes the heeling force reach a point where further flattening of the sail is not possible, increased twist to leeward near the top is a very effective way of keeping the boat on its feet, especially if it can be accompanied by an increase in sail drive from lower down.

The third reason why an elliptic lift distribution is not always desirable for a mainsail applies to fractionally rigged boats. Fig. 4.1 shows how the trailing vortex centred near the top of the mast induces a downwash velocity over the sail. This downwash is produced by the circular motion of the

vortex, so above the mast there must be an analogous velocity component which now constitutes an *upwash*. Consider now the trailing vortex from a foresail. Below the point of attachment to the mast it will have a tendency to decrease the effective angle of incidence on the mainsail. Above the point of attachment the upwash will have the effect of increasing the mainsail's effective angle of incidence. This suggests that fractionally rigged boats should have more twist in their mains than masthead or cat rigged boats.

## §4.2  Sail Interaction

A great many misleading statements have been made in yachting publications about the important question of sail interaction. The problem had its origins not with sailors but with aerodynamicists; the jib-mainsail combination has its analogue in the leading edge slot of an aircraft wing. The popular aerodynamic explanation for its operation was to invoke the Venturi effect in which the air flow, being confined to a narrow slot, speeded up thereby reducing the probability of separation and improving lift. Unfortunately this picture is quite wrong, and along with all the other information from aeronautics has found its way into the folklore of sailing. Although the aerodynamicists have long since

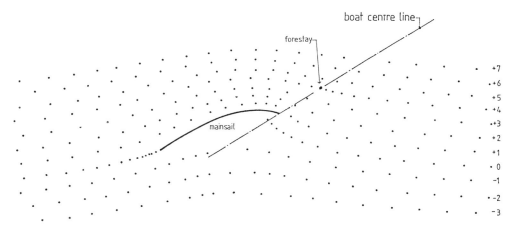

*Fig. 4.11* Two-dimensional potential flow around a mainsail alone, as measured by electric analogue plotter. This is an accurate representation of the flow for situations where separation or stall are not occurring. The sail angle of incidence is 15° and is measured by the angle between the direction of the undisturbed flow and the straight line from luff to leech. The dashed line shows the orientation of the centreline of the boat making an angle of 30° to the wind.

The solid dot shows the position of the forestay and hence the luff of the jib when it is set. An important characteristic of such flows is the extent of the upwash near the luff and the crowding of the streamlines on the leeward side, indicating a low pressure there with higher pressure on the windward side. This diagram is to be compared with fig. 4.12 where the flow lines at a large distance away are the same and have the same labels.

corrected the mistake, the misinformation still lingers in the yachting fraternity.

The question of just how sails interact is difficult to determine from actual flow measurements. A way to get consistent two-dimensional flow lines is to use an electric analogue plotter (see §8.2). Knowing the fluid flow pattern allows velocities and pressures to be determined at all points. All flow diagrams shown in this section have been obtained from the analogue plotter and therefore represent actual measurements of the ideal flow and not the artist's guess as to what the flow should look like.

Remember that 'ideal flow' means in this context the flow outside the boundary layer in situations where there is no separation. Since the boundary layer is only of the order of a centimetre thick and flow is normally attached over correctly trimmed sails going to windward in a moderate breeze, the ideal flow correctly describes the real world except for boundary layer drag. Lift and vortex drag for two-dimensional situations are given correctly.

Fig. 4.11 shows the general nature of the flow around a single sail. There are some important features of this standard flow pattern that should be noticed. The first is that there always exists a stagnation flow line which stops somewhere near the leading edge of the sail and continues off the trailing edge. The stagnation line separates the flow into that which goes to leeward of the sail and that which goes to windward. Another feature of the stagnation line is that it leaves the leech parallel to the surface. This is known as the Kutta condition (see §3.3.3).

A further characteristic is that where it approaches the sail it exhibits upwash; the flow lines start curving 'up' toward the sail even before they reach it. This is not some mysterious telepathy on the part of the wind but is the result of pressure changes due to the presence of the sail being transmitted in all directions just like sound waves. Since they travel rather faster than the wind, these pressure effects can move ahead and influence the flow before it reaches the sail. It is this upwash that allows it to fill at a smaller angle of incidence than might otherwise seem possible. As the angle of incidence increases so does the upwash. This means that the streamline which goes just to leeward of the sail must turn sharply in order to follow its curvature. A favourable pressure gradient is therefore required if the flow is to remain attached. There may be a momentary separation near the luff giving a 'bubble', followed by re-attachment.

Note that the fourth streamline up from the stagnation line passes through the forestay position.

As explained in §8.2, the Kutta condition which requires that the stagnation line come smoothly off the trailing edge is obtained by feeding electric current into the model and it can be shown that this is directly proportional to the amount of circulation in the flow. This circulation is in turn directly proportional to the lift produced. Thus the electric analogue device provides a very quick method of determining the lift. The values below for the circulation will be given in arbitrary units which have no absolute meaning but have comparative value only. In fig. 4.11 the angle of incidence is 15° and the centreline of the boat is at 30° to the undisturbed apparent wind direction. Under these conditions the measured circulation is 109 arbitrary units.

Each flow line is really an equipotential line measured at quarter-volt intervals with respect to the stagnation line. Well away from the sail the lines are equally spaced and represent a uniform flow. The stagnation line labelled zero separates the flow to go to leeward and to windward of the sail; the windward lines are labelled −1, −2 and −3 and the leeward lines +1, +2, +3 etc.

Fig. 4.12 shows the flow when a foresail is present. The dashed lines are the jib stagnation lines. Again the flow lines are measured at quarter-volt intervals with respect to the mainsail stagnation line, which means that well away from the sail the flow lines in both diagrams coincide in position and in labelling. The only exception is the dashed stagnation line to the jib. The foresail modifies the flow in a number of important ways. The stagnation streamline now approaches more smoothly into the mainsail so that the flow over the leeward side has less curvature. The value of the minimum pressure is now not so low because the streamlines are further apart, thereby reducing the adverse pressure gradient toward the leech and reducing the chance of flow separation. Thus improved flow over the lee side of the main has been obtained by a *reduction* in flow velocity near the luff rather than an *increase* near the leech which is often erroneously claimed. If you remember that the spacing of the streamlines is a measure of flow velocity, you can see that there is little change in this in the 'slot' near the leech.

The number of flow lines between the forestay and the mast is a measure of the amount of air going through the slot. In fig. 4.11 flow lines 1, 2 and 3 all pass through this region. However with the jib only line 1 passes through the slot while 2 and 3 go to leeward of it. So not only has the velocity of flow in the slot been reduced but also the quantity of air

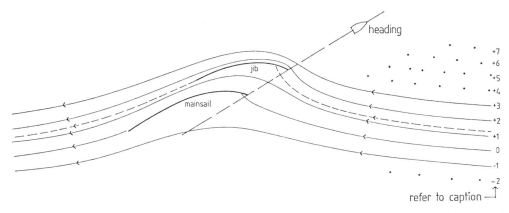

refer to caption —↑

Fig. 4.12 The course angle and mainsail angle of attack are the same as in fig. 4.11. Now, however, a jib has been added at an angle of attack of 5°. The streamlines are labelled as in fig. 4.11. Notice that the upwash on the main has been reduced but is large on the jib. Whereas with main alone streamlines 1, 2 and 3 passed between the mast and the forestay, now only streamline 1 does so. That is, the flow in the slot has been reduced. The presence of the main has redirected the flow to the leeward side of the jib increasing the velocity there and reducing the pressure. The dashed streamline is the stagnation line for the jib, which of course has no counterpart in fig. 4.11.

flowing through, as it must be. Thus the same air that formerly went between mast and forestay now flows close around the lee side of the jib.

That the flow velocity in the slot should be reduced is theoretically reasonable, can be seen in fig. 4.13 where the circulation around each sail is drawn. In the slot, the circulations oppose one another so that a *reduction* in flow speed is to be expected.

The upwash flow ahead of the mainsail causes the stagnation point on the jib to be moved more around to windward, increasing the effective angle of attack. This allows the boat to point higher while still keeping the sail filled. The reduction in air through the slot means an increase in the flow around the lee side of the jib. Put another way, the main has the effect of diverting air around the lee side of the jib, so increasing greatly the flow velocity there. This reduces the pressure on the leeward side, which combined with the lower speed higher pressure region between the two sails, produces a strong pressure differential across the foresail.

The circulations around each sail were measured for fig. 4.12 by measuring the current necessary to maintain the Kutta condition of flow off the trailing edges, giving the following results in arbitrary units which may be compared directly with fig. 4.11. For the mainsail the circulation has dropped from 109 units for main alone to 59.5, and for the jib it is 92. The combined circulation of 151.5 is considerably greater than for mainsail alone, as is to be expected on grounds of increased sail area. More significantly, the mainsail circulation has been reduced and even

though it has greater 'area' than the jib provides only about 65% of the lift. Remember that the lift is at 90° to the undisturbed wind flow and in this instance the boat's centreline is at 30° to this direction. The forward driving force component is therefore given by multiplying the lift by sin 30° = 1/2. Providing the course angle doesn't change the driving force is then proportional to the circulation. Note that the angle of attack on the main is now less so that it must be sheeted in harder with a jib. Also, because the flow

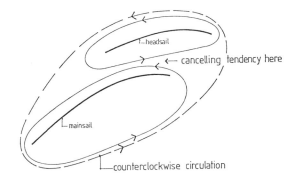

Fig. 4.13 As explained in §3.3.3, circulation is associated with the flow around a lifting surface. This does not mean that there is a bodily flow of air around sails. It means that the flow around the sail is asymmetric so that the average velocity parallel to a closed path is not zero as it would be in a uniform flow. Since both sails are producing lift in the same direction, they both have counterclockwise circulations resulting in a tendency to cancel the flow speed in the slot. The path for determining the overall circulation is represented by the dashed line and is simply the sum of the two separate circulations.

velocity over the lee side of the main near the luff is slowed when the jib is present, the pressure there is increased tending to fill the main the wrong way. This of course is the well known backwinding of the main by the jib. The expression is unfortunate since it conjures up the idea of a stream of air hitting the jib and bouncing off onto the back of the main. As I hope this analysis has made clear, that picture is very far from reality.

It should be evident from the above that it makes no sense to speak of the genoa as being the more efficient sail. The presence of the mainsail is what makes the foresail more efficient than it would be alone. Thus the proper trim and shape of the mainsail significantly affect the efficiency of the jib. For instance, if there is a tendency for separation on the main then velocities near the leech of the jib will be reduced, decreasing its effectiveness. Since mainsail trim determines the upwash on the jib, it directly affects the boat's pointing ability.

Nothing has been said about the effect of the mast. Its main effect will be to produce a separation bubble near the luff of the mainsail, after which flow becomes re-attached. Flow separation cannot be modelled by the electric analogue method which is why the mast has been left out in the flow measurements shown here. Undoubtedly the mast does impair the effectiveness of the mainsail, so the superiority of an overlapping genoa is clearly a combination of the undisturbed air flow into the luff with the beneficial interaction of the mainsail. An interesting experiment was carried out in the days of 6 metre racing by Sherman Hoyt, who built a boat called *Atrocia* which had a large genoa and small main. The argument probably was that the genoa was inherently more efficient because of the absence of a mast along its luff. The boat was unsuccessful. It could be argued that the relative effect of the mast on the small main was now greater, but most of the difference was surely due to changed sail interaction: the small mainsail was no longer capable of providing the conditions necessary to make the genoa efficient.

Although we generally think of air as impinging on sails, for the most part it is flowing parallel to the sail surface. The pressure forces which determine sail shape result from differences in flow velocities rather than air striking the cloth like a stream of bullets. Although this picture was used in the momentum change theory of lift in chapter 3, it would not be what we would see if somehow the air flow could be rendered visible. Then, instead of our ships being 'borne with the invisible and creeping wind' we would see regions of pressure change. This would have a radical effect on the way in which we sail. To get some idea of this, fig. 4.14 shows the interaction between two sails of the same camber and chord in terms of the resultant sail pressures. This diagram is worth close study as it summarises all the features of sail interaction that have just been discussed.

Possibly the most common question concerning sail interaction is: what should be the relative sheeting angles of jib and main? In fig. 4.12 the main has an angle of attack of 15° and the jib 5°. This is the angle between the direction of the undisturbed flow far ahead of the sail and the straight line drawn from luff to leech. The effectiveness of the sail

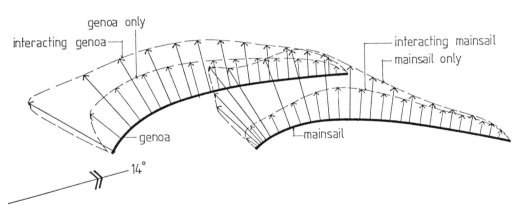

*Fig. 4.14* The net force on a sail is determined by the difference in pressures on each side. One 'envelope' of arrows represents the sail forces when only the mainsail is present: note the large forces near its luff. The other arrows indicate the forces when the two sails interact: the main's contribution is now greatly reduced whereas the genoa contribution is increased beyond that made by the genoa alone. If the pressure forces near the luff of the main get reduced to zero or are reversed in direction by interaction, then it will luff.

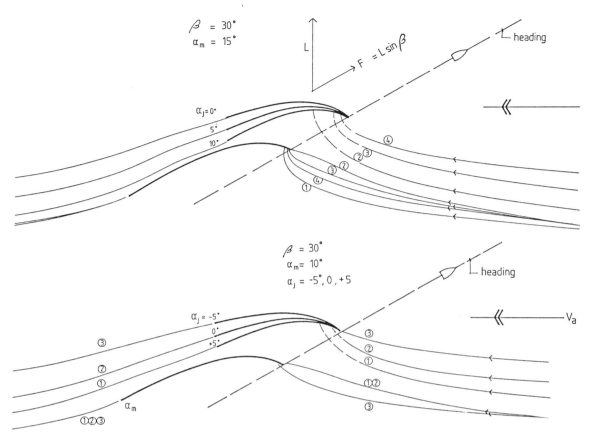

*Fig. 4.15* The effect of varying the jib sheeting angle as determined with a two-dimensional electric analogue plotter. Only the stagnation lines are shown. In (a) the main has an angle of attack of 15° and in (b) it is 10°. Sheeting the jib in causes the stagnation line to move around the luff of the main in a counterclockwise direction. Sheeting the jib in too hard causes the stagnation line to the main to approach from the leeward side resulting in backwinding. On the other hand the largest possible angle of attack is required on the genoa to maximise the lift. This is why some types of boat have been found to go to windward best with the main luffing slightly near the mast. A table in the text compares the driving forces for the situations depicted here.

combination can be estimated from the circulation as measured by the current flow into the model at the Kutta condition as described above.

In fig. 4.15a the main has an angle of attack $\alpha_M$ of 15° and in (b) of 10°; in both cases the boat's heading $\beta$ is 30° to the wind. The diagram shows the effect on the stagnation flow lines of varying the jib sheeting angle $\alpha_J$ in 5° steps. Sheeting the jib in too far causes the mainsail stagnation line to approach the sail from the leeward side producing backwinding. The effect is more pronounced with the mainsail at 10° angle of incidence than 15°, as might be expected (remember that sheeting in the main increases the angle of incidence). It is interesting to see what these jib sheeting angles do to the circulation and hence the drive; this is summarised in the table.

| $\alpha_M$ | $\alpha_J$ | $K_M$ | $K_J$ | $K_T$ | *Flow in slot* |
|---|---|---|---|---|---|
| 15° | 10° | 38.5 | 109.5 | 148 | |
| 15° | 5° | 59.5 | 92 | 151.5 | |
| 15° | 0° | 75.1 | 48.2 | 123.3 | |
| 10° | −5° | 72.3 | 36.7 | 109 | 710 |
| 10° | 0° | 48.2 | 73.7 | 121.9 | 386 |
| 10° | 5° | 39.6 | 92.4 | 132 | 212 |

$\alpha_M$ and $\alpha_J$ are the angles of attack on the main and jib respectively. $K_M$, $K_J$ and $K_T$ are the relative circulations of main, jib and total for both. 'Flow in slot' is measured by the voltage difference between the main and jib stagnation lines, a measure of the amount of air passing between the two sails. There is clearly a tendency for slot flow to decrease as the sail drive increases.

For a fixed mainsail sheeting angle the effect of changing the headsail sheeting angle is to vary the relative contribution of the two sails to the total drive. In fig. 4.16 it can be seen that the maximum drive is obtained when the jib is contributing about twice the drive, area for area, of the main.

Of course there is nothing absolute about the angles of incidence shown in fig. 4.15; they can only give qualitative information as to the kinds of effects produced. Remember that the measurements are two-dimensional only, and stall and separation effects do not show up. Furthermore the shape of real sails depends on the forces to which they are subjected. Here we have conveniently assumed that the sail shape is always the same.

Up till now we have considered sail interaction only for the windward sailing situation. On a reach the sails are effectively farther apart so we would expect less interaction. This is indeed the case, as fig. 4.17 makes clear. The total circulations for the three genoa sheeting angles of 5°, 15° and 22° are 130, 151 and 151 respectively: clearly genoa adjustment is not nearly so critical. The resulting best circulation is only marginally greater than the windward case, but now the sail drive is twice as great since the lift is now parallel to the heading.

Although a genoa angle of attack of 22° would probably result in stalling, there is no loss of drive at 15° whereas there is at 5°. This brings up a point about sail trimming on a reach, or for that matter on any point of sailing where maximum lift is required. Beginners are often told that they should sheet in the sails to the point where they just don't luff. This may be convenient for teaching beginners but since lift rises with angle of attack the latter should be maximised whenever possible, the limitations being heeling force and stall. Stall is indicated by the leeward telltales which should not be allowed to twirl, and leech telltales curling back to leeward. So better advice to the beginner would really be: sheet in as much as you can while keeping all telltales streaming.

The question is often asked whether the combination of main and genoa is more efficient than

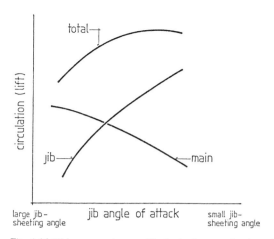

*Fig. 4.16* This summarises qualitatively the contribution of the mainsail and foresail to the overall drive as the jib's angle of attack is varied. The jib circulation or lift increases with increased angle of attack. In so doing it reduces the effective angle of attack on the mainsail and its lift drops. In most sail combinations the optimum will occur when the jib is contributing more, area for area, than the main.

main alone, as in a catboat or una rig. Within the limits of two-dimensional flow as measured by the electric analogue plotter, this can be answered providing we define what we mean by 'efficient'. If we assume light wind conditions where heeling force is not a limiting factor, efficiency could be defined as the magnitude of the circulation per unit of sail area. For a given apparent wind angle this is proportional to the sail drive per unit area of sail. In our two-dimensional analogue, area is collapsed into sail chord length. In figs. 4.12 and 4.15 the jib/main chord length ratio is 0.74. When the main alone is used, the circulation in arbitrary units is about 109 at an angle of incidence of 15°. The best main–jib combination when the main is kept at 15° gives a circulation of about 150: although the sail area has been increased by 74% the sail drive has gone up by only 30%. Not too much should be read into this. Three-dimensional effects could change the picture considerably and one would certainly expect differences between a masthead and a three-quarter rig. The effect of the mast on the performance of the mainsail is important and could well nullify the seeming advantage that the above figures suggest for the one-sail rig. Sail drag coefficients, which affect windward performance, have not been included.

Although there have been some excellent wind tunnel measurements of Finn dinghy rigs, there do not appear to be any comparable measurements of

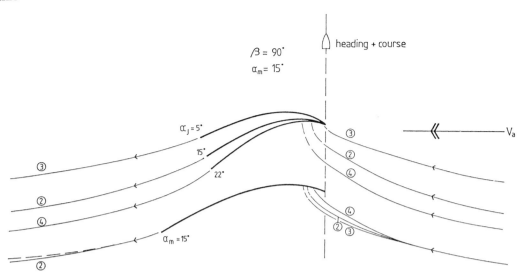

Fig. 4.17 Sail interaction when the heading is 90° to the apparent wind. Here the sails are effectively further apart and the interaction correspondingly reduced. However the practicalities of sheeting the foresail usually result in a considerable increase in sail camber compared with the windward situation.

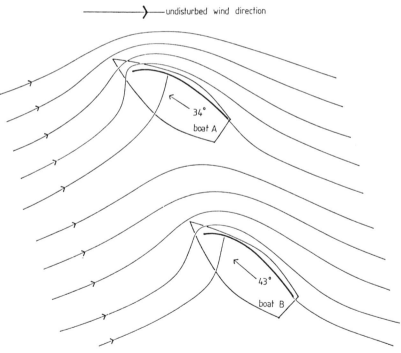

Fig. 4.18 Sailors are familiar with the 'safe leeward position', for which the flow lines have been plotted by electric analogue plotter. The angles of incidence have been exaggerated to show the effect more clearly. Just as a mainsail redirects the air around the leeward side of the genoa, so does B's sail improve the air flow over A's. The presence of B increases the upwash on A so that A can point higher. Although A is pointing 9° higher than B, the closer spaced streamlines to leeward of A show that she is experiencing greater sail forces than B and hence is footing faster and pointing higher.

jib-headed rigs from which lift coefficients and drag angles could be compared.

One final piece of interesting information that can be obtained from the analogue plotter concerns the safe leeward position. Fig. 4.18 shows two boats in close proximity sailing to windward. A is ahead but to leeward of B. Just as jib and main interact so do the sails of A and B. All the previous arguments that showed how the mainsail improved the performance of the jib can now be applied to the sails of boat B improving the performance of A. The presence of B's sail increases the upwash on A's so that A can point higher. The increased speed of flow around the leeward side of A's sail means that not only is the wind direction more favourable for her but the apparent wind is also stronger. Although A is pointing 9° higher than B she is also sailing faster since her streamlines are more closely spaced to leeward, indicating a greater sail force.

## §4.3   Full Size Measured Sail Data

Full scale measurements on yachts are notoriously difficult, yet for 40 years sail forces were calculated from the so-called *Gimcrack* coefficients determined indirectly from full-scale measurements. Probably the first truly scientific attempt to study the behaviour of the sailing yacht was that of K. S. M. Davidson of the Stevens Institute in New Jersey. In 1936 he published a now classic paper in the *Transactions* of the Society of Naval Architects and Marine Engineers. It was concerned largely with tank testing, but predicting the speed of a yacht when its resistance is known requires a knowledge of the sail forces in a given wind strength.

*Gimcrack* was a sloop of 2650 kg displacement, a waterline length of 7.25 m and a sail area of 40.3 m$^2$. A model was first tested in the tank and full scale towing trials confirmed the predicted upright resistance. Sailing trials were then conducted and boat speed, heel and apparent wind measured. Observations were recorded only when the helmsman considered that the yacht was sailing steadily on the optimum course for the existing wind.

Measurement of the speed and heel of the boat could be related, via the model tests, to the forces required to produce that speed and heel. Since sail forces are equal to hull forces when sailing steadily, this allowed the determination of the real sail forces. Fig. 4.19 summarises these *Gimcrack* coefficients. The corresponding forces in newtons may be calculated from: force = coefficient × 0.6AV$^2$,

where A is sail area in m$^2$ and V is apparent wind speed in m/sec. Or, the force in pounds is 0.00119 × AV$^2$ × coefficient, where A is the sail area in ft$^2$ and V is the apparent wind speed in ft/s. Possibly the most useful way to express this is with the wind speed in knots and everything else in metric units. We then have: force (newtons) = coefficient × 0.159AV$^2$, where A is in m$^2$ and V in knots.

When tank tests were carried out on hulls the designer wanted a prediction of boat speeds. Using these coefficients the sail forces could be calculated from the boat's sail area, compared with the scaled-up model forces as a function of speed, and hence a prediction of the speed of the full size boat obtained. Obviously the method was only valid for comparing hulls since they were all assumed to be carrying rigs scaled up or down from *Gimcrack*'s. Despite the obvious drawbacks of this method, boat speed prediction from tank tests has in the past relied heavily on the *Gimcrack* coefficients largely because of the lack of any other reliable full scale data.

The curious feature of fig. 4.19 is the rapid decrease in all the coefficients with angle of heel, which is basically because a number of other variables are involved with increasing heel. Increasing wind would stretch the sailcloth and so reduce the effective incidence. It is also probable that the helmsman would point higher and trim the sails flatter as the wind strength increased. Although a correction was made for wind gradient, it is now known that this depends on meteorological conditions and these are likely to be different in different wind strengths.

Fig. 4.20 gives a set of more modern data made by the same method as the *Gimcrack* coefficients in the early 1970s, using a more modern masthead rig and modern sailcloth. Unlike the *Gimcrack* coefficients, which apply only to sailing hard on the wind, fig. 4.20 gives values for all points of sailing. Combined with wind speed and sail area, the curves can be very useful for approximate determinations of sail forces for a sloop rig.

Apart from the *Gimcrack* method, other ways of determining actual sail forces have been tried. One idea was to put strain gauges in the rigging. This is difficult enough with a masthead rig and a stiff mast since the forces of interest are small components of the forces measured, and critically dependent on rigging angles. It is even more difficult with a bendy mast and where rigging angles vary with load. In addition, only a fraction of the rigging load is due to sail forces because of the large static tensions which

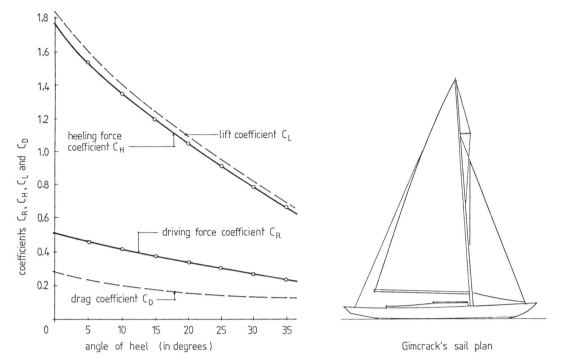

Gimcrack's sail plan

*Fig. 4.19* When a hull is tank tested the designer generally wishes to estimate the boat's windward performance. These sail coefficients allow sail forces for different sail areas and wind strengths to be calculated. Comparison with the tank-tested hull forces allows the boat speed and leeway to be determined. Since not all boats have a rig identical to *Gimcrack*'s such predictions could only be regarded as comparative rather than absolute; in this sense any reasonable values for sail coefficients would have been just as useful. The values given here are for sailing to windward and presumably correspond to a heading of about 30° to the apparent wind. The most often remarked-upon characteristic of these curves is their 'droop' with heel angle, which is commented on in the text.

must always be present. Another way of determining full scale sail forces is to measure mooring loads. Basically there are two ways of doing this. One can moor the boat at a fixed angle to the wind and measure the forces in the mooring lines, or one can mount the rig on a platform on dry land and measure the forces imparted to the platform. The first of these works reasonably well for light centreboard boats kept upright by crew weight; the sail data of fig. 3.26 were obtained in this way. With keelboats the method does not work well because of continual shifts in speed and direction of the real wind. This results in large fluctuations of the angle of heel which do not occur when a boat is moving through the water because it then has aerodynamic and hydrodynamic damping (explained further in chapter 6). Possibly the best full scale method is to mount the rig on a rotatable platform on flat land where the wind flow would be similar to that over water, and by means of a wind vane or more sophisticated feedback mechanism keep the sail's angle of incidence fixed while fast-acting instrumentation measured the sail forces as a function of wind speed and time. Although a little work has been done along these lines, modern signal processing techniques could make this more fruitful in the future.

One of the difficulties inherent in all full scale measurements is the accurate determination of the apparent wind speed and angle. Wind speed and direction are normally measured at the top of the mast, and the wind gradient must therefore be known in order to determine the mean wind speed incident on the sail. The anemometer and wind vane will be affected by the angle of heel so this must also be corrected for. Worst of all, the air flow at the top of the mast is considerably distorted by the presence of the sails in a manner not easy to predict. One measurement has shown that the masthead wind direction is rotated about 7° toward the direction of heel when sailing to windward. This will differ with

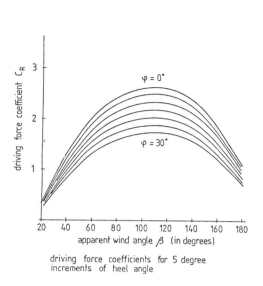

driving force coefficients for 5 degree
increments of heel angle

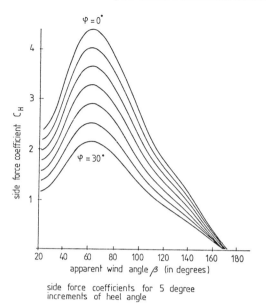

side force coefficients for 5 degree
increments of heel angle

*Fig. 4.20* Modern sail coefficients obtained in the same way as those of fig. 4.19. That is, full-scale measurements of boat speed, wind speed, heel angle were used to determine hull forces from scaled-up tank test results. From these, sail forces and hence coefficients could be derived. These graphs, for a masthead sloop, are useful for obtaining approximate values of steady sail forces for stress or stability calculations.

course sailed, angle of heel and rig but gives some idea of the size of the error. A better place from the point of view of measurement is about halfway up the mast and about 3 or 4 m away, but unfortunately equipment there is not likely to survive the foresail for very long. These variations between instrument readings and the true flow of wind or water, although disturbing for quantitative evaluations, are normally fed only into the brains of the crew where they are processed in a far from quantitative way into appropriate action. To quote C. A. Marchaj, '. . . the effective use of sailing instruments becomes itself an additional skill which can hardly be perfected by those who do not dare to brave the realms of sailing theory.' In view of these remarks it must be conceded that present techniques of full scale sail measurement have not yet reached the point where quantitative predictions of absolute performance can be estimated for anything but the smallest of boats.

## §4.4   Wind Tunnel Sail Data

A kind of Murphy's Law seems often to operate in scientific research: if an experiment is easy to perform then it is difficult to analyse, and if it is difficult to carry out then it is easy to analyse. This is

the kind of situation that exists with sail measurements. Direct measurements on real sails are directly interpretable but very difficult to make. Excellent measurements can be made with relative ease in a wind tunnel, but since real flow conditions cannot be matched, because of various scaling difficulties, 'tunnel blocking' etc, such measurements are very hard to interpret. In this section some wind tunnel data on both standard and unusual sails will be presented which might provide you with food for thought.

In chapter 2 it was pointed out that the pattern of fluid flow is determined by the flow velocity, the fluid viscosity and density as well as the shape of the object. These quantities were combined into one dimensionless quantity called Reynolds' number. The flow around two similarly shaped objects of different size will be the same providing the Reynolds' number is the same in each case. This is true providing no surface waves are involved, which is certainly the case in the study of sails.

Reynolds' number is the product of flow velocity times linear size divided by kinematic viscosity. Thus if we want to use a wind tunnel to study the flow around a 1/10th scale model we must increase the wind speed by a factor of 10 to keep the Reynolds' number the same. A 10 knot breeze on the full-scale sail now must become a 100 knot hurricane to get the same flow conditions for the model. Since the stresses in the cloth are proportional to the velocity squared times the linear dimensions, the stress in the 1/10th scale model is 10 times as great (for the same cloth thickness). This means more stretch, making the shape of the model sail different from that of the full size one. This is a fundamental problem in wind tunnel testing of scaled-down sails and for this reason many measurements have been made using 'tin' sails, metal plates bent into fixed cambers.

Another way around the problem is to measure the model sails in a water tunnel rather than a wind tunnel. Because of the different densities and viscosities of water and air it is possible to scale down by a factor of 20 or so while retaining approximately the same Reynolds' number and fabric tension. Another problem with wind tunnel measurements is an effect called blocking. Even at some distance from a sail the air flow will move along some sort of curved path, yet the flow along the walls of a wind tunnel is constrained to be straight. Unless the cross-section of the tunnel is several times greater than that of the model a distortion of the flow will result.

Despite all these difficulties wind tunnel measurements have so far supplied the most useful information about sail characteristics from the point of view of the sailor. Useful results have been obtained with soft sail models in wind tunnels; although the fabric tension is not scaled properly it is still possible to get some idea of trends in aerodynamic properties when the sail shape is modified to change camber or twist. Knowing the absolute values of lift or drive coefficients is of little value to the helmsman while sailing, especially in a small boat without instruments, but what is of value is the knowledge of how these quantities *change* with change of sail shape.

## §4.4.1 Sail Facts from Wind Tunnel Measurements

Fig. 4.21 is taken from data representing one of the few reliable measurements of the effect of a mast on the characteristics of a mainsail.

Aerodynamic theory teaches us that the higher the aspect ratio the higher the resulting lift to drag ratio, but such predictions ignore the effect of the mast. To calculate the effect is difficult, but wind tunnel measurements give interesting results. Curve 2 in fig. 4.21 is for the highest aspect ratio and yet its

performance, both with regard to maximum lift and maximum lift/drag ratio, is inferior to curve 1 for the lower aspect ratio. Even curve 3 for the lowest aspect ratio would be a superior sail under most conditions to that of curve 2 with nearly twice the aspect ratio.

These apparently anomalous results are of course due to the presence of the mast. A major effect is to produce a separation bubble just aft of the mast followed by re-attachment of the flow. For a tall narrow sail a greater relative area will be affected by this lift-squandering bubble. The mast in these tests was circular in cross-section and had a diameter of 1.3% of its height. Aspect ratio is here defined as luff length divided by foot length.

A careful set of measurements has been made on a model of a Finn class rig in which the effect of the boom vang tension was studied. Every sailor now knows that increasing the boom vang (kicking strap) tension flattens the sail, especially if it is coupled with

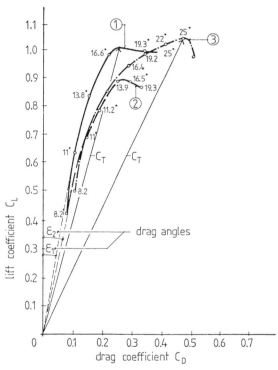

Fig. 4.21 Wind tunnel measurements of model sails, all with a mast of the same dimensions. Curve 1 is a sail of aspect ratio 2.3, curve 2 has AR = 3 and curve 3 has AR = 1.6. Although the lift/drag ratio should improve with aspect ratio, the relatively greater effect of the mast on a tall narrow sail means that these two opposing effects must result in a best compromise, as is the case with most aspects of yacht design.

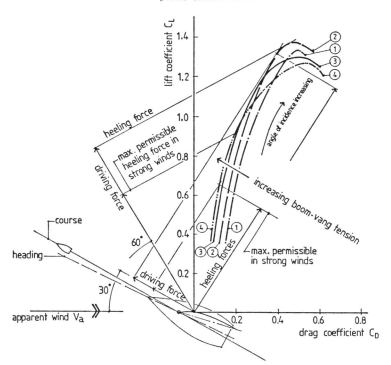

*Fig. 4.22* A set of polar curves for a Finn dinghy sail as a function of increasing tension on the boom vang (kicking strap) from curve 1 through curve 4. Sail driving forces are shown when hard on the wind and when on a reach. These are determined in the same way as in fig. 3.27, in light winds when heeling force is not a problem. The reader can draw his own conclusions as to vang tension and how sensitively performance will depend on it. In strong winds the sail must be operated at a lower heeling force coefficient in order to remain upright, and this changes somewhat the conclusions about vang tension.

a bendy mast. What is interesting is to see how the aerodynamic properties of the sail are affected and thereby predict how a particular change in sail shape will affect performance. Although these measurements do not scale cloth tension properly the trends are not likely to be changed by this. Though they were made for a particular rig, again the trends will be applicable to all rigs even those with little mast bend.

Fig. 4.22 summarises the results in the form of sail polar diagrams. Increasing the boom vang tension increases the maximum lift to drag ratio, but as we saw in §3.5 this is not of fundamental importance. Four sailing situations are shown: closehauled at an apparent wind angle of 30° in light and in strong winds, and reaching at 60° also in light and strong winds. For light winds we can determine the maximum driving force as in fig. 3.27 which shows clearly that not much boom vang tension is required. However the driving force does not depend too sensitively on vang tension as curve 3 gives almost the same drive when on the wind. If the winds are strong we must reduce the angle of incidence and luff a little to reduce heeling. The sail is now operating at a lower lift coefficient to compensate for the increased wind speed. Now, for a given heeling force the maximum drive is given by curve 4, maximum boom vang tension. Note that under these conditions the drive is quite critically dependent on the vang tension.

On a reach in light winds curve 2 is again best. Unlike the windward case, vang tension is now quite critical. Good sailors will often state that, contrary to general belief, sail trim can be more important on a reach than on the wind. Their reasons are rarely stated but fig. 4.22 certainly shows one of them.

In strong winds the heeling force limitation is the same as when on the wind and now curves 3 and 4 both produce about the same driving force; but a reduction in tension to curves 1 or 2 has a quite deleterious effect.

These results are in accord with what has been empirically determined over years of competition.

The new dimension added by the wind tunnel measurements is the knowledge of *why* a certain tension is necessary. The enjoyment and skill of sailing are always enhanced by this added dimension.

The effect of sail shape or camber is discussed a great deal in the sailing literature. Unlike the invisible antics of air flow, it is probably observed, discussed and argued about more than any other boat parameter. Our fundamental knowledge of how shape influences the aerodynamic characteristics of sails and hence performance comes from wind tunnel measurements. Discussions of sail shape usually jump directly from describing it in some way to saying what effect it will have on boat performance under various conditions. Instead, we will look at the aerodynamic characteristics produced by various shapes and from these infer the effect on performance.

A difficulty that arises immediately is how to characterise sail shape. If the sail were shaped so that it fitted snugly around a cylinder, then its curvature could be expressed by one number, the radius of the cylinder. Usually, however, the radius of curvature of a section of sail will vary continuously from luff to leech along a horizontal line, and also as we move up the sail so that it has curvature along vertical sections which in turn vary from luff to leech. To fully specify a sail shape would require an infinite set of numbers, an inconvenient state of affairs circumvented by approximating, rather grossly. The most detailed practical description of sail shape then usually consists of the following information: for each of about ten horizontal sections equally spaced up the mast, a specified maximum camber, the position of this maximum point and the twist (fig. 4.23). If these parameters are all specified for eight or ten stations equally spaced up the mast together with the sail plan form, then the aerodynamic characteristics would be reasonably well defined and could be measured. At least they could be measured for a uniform air flow, one in which there is no wind gradient. But this is far from reality, and besides, one of the causes of sail twist has to do with the wind gradient as already explained. To get useful information from wind tunnel measurements we must further simplify and study the effect of the amount of camber and position of maximum camber for untwisted sails in uniform flow, which has been done using metal plates bent to shape and supported without a mast. Hopefully these considerations will have emphasised what a complex device a sail really is, and help to put into context the qualitative nature of the conclusions that follow.

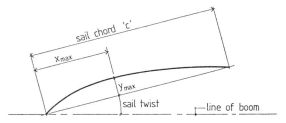

Fig. 4.23 The shape of a sail is reasonably well defined by specifying the parameters shown here at a number of stations at equal intervals up the mast. The sail chord is the straight line from luff to leech, of length c. Twist is the angle between the boom and the chord. $y_{max}$ is the maximum value of the camber. $y_{max}/c$ is the maximum value of the camber ratio. $x_{max}$ is the position of the maximum camber from the leading edge, sometimes expressed as a ratio $x_{max}/c$.

Fig. 4.24 shows the effect of changing only the value of the maximum sail camber. The inset summarises the variation in the two main aerodynamic characteristics with camber; these are the maximum force developed by the sail, represented by the coefficient $(C_T)^{max}$, and the maximum lift/drag ratio $(L/D)_{max}$. There is an increase of about 50% in sail drive in going from the flattest to the fullest sail. Changing the sail camber thus has a strong effect on the horsepower available.

Whether the full sail force can be used or not depends of course on the point of sailing and wind strength, as is shown in fig. 4.24 in the same way as in fig. 4.22. Sail drive is the component of the sail force along the course; heeling force is the component at right angles to the course and is represented by the length of the dashed lines in the diagram. In light winds where heeling force is not a problem it is clear that maximum drive will be obtained with the full sail of 14% camber. When sailing on the wind the heeling force is reduced by nearly 60% when the sail is flattened from 14% to 5% camber. On the other hand, if one pays off from 30° to 60° without changing the sail shape there is a 40% reduction in heeling force. This then is the origin of the rule which says that one should use the maximum sail camber consistent with stability. There is, however, one important exception to this which was discussed in §3.6: when the wind is very light, less than about 3 or 4 knots. Under these conditions separation of the flow may occur around a full sail because of the adverse pressure gradients on the leeward side. Then one must flatten the sail and be satisfied that less drive is better than none at all.

We have just discussed the effect of changing the

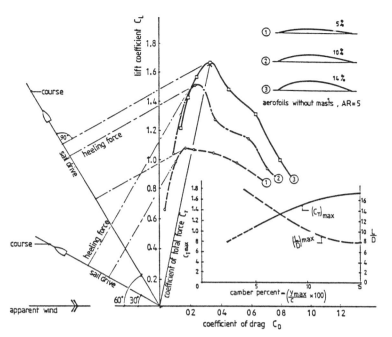

*Fig. 4.24* There are well known methods for changing sail camber and these change the aerodynamic characteristics of the sail, as shown in this partial polar diagram for maximum camber ratios of 5, 10 and 14%. For reaching at 60° to the apparent wind, maximum drive is obtained with the fullest sail section. Heeling force is proportional to the length of the dashed lines. For windward sailing at 30° the heeling force is greater, as expected. Notice that there is little difference in the sail drive between 10 and 14% camber ratio when on the wind, but there could be a worthwhile saving in reduced heeling force with the flatter sail. The inset graph shows how the maximum total sail force coefficient and the maximum lift/drag ratio vary with camber. A greater total sail force is always associated with a greater drag.

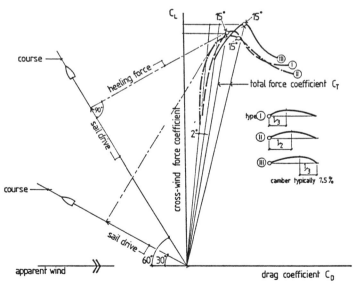

*Fig. 4.25* For a fixed camber of 7.5%, this shows the aerodynamic effect of changing the position of maximum camber. For windward work when heeling is not a problem, maximum camber at the mid-chord position is definitely best. Driving force on a reach is marginally better with the maximum camber at the one-third chord position. Unless brought about by sail interaction there seems little reason to want to have the maximum camber at the two-thirds point.

camber when the point of maximum camber was fixed at about 45% of the chord length back from the luff. So the question naturally arises: for a fixed maximum camber, what effect does changing the position of the maximum have? This is shown in fig. 4.25 where the camber is 7.5% but the position of the maximum is varied from one-third of the way back from the luff to one-half to two-thirds. Again driving forces are shown for sailing angles of 30° and 60° with respect to the apparent wind direction. Although the differences are not great, maximum camber at the mid-point of the chord clearly seems best for windward sailing. For reaching there is little between them as far as drive is concerned, but maximum camber at the halfway point is still to be preferred since the heeling force is somewhat less than when it is two-thirds of the way back from the luff. Thus, as a general rule the most useful sail will have its maximum camber at the mid-point between luff and leech.

Two points should be remembered, however. We have been speaking of light wind conditions. A maximum camber at the one-third position helps slightly to reduce sail force for strong wind conditions, and sail interaction could change the picture. For instance when the mainsail is backwinded by the genoa the point of maximum camber will certainly move well aft regardless of sail adjustments simply because of the distribution of pressure across the mainsail.

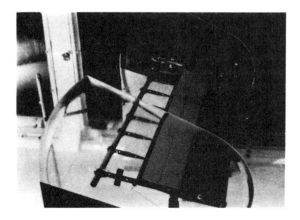

*Fig. 4.26* The apparatus for testing double surface sails in a wind tunnel. The large flat plates ensure two-dimensional flow.

## §4.4.2 Wind Tunnel Measurements on Double Surface Sails

Double surface sails are formed by wrapping the cloth around a circular leading edge (mast) and bringing the two sides together to form a sharp trailing edge. They are used for hang gliders, yacht sails on unstayed masts, and have been proposed for a form of vertical axis wind turbine. Two-dimensional measurements have been made by placing large flat plates at each end of the aerofoil to prevent crossflow effects (fig. 4.26).

There are many attractions in using unstayed masts: the quite considerable parasitic drag due to the rigging is completely eliminated, sail handling can be made simpler, and the possibility of using a double sided sail could give a performance bonus. In the wind tunnel measurements given below the sail model was formed by wrapping 1.23 oz woven nylon around a tube whose diameter was about 7% of the

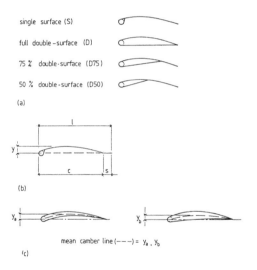

*Fig. 4.27* (a) The lift and drag curves given in fig. 4.28 are for the four types of sail shown here. S is a normal single surface sail. D is a full double surface sail, and D75 and D50 are three-quarter and half double surface sails respectively. The sails are inflated by static pressure and are constructed so that they can slide around the leading edge spar as different forces appear on the separate sides. (b) explains the definition of the slackness, s. The quantity s/c is called the slackness ratio. The maximum camber y is of course related to s. The diameter of the leading edge is about 7% of the chord length. (c) shows typical sail shapes at high (left) and low (right) angles of attack. At small angles of attack the windward pressure is almost the same as the internal pressure whereas the leeward pressure is much less than the internal pressure resulting in much greater curvature. Such an aerofoil has lower drag than the thin single surface one.

chord length. The trailing edge was held rigidly in place but with varying amounts of slackness so as to control the camber and measurements were made on the four different sail profiles shown (fig. 4.27). All the sails were free to rotate about the leading edge and the double surface sails were inflated by stagnation pressure air through a vent in the leading edge. At high angles of attack (c) the two layers of fabric tend to equalise in cordwise length to form an aerofoil with little thickness; however at small angles of attack the pressure difference across the windward surface is less than across the leeward surface. This is because the windward pressure is only slightly greater than the internal pressure, whereas the leeward side pressure is considerably less than the internal pressure producing a greater curvature on that side. The mean camber line shown dashed in (c) is the line midway between both surfaces and this remains constant as the angle of attack is varied for given fixed leech positions.

Fig. 4.28 displays some of the measured results, all

made at a Reynolds' number of $1.5 \times 10^5$. These graphs are not the usual polar form but are plots of lift and drag coefficients against angle of attack. In all cases the maximum camber ratio is 10.6%. The clear superiority of double surface sails under these conditions is obvious. The half and three-quarter double sails have higher lift coefficients at all angles of attack than the single surface sail. All the sails have about the same drag at very small angles of attack up to about 5°. At larger angles there is little difference between the three double surface sails, but the single surface one exhibits much greater drag. Whereas the single surface sail starts to stall at an angle of incidence of 15°, the double sided sails show no fall in lift up to 21°. The maximum lift/drag ratios of the half and three-quarter double surface sails are much greater and occur at much higher values of the lift. This is very important for sailing in light airs where large lift is required, and if this is also accompanied by low drag there is a substantial improvement in performance.

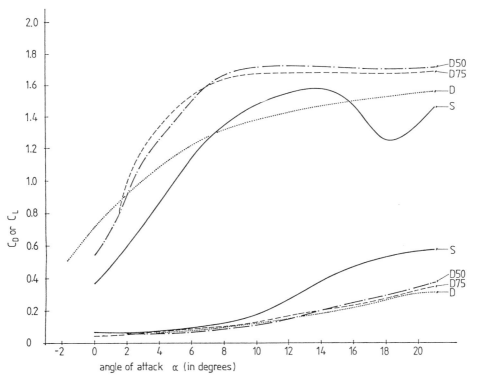

*Fig. 4.28* The upper four curves give the lift coefficient as a function of angle of attack for the four types of sail labelled. S is single surface, D is full double surface, and D75 and D50 are three-quarter and half double surface sails respectively.

The lower four curves give the corresponding drag coefficients. The higher lift and lower drag of double surface sails under these conditions is evident.

Fabric porosity was found to affect the lift, drag and position of the centre of pressure. Non-porous sails made from Mylar-coated Dacron gave higher lift and less drag when compared with the slightly porous sails of the same camber. The centre of pressure also moved further forward with non-porous sails. An unusual characteristic of double surface sails is that instead of fluttering unstably at zero angle of attack they would inflate because of the ram air pressure inside, forming a stable symmetrical streamlined aerofoil. Because the sailcloth could move around the leading edge, the leeward surface would camber leaving the windward surface nearly flat, giving a shape similar to an efficient asymmetric aerofoil. This effect was paramount in reducing the high drag normally associated with single surface sails.

Many of the trends in yacht design have been facilitated by the development of new materials. Such developments will undoubtedly result in the increased use of unstayed masts. As a corollary to this, double surface sails would be a natural way to go.

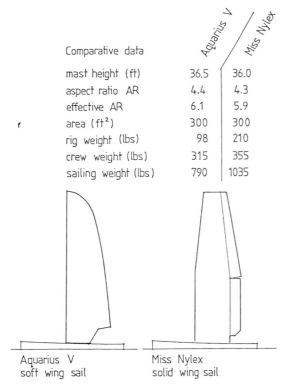

| Comparative data | Aquarius V | Miss Nylex |
|---|---|---|
| mast height (ft) | 36.5 | 36.0 |
| aspect ratio AR | 4.4 | 4.3 |
| effective AR | 6.1 | 5.9 |
| area (ft²) | 300 | 300 |
| rig weight (lbs) | 98 | 210 |
| crew weight (lbs) | 315 | 355 |
| sailing weight (lbs) | 790 | 1035 |

Aquarius V
soft wing sail

Miss Nylex
solid wing sail

*Fig. 4.29* The plan form and other particulars of the soft sail rig on the left and the wing sail on the right, whose performances are compared in the text.

If this glimpse into the future is a little blurred, the same advances in strong light materials may bring the wing sail out of the laboratory of C Class catamarans into the more demanding world of cruising and racing yachts where reliability is as important as lift to drag ratios.

The 1976 Little America's Cup challenge for C Class catamarans saw, for the first time, a unique opportunity to compare a modern wing sail with an 'old-fashioned' soft sail rig. Both boats had very similar hull forms so that any performance difference was considered to be due to the rigs only. Very complete performance data were recorded during the racing between *Miss Nylex* with the wing sail and *Aquarius V* with the soft sail. These measurements were then compared with performance predictions of the boats based on the aerodynamic properties of their rigs and their hull hydrodynamics. As well as being a fascinating comparison between two very different rig types, the excellent agreement between the measured and predicted performance gives us some confidence in all this physics, which you, the reader, may well be coming to doubt by now!

Fig. 4.29 compares the two rigs. Such things as area and aspect ratio are virtually the same. $AR_{eff}$ means effective aspect ratio and this differs slightly between the two rigs because of a more effective seal between the sail and the deck in the case of *Aquarius* (see §4.1.1). Only in one parameter is there a really big difference: the rig weights. The wing sail weighed 210 lb (95 kg) and the soft sail rig somewhat less than half that at 98 lb (44 kg).

Although I have described the soft sail rig as old fashioned, it was in fact as high-tech as the times allowed. It was lightweight and boomless; a bendy rig of essentially elliptic planform; high aspect ratio with deck seal; capable of controlling sail twist and mast rotation and able to vary camber up to 20%. The wing sail could also control twist and camber by means of the 25% chord full-span flap system shown. The wing section was NACA 0015 and maximum flap angle 40°.

From two-dimensional measurements of the lift and drag of these sails, the three-dimensional polar curves were derived by calculating the vortex drag from the aspect ratio and adding parasitic drag (fig. 4.30). For small cambers and small angles of attack corresponding to small aerodynamic forces the differences are not great, but for large values of the force the wing sail drag is much less. Furthermore, the wing sail has the advantage both in magnitude of thrust available and ratio of thrust to heeling force for

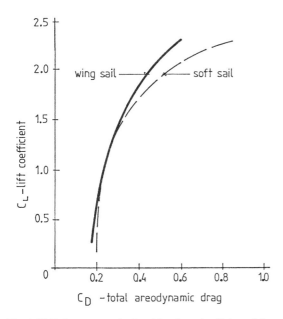

*Fig. 4.30* Polar curves calculated for the soft sail rig and the wing sail described in fig. 4.29. These curves include the effects of parasitic drag from hull and rigging of the C Class catamarans on which they were used.

both the low and high ranges of aerodynamic force. Only in a narrow region in the mid-range are the two rigs equal. Clearly the wing sail is aerodynamically superior to the soft sail rig. Despite this, the testimony of history is that the 1976 Little America's Cup was won by the boat carrying the 'old fashioned' soft sail!

The aerodynamic advantage alone is insufficient even though the hulls were essentially the same: a correct prediction of the overall performance must include the effects of weight. Apart from the different sails the main difference between the two boats was in weight and this originated almost entirely with the rigs. Full performance prediction calculations were then carried out which allowed for weight, giving the results in fig. 4.31a. This shows that the lighter boat, *Aquarius V*, has the advantage on all points of sail in light airs, $V_T = 5$ knots, despite her less efficient rig. In winds above 15 knots the heavier *Miss Nylex* has the advantage except for broad-reaching and tacking downwind. At 10 knots both boats perform equally well.

From fig. 4.31a it is easy to predict the outcome of the racing if we know the prevailing wind strengths and the times spent on each point of sailing. Fortunately for *Aquarius V* the wind was

predominantly light, leading to the inevitable result in agreement with the predictions. Fig. 4.31b shows the predicted relative performance curve with the triangles giving the minutes advantage of the winner in each of the seven races. This is a very gratifying vindication of prediction methods and our understanding of how sailboats work.

We have, of course, assumed that races are won on boat speed alone which is not by any means universally true. C Class catamarans are not very manoeuvrable so tactics probably did not play a significant part. Furthermore the races were sailed in flat water so the steady state aero- and hydrodynamic properties measured by the wind tunnel were completely applicable.

If nothing else is learned from this exercise, the one message that should be received loud and clear is that the ultimate performance of a boat depends on many interacting factors, even the weather.

## §4.5 Theoretical Sail Design

Most of our fundamental understanding of how yachts work has been adapted from aeronautics where the greater potential for economic gain has been a concentrated catalyst. This is perhaps a little unfair to the early pioneers for whom the solution of the age-old 'secret of flight' was paramount. The driving force behind a better fundamental understanding of yacht sails is still largely the pioneering problem of how to extract the most energy from nature. One suspects, however, that a growing element in this process has to do with financial spin-off, as observation of any America's Cup campaign makes clear.

The application of rigid wing theory to sails is probably as big a step as the original development of the wing theory itself. The basic problem is that, whereas aeroplane wings are rigid so that their shape is not affected by air flow, sails are flexible so that their shape depends on the air flow which in turn depends on their shape!

Nevertheless a first step toward understanding sails theoretically is to try to calculate the lift and drag from a given sail shape. For a particular wind strength and angle of incidence the shape will be fixed and could be determined by photography, for instance. In §4.1.1 and §9.1 it is pointed out that lift is related to circulation, and that the circulation varies as we move up the sail. If the circulation at each height can be calculated, it is therefore merely necessary to add up all these contributions to the total circulation in order

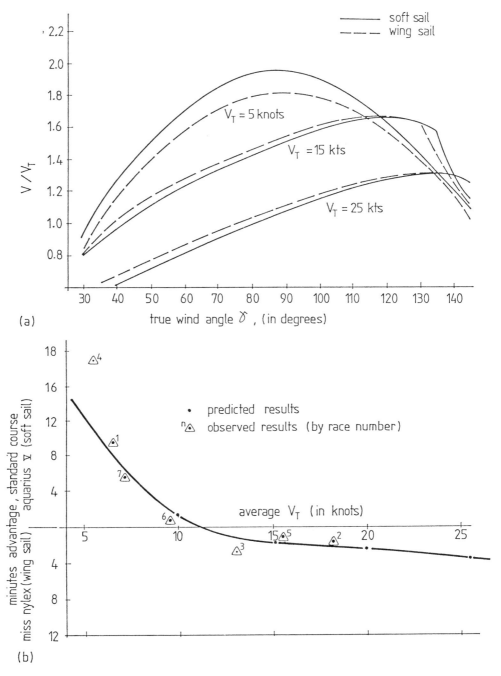

Fig. 4.31 (a) The ratio of boat speed to true wind speed as a function of true wind angle, for C Class catamarans using a soft sail rig as opposed to a wing sail. These curves are from performance prediction calculations based on rig aerodynamics, hull characteristics and overall weight.
(b) The smooth curve has been calculated from (a) and shows the advantage of the lighter boat in light airs. The triangles represent the measured results of races run in different wind strengths. The agreement between the predicted performance and the measured results is very gratifying.

to determine the lift. In a similar way the vortex drag may be determined from the circulation and the downwash velocity (some further details of how this is done are given in §9.5). This kind of calculation is known as lifting line theory and it agrees moderately well with wind tunnel measurements made on sails of fixed shape without masts. It predicts that drag increases considerably with sail twist, but as previously stated this must be treated with caution because of the wind gradient. This theory, being purely an adaptation of rigid aerofoil theory, adds little to our understanding of sails.

The first steps away from aeronautics and toward flexible sails were made in the early 1960s. It began with a theoretical investigation of 'the two-dimensional flow of inviscid incompressible fluid past an infinitely long flexible inelastic membrane'. You may never have thought of a sail in these terms, but such is the formal language of the scientific literature. If the distance between the luff and the leech is c, and if the length of the material of the sail between luff and leech is c + s, then the problem is to determine the flow when the angle of incidence and the ratio s/c are both prescribed. s is called the *slackness* and s/c the *slackness ratio* (fig. 4.32). Sail shape depends on the equilibrium between the aerodynamic pressures and the tensile loads in the sail fabric. By using an approximation known as *thin aerofoil theory* a mathematical result known as the *sail equation* was derived which relates sail shape to the angle of incidence and the slackness. Unfortunately the sail equation cannot be solved analytically, which means one cannot find a mathematical expression that will

describe the results. Instead one must resort to numerical computation for each specific situation, for which computers are required.

It is found that to a first approximation the effects of angle of attack and camber can be separated (see fig. 4.33a). For an uncambered aerofoil such as the symmetric sections used for rudders and keels, zero lift occurs at zero angle of attack. For a normal cambered aerofoil some lift occurs at zero angle of attack and can be said to be caused by the camber. The amount of lift and its location depends on the shape of the camber. The lift force due to camber is generally located near the point of maximum camber. The lift due to angle of attack is always located at the quarter chord point. The total lift is the sum of that due to angle of incidence and that due to camber, and its location is at a point between the two lift forces appropriately weighted by their relative magnitudes.

Even for a sail it is possible to have camber at zero angle of attack, although the situation is rather unstable and can easily switch to a camber-flutter mode. Fig. 4.33b shows how this is possible. The sail distorts the flow pattern causing upwash into the luff

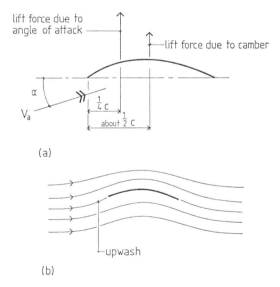

(a)

(b)

Fig. 4.33 (a) Cambered aerofoils in general exhibit lift at zero angle of attack. This lift force due to camber alone occurs at about the mid-chord point. The lift due to angle of attack, on the other hand, acts through the quarter chord point. (b) Lift at zero angle of attack is in principle feasible with a sail because of the upwash at the luff, even though the angle between the chord line and the undisturbed flow far ahead of the sail is zero.

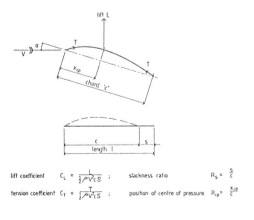

Definition of two-dimensional sail parameters

Fig. 4.32 Definitions of the quantities used in the theoretical description of sails. T is the tension in the cloth.

119

so that it can fill even when the angle between the chord and the undisturbed flow direction is zero. Here the lift is due solely to camber and so depends only on the slackness. The theory then shows that the lift due to camber at zero angle of attack for a flexible aerofoil is about 8% less than that for a rigid circular arc aerofoil of the same camber. For all aerofoils the lift depends on the angle of attack; however, the approximate theory of flexible aerofoils shows that it increases more slowly with angle of attack than is the case for a rigid aerofoil. Furthermore the effect is greater as the slackness ratio increases. The analysis further shows that the centre of pressure is at the quarter chord position for high values of the sail tension and moves back toward the mid-chord position as the tension is reduced, while at the same time the position of maximum camber moves from the 40% chord position back toward the 50% point. Theoretical sail profiles for an angle of attack of 6° and two different sail tensions exhibiting these effects are shown in fig. 4.34.

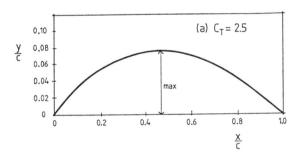

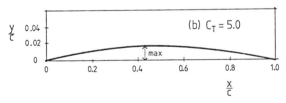

theoretical sail profiles at 6° angle of attack

when (a) $C_T = 2.5$

(b) $C_T = 5.0$

Fig. 4.34 Theoretical sail profiles calculated from the 'sail equation', an approximate calculation accurate at small angles of attack and large sail tension. Here the profiles are calculated for two values of the tension coefficient which is related to the tension and the slackness ratio as shown in fig. 4.32. Note that the vertical scale is expanded by a factor of 10 compared with the horizontal one, so these curves represent very flat sails indeed.

The theory just described contains approximations which mean that it is accurate only for small cambers and low angles of incidence. A better method is to start from a sail profile given by the approximate theory and calculate the velocity distribution over the sail from which, with the aid of Bernoulli's theorem, the pressure distribution may be determined. The profile shape is then re-calculated so that the sail tension and pressure forces are in equilibrium. The pressure distribution is then re-determined and the process repeated until there are no further changes. This exact theory gives a greater sail camber under the same conditions and the difference increases with increasing angle of attack. Although the camber is increased the position of the maximum camber does not change. The increased camber does, of course, produce increased lift. The exact theory also shows a small change in the position of the centre of pressure with angle of attack which is not predicted by the approximate theory.

The balance between aerodynamic forces and sail tension has been alluded to several times in this section. It is therefore worth looking a little more closely at the relationship between the total sail force and the tensions in the cloth. Consider a section of sail experiencing a total force $F_T$ (fig. 4.35). This force must be transmitted through the sail to the boat. A sail can sustain only tension forces. If you fasten a piece of string at each end so that it is horizontal and pull down on it near the middle you know intuitively that the tension in the string will be much greater than your downward pull. In fact a piece of string which was too strong to be broken by pulling horizontally could easily be broken in this way. The situation for a sail is, as fig. 4.35 shows, exactly the same. Just as the tension in the piece of string must be the same along its entire length, so the tension in the sail fabric is the same from luff to leech. If the tension were not constant the sail would either buckle or stretch. The points where the tension forces are applied to the boat are at the luff directly to the mast and at the leech to the boom and mast via another sail tension running from the clew to the headboard. It is clear from the diagram that for a given sail force the tension in a full sail is much less than that in a flat one.

Since the tension forces at luff and leech are tangential to the sail at those points and since they must add up vectorially to the total sail force, it should be clear how the magnitude and direction of the total sail force determines the angle of the sail at luff and leech. It is also clear from fig. 4.35 that since

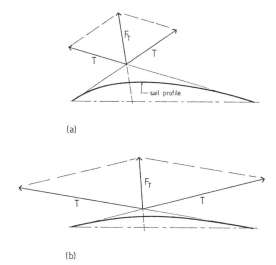

(a)

(b)

*Fig. 4.35* The tension components in the plane of a horizontal section of sail must have the same magnitude from luff to leech, just as the tension in a piece of stretched string is the same along its length. The sail force is transmitted to the boat via the tension at the luff and leech. These two equal tension forces must therefore add up vectorially to the total sail force. This diagram shows a full sail (a) and a flatter one (b). In both cases the magnitude of the sail force is the same. Because of the angle of the tension forces their magnitude must be very much greater in the case of the flat sail.

the two tension forces are equal the total sail force $F_T$ must bisect the angle between them (this is a geometric property of all quadrilaterals). So we see that there is a direct connection between the magnitude and direction of the total sail force and the centre of effort on the one hand, and the tension forces in luff and leech on the other.

Remember that what we have been discussing here are the tensions in the plane of a sail section. More obvious to the observer are the effects of tension in the plane of the sail. The tension forces in the neighbourhood of the leech are very considerable because it is being pulled forward horizontally by the sail forces, so to prevent excessive twist the leech must be tensioned with the mainsheet or boom vang (kicking strap). Wrinkles radiating out from the clew and head of the sail are a sure sign of the tension in these regions.

Computers have influenced sail design in recent times almost as much as the development of new materials. There are three aspects to the use of computers in the production of sails. The first is to determine the flow and pressure distribution around

a sail and hence predict its polar curve. The second is to design the shape of the cloth panels which when stitched together will give a sail with the desired aerodynamic characteristics. The third is to automate the cutting so that the design shape can be exactly reproduced.

One of the problems in sail development is specifying the shape. Whereas a set of line drawings exists for almost every hull built, the same detailed shape specification for the sails rarely exists. Unless the shape of a sail is properly specified it is difficult even for the slow 'Darwinian' developmental process to be very effective since vague qualitative statements about the shape of a successful sail are not much help to the designer who wishes to make just a small change in the hope of improving it, or simply to make an identical replacement. In the discussions so far, a sail has been specified by its maximum camber, the position of this maximum and the twist at each of ten stations up the mast, giving 30 numbers altogether. A fuller specification from which a set of curves like fig. 4.36 could be drawn is really required. Tradition has established that lines be painstakingly drawn for hulls but the same tradition does not exist for sails. Instead we look to the computer for assistance, just as hull designers are doing to remove much of their drafting drudgery. If a few thousand numbers are required to specify a sail, these are just as easily stored and used by a computer as three or four numbers are by a human.

The determination of flow and pressure distribution by an approximate method has already been discussed. Improved methods are constantly being developed, but the problem is much more difficult than that for pressure distribution over aircraft wings. Sails are generally much more highly cambered and twisted than wings; they have low aspect ratios and elastic properties which depend on direction; they are porous and leak air through when under load. As if this were not enough, there are the added complications of strongly coupled interactions with other sails, very large angles of attack in some downwind sailing situations, variable parasitic drag and non-steady conditions when pitching and rolling, and of course the inclusion of the appropriate wind gradient.

Although this first aspect of computer use in sail design is in an ongoing state of development, the second aspect is inherently simpler yet still a complex computational problem. The question is: given a certain required sail shape how should the panels be cut to realise this? Computer codes which make

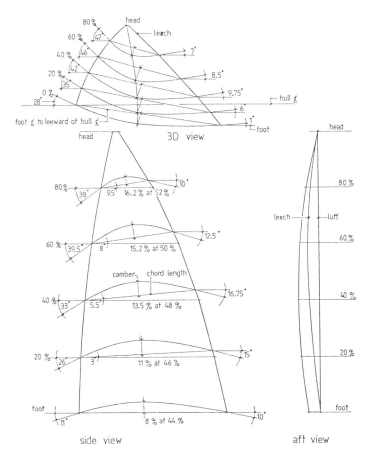

*Fig. 4.36* Specification of sail shape for a cat rigged boat sailing upwind in smooth water. Notice the large number of parameters needed to give a reasonable specification. The 'symmetry of sailing' tells us of the equal contribution of the underwater body and the sails to the progress of a yacht, yet although the designer draws up detailed underwater lines, rarely does he draw up 'sail lines' with the same detail.

extensive use of graphics can allow the sailmaker to produce a 'lines drawing' of his sail, view it on a screen from any desired angle, and amend it if necessary before setting in motion the program that computes the panel shapes. Computer graphics can be a very powerful tool in translating human thoughts into the cold digits of computer calculation. The sailmaker needs no special knowledge of computers; once a few basic dimensions of the sail have been entered via the keyboard the entire design process is accomplished by interacting with the screen display. For the system shown in fig. 4.37 a pointer on the screen can be positioned by moving the 'mouse' on the table next to the computer. A 'menu' appears on the screen listing a number of tasks the program can

perform. If the camber is to be adjusted the pointer is simply positioned over 'chord adjustment' and a display like that in fig. 4.37 appears. Associated with each chord is a set of control points which can be used to determine the cross-sectional shape of the sail: camber can be changed by moving these points with the mouse. At any time while shaping, the sail can be viewed in three dimensions from any point around the yacht and from any height above or below it. Fig. 4.38 gives a three-dimensional view of a mainsail showing panel layout, battens and reef points. In effect a sail can be designed, viewed from any angle, changed, viewed again, and only when the sailmaker is completely satisfied are the offsets for the various panels printed out or plotted directly on the sailcloth.

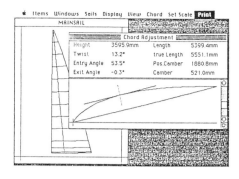

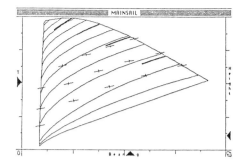

*Fig. 4.37* Putting the correct shape into sails is as important as it is with hulls. Sail design is very much a visually oriented procedure for which computer graphics are ideally suited. This is part of a computer display in which sail shape is being adjusted. The sailmaker interacts directly with the display without the need to type into a keyboard. *(Reproduced by permission of Sails Science Ltd, Box 47–234, Auckland, NZ)*

*Fig. 4.38* By means of computer graphics a sail designer can 'build' a sail in the computer and view his handiwork in three-dimensional views such as this from any desired direction. When he is satisfied with the shape he can print out the offsets necessary to cut the panels to shape or even draw them directly on the sailcloth. *(Reproduced by permission of Sails Science Ltd, Box 47–234, Auckland, NZ)*

The third aspect of computer use is the simplest problem to solve. Having computed the panel shapes it is necessary to mark them or cut them out accurately. Marking out is done with a very large version of the familiar plotter associated with most computers as a peripheral device for producing hard copies. This allows much faster and more accurate marking out of the cloth than was possible by former techniques. For production of a large number of sails of the same design computer driven cutting machines are also used.

## §4.6 Some Practicalities of Sail Design and Use

Purely analytical methods cannot as yet be used alone. Apart from the difficulties already mentioned, flow separation and the effects of pitching on the air flow cannot be handled properly. The analytical approach is undoubtedly a valuable tool for indicating trends in proper sail shape selection and for determining how sail shape influences aerodynamic characteristics. It is clear from the problems mentioned that the use of analytical methods is going to be most successful for moderate air and smooth water where flow separation is unlikely and the flow is fairly steady. In heavy air as well as in the very light, and also when pitching is pronounced, a more empirical approach should be taken.

A glance at fig. 4.36 shows that cross-sectional shape of a real sail changes considerably from foot to head. Here the maximum camber increases from 8% at the foot to 16.2% at the 80% chord, in an effort to minimise the vortex drag which is large at large lift. Since the sail is triangular the loading (the lift as a function of height) falls off more rapidly than an elliptic loading distribution requires. Thus increasing the section lift coefficients tends to correct for this. Reducing vortex drag is important in smooth water sailing since it is by far the greatest contributor to the total drag in light weather when the highest possible lifts are being used. When sailing to windward in smooth water pointing ability is very important and could be improved by flattening the sail lower down, thereby improving the lift/drag ratio at the expense of a slight reduction in drive. Although this will raise the centre of effort, the increased heeling moment should not be a problem in light weather.

Flow past the mast produces eddies downstream on both sides of the sail. The extent to which they disrupt flow over the sail depends on the ratio of the mast diameter to the chord length. Unless the mast is tapered the effect will obviously be much worse near the head than the foot. Flow around the mast will produce a separation bubble on the leeward side of the sail followed by re-attachment of the flow. Probably this re-attachment should occur before the point of maximum camber to minimise the possibility of complete separation. For this reason the position of maximum camber should move aft with height up the sail. Because of the possibility of stall at the trailing edge this shift aft should be limited to about the 55% chord point.

Sailing in waves calls for a different approach that has been gleaned from experience. Pitching greatly increases the hull drag. Maximum sail drive is then required in order to maintain enough boat speed for the keel to supply the required lift without too much leeway, and in fact this must often be obtained by paying off slightly. Although vortex drag will be increased and the lift/drag ratio reduced, experiment has shown that this is the better compromise. Thus sails designed for choppy conditions must be fuller throughout and especially near the foot. Greater twist is now required to minimise stalling in the upper part of the sail when the mast pitches aft. It has been found better in rough water to have the upper part of the sail luffing when the mast pitches forward than to have it stalling when it pitches aft.

Since it is rarely convenient in practice or for the pocketbook to change sails for different sea conditions, we would like to transmute our sails into a different shape by pulling the appropriate strings. Remember that sail adjustments are edge constraints only and can do little to change the shape that is cut into the seams. So if a single sail is to be used it must always be a compromise. Since we now know the aerodynamic properties required for the prevailing conditions and the sail shape necessary to create those properties, there remains only the practical problem of how to trim for the wanted shape.

### §4.6.1 Genoa Trim

The relative size of a genoa is commonly specified by its overlap, determined as follows. Draw a line from the clew perpendicular to the luff (fig. 4.39), called the longest perpendicular, LP. The base of the foretriangle, from the forestay to the front of the mast, is labelled J. The overlap in per cent is then given by (LP/J) × 100. A no. 1 genoa will usually have an overlap of about 150%, a no. 2 about 125% and a working jib about 90%. Wind strength of course dictates which sail to use and the important indicator here is the amount of heel (see §3.5.2). Within the operating range of each sail there are a number of adjustments that can be made to fine tune drive and heeling. For optimum windward ability in smooth water the sheeting angle should be as small as it can be made without excessively backwinding the main (see §4.2). In rough water windward ability is inherently worse but can be optimised by both increasing the genoa sheeting angle and paying off. This keeps the angle of attack about the same but rotates the total force vector so as to increase the

Fig. 4.39 Genoa overlap is measured in per cent and is defined as (LP/J) × 100. LP is the longest perpendicular from the luff to the clew and J, the base of the foretriangle, is the distance from the forestay to the front of the mast.

driving force. It also reduces the interaction with the mainsail and lessens backwinding. On a genoa trimming the sheet has the effect of changing the twist, the camber and the sheeting angle. Changes of genoa sheeting angle are easiest done by means of a Barber hauler. From the earlier discussion of the effect of camber and twist you should be able to decide on the appropriate sheet tension.

Unfortunately there are other adjustments that affect twist and camber. The first is the fairlead position; moving the lead forward increases the camber, especially at the foot, and decreases the twist. The second is the forestay sag. For masthead rigs this is adjusted by the backstay tension and the sag is at a maximum about halfway up the sail. A sagging forestay increases the camber and to some extent reduces twist. Increasing backstay tension flattens the sail and moves the point of maximum camber aft. The final adjustment available is the halyard tension or Cunningham hole: greater tension moves the point of maximum camber forward and less tension moves it aft. Adjustment for good windward performance is less critical with the draft forward, but best windward performance in smooth water will be obtained with the maximum camber close to the 50% chord point. Needless to say, all these adjustments must be made with the aid of telltales as explained in chapter 3.

## §4.6.2   Mainsail Trim

As with the genoa, we are concerned with controlling twist, camber and position of maximum camber. A mainsail also has an additional function. As its angle of attack can be readily varied one has easy control over the total sail force. Varying its contribution to the overall rig force moves the centre of effort forward or aft and so enables the boat to be kept in balance.

When sailing on the wind mainsheet tension controls directly the leech tension which in turn controls the amount of twist. More tension means less twist. As discussed earlier some twist is normally desirable. Slackening the mainsheet produces a large amount of twist very quickly, which luffs the upper part of the sail. This not only reduces the heeling force but rapidly brings the centre of effort down thereby quickly reducing the heeling moment. Under normal smooth water conditions the standard rule of thumb for mainsheet tension is to adjust it so that the top batten is parallel to the boom. The kicker or boom vang also controls twist but in a sense it is supplementary to the mainsheet, giving control at large sheeting angles where the mainsheet's downward component is small.

If your boat allows control of mast bend you have a very effective way of changing the amount of sail camber. The effect is greatest halfway up the sail. When the mast bends in the fore-and-aft plane the distance between the luff and the leech is increased pulling the sail flatter. Since the lower part of the sail is little affected by mast bend, flattening there is best accomplished via the clew outhaul or flattening reef. Control over the position of maximum camber is best obtained by use of the Cunningham: increasing tension moves the point of maximum camber forward.

The final adjustment normally available is the mainsheet traveller position. Basically this is a device for controlling the angle of incidence of air on the mainsail. In this role it is a sail force varying device and has a direct influence on boat balance. Unfortunately the geometry of traveller design is usually such that as the boom swings out the mainsheet tension changes, so both twist and angle of incidence are affected. As with the genoa all of these controls are to some extent interdependent. Adjustment to the final desired shape must then be a process of successive approximations where all the adjustments must be cycled through several times because of the way they interact on each other.

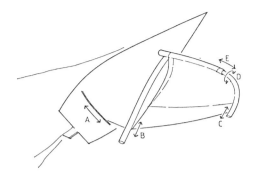

*Fig. 4.40* Bird's eye view of a soft sail rig design employing features that would make its aerodynamic performance come close to that of a solid wing sail and yet would be lighter and more practical to handle. A is a track that allows control of the angle of incidence. B and C are clew and 'head' outhauls respectively for controlling the sail camber. D is a rotatable upper mast section which controls sail twist; vertical movement of this section could allow for reefing. The mast is self-supporting and the sail 75% double sided. For large rigs, strength of materials poses the main design problems.

I would like to conclude this chapter with a little flight of the imagination. We saw earlier that wing sails, although clearly superior aerodynamically, have a weight disadvantage, not to mention a few practical handling difficulties when not actually sailing. Since it is unlikely that one will ever be able to build a wing sail as light and convenient as a soft sail, it is clear that one should attempt to endow the soft sail with wing sail qualities.

We saw in §4.4.3 how 75% double surface sails have lift and drag coefficients approaching those of solid aerofoils. Such sails need unstayed masts. The technology for building large unstayed masts already exists and doubtless will improve with time. All that is needed is a method of camber and twist control to match that of the wing sail. Fig. 4.40 shows a possible way of achieving this. The top curved section of the mast passes down inside the mast and can be rotated from the bottom. The boom moves in a curved track and stays in the horizontal plane. Rotating the upper section of the mast independent of the boom can produce any desired twist. Camber could be controlled at foot and head by means of clew and head outhauls. Because of the sail planform, near-elliptic loading could readily be obtained.

These features, combined with the use of a double-sided sail and the absence of parasitic drag due to stays, could produce a rig matching a wing sail in aerodynamic performance yet lighter and more practical.

# Chapter Five

# THE HULL

## §5.1 Introduction

Unlike the cars we drive or the houses we live in, the boats we sail can trace their pedigree back about 6000 years. Our earliest knowledge of boats comes from Egypt, where about 4000 BC they had already advanced from mere rafts of reeds.

However the true ancestors of our present day hulls were the vessels built by the Greeks. They appear to have been the first to be built with a keel, stem and stern post and with internal framing with planks attached edge to edge. This Mediterranean method was to dominate European boatbuilding until the emergence of synthetic materials in the second half of the twentieth century. As the centroid of European culture and influence moved north the southern boatbuilding techniques moved with it, and as is so often the case with a change of environment, new ideas emerged. The most important was the improvement of the steering gear. The first step in this process is seen in the very beautiful hulls built by the Vikings in the 9th and 10th centuries. Unlike Mediterranean ships which were steered by two oars near the stern, the Viking ships had a single high aspect ratio steering oar fixed so that it would rotate about its long axis and controlled by a short tiller projecting athwartships into the boat. Since Viking helmsmen were apparently right handed this rudder or 'steering board' was mounted on the right hand side of the boat. The word for 'steer board' in all Germanic and Nordic languages is almost identical and it is from this that our modern English word 'starboard' has been derived, though of course it now refers to the right hand side of the ship and no longer has any association with steering.

Fig. 5.1 shows the elegant lines and construction of a hull of this era. Although primarily powered by oars, this ship was also equipped with a squaresail for downwind sailing. Viking craft are often described as long narrow vessels, yet as the picture shows they have considerable beam and must have good form stability. It must have made them seaworthy in the open ocean and quite capable of journeys across the North Atlantic as they are alleged to have accomplished. The beam/length ratio of the ship pictured is 0.22 which for a boat 22 m overall would be comparable to a modern yacht of similar size.

The great step in European hull development came at the very end of the 12th century when someone, probably in the Netherlands, hung a rudder over the stern and put the tiller along the centreline of the boat. This step was important, not only because of the improvement in handling but also because it necessarily gave rise to a differentiation between bow and stern.

In order to appreciate why a hull is shaped the way it is, one must first ask what functions it is required to fulfil. The criteria for a power driven vessel are the simplest. First, it must provide a volume in which to house the accommodation, the engine, the supplies and the cargo. Second it must be seaworthy enough

*Fig. 5.1* In ancient times ships were probably the most highly developed devices created by humans. This is the *Gokstad Ship* built in Norway in the 9th century. It is a fine example of form and workmanship. By slow development in the manner of the quotation at the beginning of this chapter, the Vikings had found the ultimate shape for a rowed vessel. The fine ends combined with considerable amidships form stability must have made these vessels easy to row and stable in a seaway. (*Reproduced by permission of the University Museum of National Antiquities, Oslo*)

for the routes envisaged and provide the lowest possible resistance consistent with this and the volume required. The criteria for the hull of a sailing yacht include these and a third, the need to resist heeling forces and have a large resistance to sideways movement and yet a small resistance to forward movement. It is this added criterion that makes yacht design so complicated, so fascinating and even still an art to some extent.

## §5.2  Hull Parameters

Just as sails are complicated three-dimensional shapes and cannot be described fully by just a few parameters, so a hull is really only fully described by its lines drawings. A representative set is shown in fig. 5.2. *Sections* are the outlines formed on planes perpendicular to the centreline of the hull. *Buttock* lines are the outline shapes obtained on vertical planes cutting the hull parallel to the centreline. *Waterlines* are formed by planes parallel to the water surface.

Just as sail shape can be characterised approximately by a few numbers such as maximum camber ratio and twist, so there are a set of conventional numbers for describing approximately the hull shape; they are its overall length, waterline length, beam, draft and displacement. Another quantity, not so familiar but a very useful measure of the shape of a hull, is the *prismatic coefficient*, $C_p$. Mathematically it is defined as $C_p = \nabla/(A_0 \times L)$ in which $\nabla$ is the immersed volume, $A_0$ is the maximum section area and L the waterline length. The product of $A_0$ and L is the volume of a prism whose cross-section is the largest hull section (fig. 5.3). So what the prismatic coefficient measures is the ratio of the actual underwater volume to the volume of the larger prism. Thus, for a box-shaped barge $C_p$ could be 1.0. A boat with very fine ends will have a low prismatic coefficient. Of course, a fine bow with a full stern could give the same prismatic coefficient as a full bow and a fine stern. In calculating $C_p$ most designers do not include the cross-sectional area of the keel, although this practice is not completely universal.

One of our hull requirements concerned accommodation. Clearly a large prismatic coefficient is needed to achieve a roomy boat, but unfortunately, as intuition tells us, the fineness of the ends is related

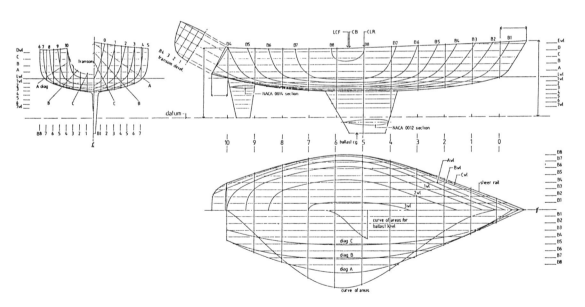

*Fig. 5.2* An example of a set of yacht lines. B1 to B8 are the buttock lines. Their shape is shown on the upper right and their position at the lower right. Waterlines are formed by horizontal planes and are labelled 1WL to 7WL below the load waterline (LWL), and AWL to EWL above. Sections are shown on the left. The diagonals A, B and C are arbitrarily placed in order to increase the number of offsets (co-ordinates) specifying the section shape. CLR is the centre of lateral resistance which is traditionally calculated in a geometric way and does not necessarily coincide with the actual hydrodynamic CLR. CB is the centre of buoyancy and LCF is the longitudinal centre of flotation.

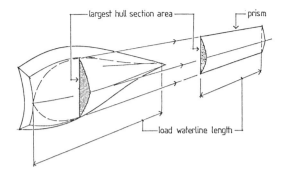

$$p.c. = \frac{\text{displaced volume}}{\text{l.w.l length} \times \text{largest hull section area}}$$

*Fig. 5.3* The prismatic coefficient is a measure of the fineness of the ends of the hull. It is the ratio of the actual underwater volume to the volume of a prism whose cross-section is the largest hull section. Clearly the maximum value of the prismatic coefficient is 1.0 and this corresponds to a barge-like hull with no bow or stern.

to the hull drag. Towing tank tests on yacht hulls have established a relationship between the prismatic coefficient for minimum drag and the speed (fig. 5.4). The prismatic coefficient which gives the least drag is plotted vertically and the speed to length ratio is plotted horizontally. The reason for using speed/length rather than just speed is so that the graph will be reasonably valid for boats of all sizes. This stems from the fact that the maximum speed for a displacement boat is about that of a water wave in deep water which has a wavelength equal to the waterline length of its hull. The speed/length ratio, which is really the ratio of the speed to the square root of the length, for such a water wave is given by the expression $V/\sqrt{L} = \sqrt{g/2\pi}$.

The value of the quantity on the right hand side depends on the system of units used. In metric units $V/\sqrt{L}$ has the value 1.25 for a water wave. To the extent that the wave speed is synonymous with the maximum speed of a displacement boat, this means that a speed/length ratio of 1.25 will be hard to exceed. Taken out of context this last sentence would be very loose and confusing; it has been written so because this is how it appears in almost every yachting publication. People so often speak of speed/length ratios as though they were dimensionless numbers. The ratio quoted here for a water wave is 1.25, but it only has that value if the speed is measured in metres per second and the wavelength in metres. In other words the speed/length ratio is *not* a dimensionless number like Reynolds' number. If the

wavelength is measured in feet and the speed in knots, then the speed/length ratio of a wave is 1.34. If knots and metres are used it is 2.43. Thus one must always specify the units of measurement if the figure is to make any sense.

With that diversion, we see that fig. 5.4 gives the optimum prismatic coefficient for various boat speeds relative to hull speed. Unlike power boats, yachts sail at all sorts of speeds so further compromise is necessary. If it is expected that most sailing will be done upwind in light weather, implying low speeds, then $C_p$ ought to be about 0.53. If it is going to be driven hard downwind it will spend much of its time at hull speed so a $C_p$ of around 0.6 would be more appropriate. Note that these figures are really only applicable to a 'normal' form where the bow is somewhat finer than the stern. To give a feel for the prismatic coefficient, note that the lines shown in fig. 5.2 have $C_p = 0.542$. For speed/length ratios less than 2.0 knots per metre$^{1/2}$ (or $\sqrt{m}$) the hull resistance is mainly due to surface friction drag which depends on wetted area rather than hull form. At higher speeds wave-making becomes the predominant drag component and it is here that the prismatic coefficient is more important than the wetted area. For this reason it is better to err on the side of a too high

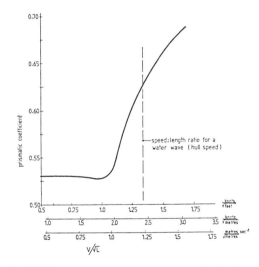

*Fig. 5.4* Towing tank measurements on upright hulls making no leeway give an approximate relationship between the prismatic coefficient which gives the least resistance and the speed/length ratio. Here the speed/length ratio is shown in three systems of units. The vertical dashed line is the ratio for a water wave in deep water. This corresponds to the so-called hull speed when the length of the wave produced is equal to that of the hull.

prismatic coefficient at low speeds rather than too low at high speeds.

Another number which helps in the characterisation of a hull is the displacement/length ratio defined as: $\Delta/(0.01L)^3$ where $\Delta$ is the displacement in tons (2240 lb) and L is the waterline length in feet. Again this quantity is often quoted without mention of the units, though the use of tons and feet is widespread. Naval architects missed the opportunity to use a dimensionless quantity here. If, instead of displacement, the volume displacement had been used, then a dimensionless number would have resulted, having the same value whether the boat was measured up in feet, metres or cubits. Although throughout this book I have attempted to familiarise the reader with the modern International System of units, I intend to make this one exception and quote some displacement/length ratios in old fashioned tons per cubic foot. But before doing that we must ask: what is the significance of this number?

From a negative point of view it is a measure of the wave-making resistance. From a positive point of view it is a measure of the carrying capacity and accommodation in the hull. The latter is obvious, but why is it also a measure of the wave-making resistance? Whenever a boat moves a distance equal to its waterline length it pushes aside a mass of water equal to its own displacement. Just as Archimedes' bath water rose when he got in so does the sea in the immediate vicinity of a boat. Energy is required to raise a mass of water above its natural level and it must come ultimately from the wind on the sails. A heavy displacement boat moving slowly may push up the same amount of water each second as a light boat moving fast. Given the same sail area, it is clear that the light boat will be faster. Of course boats are not designed this way; the 'power to weight ratio' is kept fairly constant. This is measured by the ratio of the sail area to the two-thirds power of the displaced volume. This is a true dimensionless quantity and in most designs lies between about 14 and 19.

Returning now to the displacement/length ratio: values range from around 150 tons per $(0.01 \text{ ft})^3$ for a very light displacement boat to about 400 for a heavy displacement type. The boat shown in fig. 5.2 has a value of 195 tons per $(0.01 \text{ ft})^3$.

## §5.3 Static Stability

Before talking about stability it might perhaps be a good idea to talk first about why your boat floats in the first place. If you feel that this is taking explanations too far then skip the next few paragraphs, but if you are not *absolutely* sure how boats carry out their most primitive function, then read on.

A boat which is floating is under the influence of two forces which are in equilibrium; they are equal and opposite so that there is no net force. Why then bother to talk about two forces if they cancel each other? First because we know they are there, and second because if one of them changes we would like to be able to predict what will happen. These two forces are, of course, the weight of the boat which acts downwards and its buoyancy which acts upwards. If the weight of the boat is increased by taking on stores, it must move down because the weight, being greater than the buoyancy, produces a net downward force and hence an acceleration in that direction. If the buoyancy force didn't change the boat would sink to the bottom like a stone. It is clear therefore that as a greater volume of hull is submerged the buoyancy increases. The hull keeps on sinking until the buoyancy force has just built up sufficiently that it exactly cancels the weight. We are then in equilibrium again.

Thus far, you might be thinking, very little that is new has been said. You have heard the words before – weight, buoyancy, equilibrium. They are familiar, but familiarity does not imply understanding, although it helps. The concept I am concerned about is buoyancy. What produces it? The standard answer parroted from school is: it is equal to the weight of the water displaced. Correct, but weight forces act downwards and buoyancy acts upwards. A closer look is required.

The fundamental origin of buoyancy is the change of pressure with depth. The pressure at a depth h below the surface is determined by the weight of water above. Fig. 5.5a shows a figure in the unenviable situation of being inside a tall cylinder with a smooth fitting piston, above which there is a column of water of height h. The weight of water he is holding up can be calculated if its volume is known: it is simply density × volume × acceleration of gravity $= \rho g A h$, where A is the cross-sectional area of the piston. Since pressure is force divided by area, the pressure on the piston is just $\rho g h$.

The stick figure in fig. 5.5b is in an even worse predicament. The piston above him is pushing down with a force $\rho g A h$ and the piston he is standing on is pushing *up* with the same force. The weight of the column of water of height h produces a pressure which is transmitted through the incompressible water and

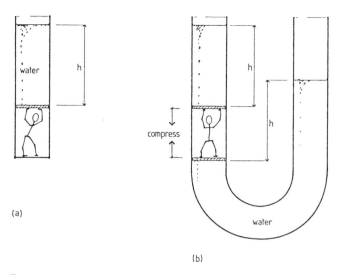

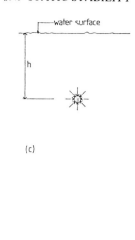

(c)

(a)

(b)

*Fig. 5.5* The pressure on the piston is equal to the weight of water being held up divided by the area of the piston. The weight of the water is its volume × the density × the acceleration of gravity: $W = Ah\varrho g$. Pressure is the weight divided by the area, so $P = \varrho gh$. A column of water bent round to a lower piston as in (b) will transmit its weight force through the incompressible fluid and push upwards. Thus an object surrounded by fluid as in (c) experiences pressure from all directions whose origin is the weight of water of height h above.

appears as an *upward* force on the lower piston. This is analogous to the arms of a balance: push down on one side and the other side pushes up. If we extend this idea further to a cylinder with movable sides, you see that our unfortunate stick man is squeezed from every direction. This is precisely the situation for a ball immersed in a fluid. The pressure at the depth of the ball is $\varrho gh$ and the direction of the pressure is always perpendicular to its surface as shown in fig. 5.5c.

Now when a boat is put into the water, different parts of it are at all depths from the surface down to the maximum draft. Since pressure depends on depth the pressure forces on the hull increase from the

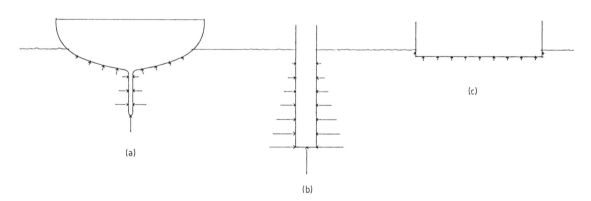

(a)

(b)

(c)

*Fig. 5.6* Since pressure increases with depth the pressure forces on a hull depend on position and act always perpendicular to the surface. The vector sum of all these forces is the buoyancy. In (b) the forces on opposite sides cancel and do not contribute to the buoyancy, which can come only from the upward pressure at the bottom acting over a small surface area. For a given buoyancy this object must be deeply immersed to acquire a sufficiently high pressure. The other extreme is (c): a small pressure force acting over a wide area produces the necessary buoyancy without the need to sink very deeply.

surface down (fig. 5.6). The total buoyancy force is the vector sum of all such elementary contributions. From the symmetry it is clear that this will be vertically upwards. The relative carrying capacity of two very extreme hull forms is shown in (b) and (c). The plank on edge has equal and opposite pressure forces on each side which cancel and contribute nothing to the buoyancy, which comes only from the pressure at the bottom multiplied by the small area of the edge of the plank. Only by allowing the plank to go very deep will the buoyancy force be reasonably large. The other extreme in (c) is the barge; although its draft is small and hence the pressure on its bottom

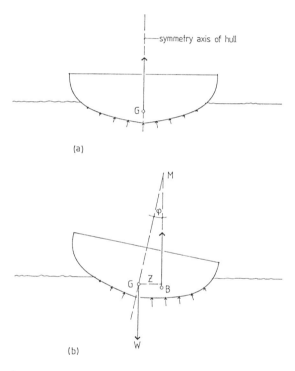

(a)

(b)

*Fig. 5.7* In (a) the vertical dashed line is the symmetry axis of the hull. The pressure forces producing buoyancy are symmetrical with respect to this axis, which means that the resultant buoyancy itself possesses the same symmetry. This is only possible if the buoyancy B acts through the centre of gravity G along the axis of symmetry. When the hull heels, as in (b), the symmetry of the pressure forces is destroyed and the buoyancy force moves out to the right. The point M where the line of action of the buoyancy crosses the hull symmetry axis is known as the metacentre. B is the centre of buoyancy, W is the weight of the boat and Z is the base of the perpendicular from G on to BM. The magnitude of the righting moment is given by the product of the weight and the distance, GZ.

small, it acts vertically over a large area giving it considerable buoyancy.

You should now be in a position to understand better 'form stability' which is stability due only to the shape of the hull. Fig. 5.7a shows a hull on an even keel. Because of its symmetrical shape the pressure forces on the hull are also symmetrical, increasing with depth as they should. Since these pressure forces are arranged symmetrically on each side of a vertical axis through the centre of gravity then their resultant force, the buoyancy, must possess the same degree of symmetry. This is only possible for a single force if it is in the plane of symmetry, so the resultant buoyancy is vertical and passes through the centre of gravity G. This is, by the way, a simple example of the *symmetry principle*, which states: the effect is at least as symmetric as the cause.

Now look at fig. 5.7b. The hull is heeled and the pressure forces, as always, are perpendicular to its surface and increase in magnitude with depth. Clearly the symmetry is now lost; the buoyancy need no longer pass through the centre of gravity. In fact it is clear from the distribution of pressure forces that the buoyancy will be displaced off to the right. Remembering that the weight and the buoyancy are always equal in magnitude, we see that a pure couple is being applied to the boat. This has a tendency to rotate the hull in a counterclockwise direction in this case. The righting moment produced by this couple has a magnitude given by the product of one of the forces and the perpendicular distance between them. If the buoyancy force is extrapolated upward until it cuts the symmetry axis of the hull, it is found that the point of intersection M remains fixed for small amounts of heel up to about 10°. If this is so, then the righting moment which is W × (GZ) in fig. 5.7b may also be written: W × (GM)sin φ. Since the weight or displacement of a boat is fixed, the righting moment is often thought of as being simply proportional to GZ which is a measure of the amount by which the centre of buoyancy moves on heeling. Our second formula for righting moment shows that it is proportional to the sine of the angle of heel providing this angle does not exceed about 10°. We will see later what happens at larger angles of heel. Point M is called the *metacentre* and the distance BM the *metacentric height*.

As well as a righting moment against heel, a hull will also right itself against pitching. This kind of righting moment is sometimes called the *longitudinal bending moment*. Fig. 5.8 shows a hull trimmed by the bow. If there is no change in displacement the

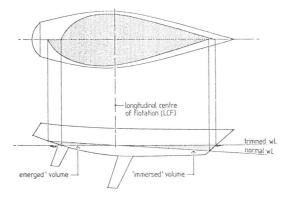

Fig. 5.8 A hull trimmed by the bow. The darkened wedge shape at the right is the 'immersed' volume and that at the left the 'emerged' volume. These two volumes must be equal since the weight of the boat and hence the volume of water it displaces is assumed unchanged. Thus the axis about which the boat pitches must be located so that it divides the area of the waterline plane exactly in half, as shown in the lower half of the diagram where the waterline plane has been darkened. The dashed line shows the position of the longitudinal centre of flotation.

volume of the immersed wedge must be equal to the volume of the emerged wedge. This means that for small angles of trim the boat rotates about an axis which divides the waterline plane into two equal areas. This axis is known as the *longitudinal centre of flotation*. An extra weight added at this point will sink the boat without trim. If it is added forward of this point it will trim by the bow and if added abaft it will trim by the stern. A longitudinal metacentre may be defined in the same way as the transverse metacentre of fig. 5.7; in this case the metacentric height is much greater since hulls normally have a much greater righting moment against pitching than against heeling.

The amount of sinkage when extra weight is added to the boat is easily calculated from the area of the waterline plane. Assume the extra weight is added in such a way that it does not change the trim. If the boat sinks by an amount x into the water the change in displacement is determined by the extra volume immersed. For small changes this is just the area of the waterline plane multiplied by x, which gives the change in displacement volume. The weight change associated with this is the weight of this volume of water which is obtained by multiplying by the density: $W = \varrho A_{WL}x$ where $A_{WL}$ is the area of the waterline plane. Any consistent set of units may be used for such calculations, but remember that it is an

approximation for small changes in displacement and is most accurate when the topsides are vertical.

## §5.3.1 Static Stability Curves

A full static stability curve is a graph of righting moment or righting arm against the angle of heel. For monohulls there are two generic forms for these curves, depending roughly on whether the centre of gravity is above or below the centre of buoyancy (fig. 5.9a, b). The dashed curve corresponds to the effect of increasing beam. An angle of heel for which the righting moment is negative means that the static stability is working toward *increasing* the angle of heel rather than *decreasing* it, as is the case when it is positive. Thus when the centre of gravity is high, as is

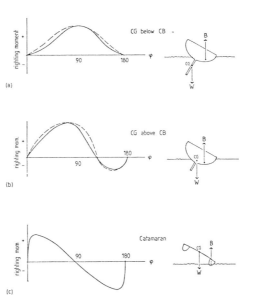

Fig. 5.9 Three generic forms of the static stability curve. In (a) the centre of gravity CG is always below the centre of buoyancy CB. At all angles of heel the hull has a positive righting moment tending to bring it back upright. The dashed line shows the modification that would occur for increased beam. (b) shows the situation where the CG is above the CB, which is common in many modern designs. There is a region of negative righting moment, implying that the boat has some stability in the upside-down position. The dashed line shows the effect of increased beam. A more extreme case of this kind of curve is shown in (c). A catamaran relies almost entirely on form stability. The symmetry of its hulls produces complete symmetry of the righting moment curve, where it is seen that the boat is just as stable upside-down as right way up.

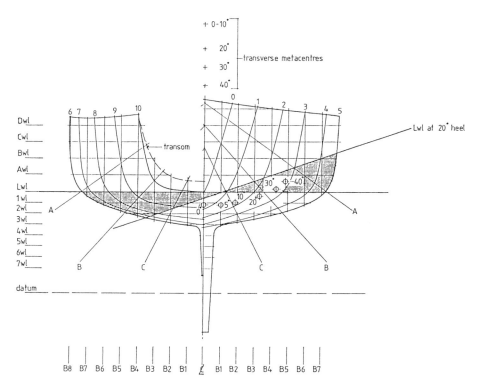

*Fig. 5.10* It is possible to calculate from a set of lines the positions of the centre of buoyancy and the metacentres for various angles of heel. Traditional methods of calculation are long and tedious. To within the accuracy of the method, the metacentre is fixed, up to 10° as theory predicts. For greater angles of heel there is a reduction in the metacentric height. This is typical of a light displacement beamy hull. The cross-hatched areas are the equal immersed and emerged wedges corresponding to a heel angle of 20°. The centres of buoyancy are shown by the symbol ⊕.

common in many modern designs, the boat may have a tendency to be stable when upside-down. This situation reaches an extreme with a catamaran hull which relies entirely upon form stability (c). The obvious symmetry means that the stability curve for a catamaran is the same whether right way up or upside down. A monohull, by contrast, has positive stability at 90° because of the ballast and hull shape, the combination of which keeps the righting moment of the monohull positive beyond 90°.

It is possible to measure stability curves, at least to 90° for boats that are not too large, or calculate them from a lines drawing. Fig. 5.10 shows the calculated positions of the centres of buoyancy and the transverse metacentres for angles of heel of 0°, 5°, 10°, 20°, 30° and 40°; also the waterline when the angle of heel is 20°. Since the weight of the boat remains fixed the total displacement volume must remain fixed, so the loss of volume due to the

emerged wedge must be compensated by the increase due to the immersed wedge. As can be seen the metacentric height is fixed for angles of heel up to 10°; beyond that there is a rapid reduction in its value. This means that the righting moment will be proportional to the sine of the angle of heel only up to a few degrees. For larger amounts of heel the righting moment will be considerably less. This is shown in fig. 5.11 which plots the righting moment against the angle of heel. The solid curve is the correct righting moment calculated from the positions of the CB and metacentre in fig. 5.10. The dashed curve is calculated from the approximate formula for righting moment which is GM sin φ, where GM is the distance from the CG to the metacentre (see fig. 5.7). Note that the two curves agree approximately only for small angles of heel less than about 10°.

Because of the beamy and light displacement form of the hull of fig. 5.10, it also has a tendency to rise

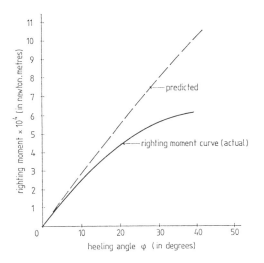

*Fig. 5.11* The solid line is the actual righting moment calculated from fig. 5.10. The dashed line is the prediction of the simple theory for small angles of heel. In this case the two agree to within about 10% only up to a heel angle of 10°.

when heeled. This effect is shown in fig. 5.12 and is characteristic of hulls of this type. Also shown is the variation in metacentric height with heel.

In practice it is difficult to know the true position of the CG, so that even if you have a set of hull lines and the patience to do the calculations it is difficult to come up with a reliable righting moment curve. By far the best and most convincing procedure is to *measure* the static righting moment. Measurements up to 90° are quite practical in boats up to about 8 m but rapidly becomes impractical with increasing size.

In §10.1 details of a measurement procedure carried out on a 6 m trailer yacht are given. Just the results are given in fig. 5.13. Notice that the righting moment reaches a maximum at about 50°, and from 60° to 90° it falls off linearly. This means that one can easily extrapolate the graph a further 20° beyond the point where measurements can conveniently be made. This is necessary since after a capsize a boat will often end up with the sails in the water and the mast at about 10° below the horizontal: a positive

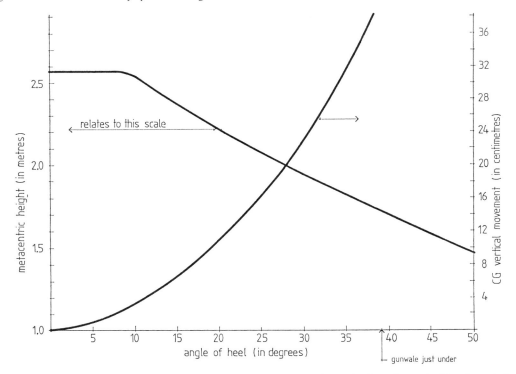

*Fig. 5.12* A plot of the metacentric height for the hull shown in fig. 5.10. Values are to be read from the left hand side of the graph and are seen to be reduced on heeling. The metacentric height is defined as the distance from the centre of buoyancy to the metacentre. The position of the metacentre in turn is where the vertical line through the CB cuts the hull symmetry axis. The other curve with its ordinate on the right shows the amount by which the centre of gravity rises when heeled.

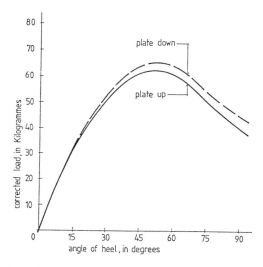

*Fig. 5.13* Although righting moments can be calculated from a lines drawing, it is undoubtedly better to measure them. This shows the results of a measurement on a 6 m trailer yacht. The solid line is with the centreboard up and the dashed line when it is down. The steep rise of righting moment at small angles of heel is due to form stability and the righting moment at 90° is due to ballast stability.

righting moment from this position is therefore a very desirable safety feature.

Although every owner should know his boat to the extent of having a measured or calculated static stability curve, it is often more useful to be able to quote a single number which is representative of its stability. Certainly if one wants to compare the stability of various boats a single number would be very useful. This is the same old problem that we had in trying to characterise the sail shape or the hull shape: we would like one number to be representative of a continuous function, which is clearly impossible. We therefore choose our number in such a way that it characterises some aspect of the stability that we consider is important. Most such numbers are concerned with the initial stability, that for small angles of heel.

Probably the best known in this category is the International Offshore Rule (IOR) inclining test (see also §10.1). Essentially the test measures the righting moment for 1° of heel. As was shown earlier the righting moment for small angles of heel can be represented by the equation: $RM = W(GM)\sin\varphi$. The difficulty with using the righting moment for 1° heel as a measure of stability is that it varies greatly with the size of boat and so is not a very useful

number for comparing different boats. Reference back to fig. 5.7 will remind you that the centre of buoyancy and hence the metacentric height BM is determined solely by hull shape. It is a measure of form stability and does not depend on where the centre of gravity is located. But stability depends also on the position of the CG: if it were above the metacentre the hull would be completely unstable. Thus the position of the metacentre is determined by form stability and the position of the CG by ballast stability. The IOR 'tenderness ratio' is a factor which compares the form stability with the ballast stability and is therefore independent of the size of the yacht. Referring again to fig. 5.7, the tenderness ratio is proportional to the ratio BM/GM. If the centre of gravity coincides with the centre of buoyancy then BM/GM = 1. If the CG is high, GM becomes small and the tenderness ratio becomes large. When a boat is measured for the IOR, BM and GM are not determined explicitly; instead GM is replaced by the righting moment for 1° which as we have seen depends directly on GM and BM is replaced by a product of the length and the cube of the waterline beam. It can be proved that these latter two are equivalent by consideration of the immersed and emergent wedges, but it is not appropriate to do that here. The tenderness ratio enters the IOR through the CG factor which is one of the multipliers converting the measured rating to the final rating. This is done in such a way as to discourage extreme values of the tenderness ratio. This is to avoid the development of lightly built yachts with a great deal of ballast low down on the one hand or very tender yachts on the other.

Another IOR stability test is the 'screening test' which is meant to give a measure of the ability to recover from a 90° knockdown. The 'screening value' is basically a quantity which depends on (BM − GM) divided by the displacement. Although the actual formula contains a number of correction factors, it is clear that a screening value close to zero or negative results when the centre of gravity is close to or below the centre of buoyancy.

Finally there is an IOR 'stability test' which requires that the yacht be heeled to 90° where it must support a weight attached to the top of the foretriangle. The weight depends on the yacht's dimensions.

So far these numbers which characterise stability have depended only on the hull design. In practice the danger of excessive heel obviously depends also on sail area. As a result there are a number of measures of stability which include sail area.

Probably the best known of these is the *Dellenbaugh angle* which compares the heeling moment due to the wind on the sails with the righting moment at small heel angles, W(GM)sin φ. The heeling moment is taken as sail area multiplied by the vertical distance from the geometrical centre of the sail area to the centre of lateral resistance of the hull. This distance is called the heeling arm. The Dellenbaugh angle is then given by:

$$\text{Dellenbaugh angle} = \frac{57.3 \times SA \times \text{heeling arm}}{GM \times W}$$

SA is sail area, W is displacement and the answer is in degrees. Dellenbaugh angles typically vary from about 10° to 20°.

Another quantity which is almost the inverse of the Dellenbaugh angle is the *wind pressure coefficient*, the ratio of the righting moment at 20° to the upright heeling moment due to the sails. Thus wind pressure coefficient is defined as:

$$WPC = \frac{(\text{righting arm at } 20°) \times W}{\text{heeling arm} \times SA}$$

Values of WPC between 1 and 2 are typical.

None of these formulae except the IOR screening value attempts to give a measure of the safety of a yacht under the extreme circumstances of a complete knockdown. A stability measurement under these circumstances is therefore desirable especially for trailerable boats which carry a minimum of ballast. What is needed is a number which is a measure of the yacht's ability to right itself from about 95° of heel in a 30 knot wind with the spinnaker still filling. This is certainly a dangerous but common predicament. A formula has been developed by the New Zealand Trailer Yacht Association which attempts to do this (details of the physical reasons for its various terms are given in §10.1).

This self-righting index is called SRI. The formula is:

$$SRI = \frac{(3T_{90} - T_{75})\,(I_s + 0.5FML)}{6B^2L + 3B^2I_s + (40L \times FML)}$$

Lengths are in metres and loads in kilogrammes. B = maximum beam, L = overall length, FML = freeboard at mid-length, $I_s$ = slant height of spinnaker halyard exit on the mast above the gunwale at deck level and mid-length point. $T_{75}$ and $T_{90}$ are the loads required on the spinnaker halyard acting at right angles to the mast to maintain angles of heel of 75° and 90° respectively.

The various components of the formula have the following significance.

$(3T_{90} - T_{75})/2$ estimates approximately the load required to maintain an angle of heel of 95° and provides an allowance for wave action.

$(I_s + 0.5FML)$ is the assumed lever arm at which a load on the spinnaker halyard acts. The numerator of the equation is therefore twice the estimated righting moment at 95° heel.

$3B^2L$ is the estimated overturning moment due to wind of about 30 knots acting on the hull when heeled to 90°.

$1.5B^2I_s$ is an estimate of the overturning moment due to wind of about 30 knots acting in the head of the spinnaker which is dragging the yacht sideways downwind while heeled to 90°.

$20L \times FML$ is the estimated overturning moment due to the weight of the crew.

The denominator of the equation is twice the estimated overturning moment due to the assumed wind and load conditions. The SRI is therefore the ratio of the righting moment to the sum of a number of overturning moments. Measured values of the SRI for trailerable yachts have been found to be in the range 0.01 to 1.7. This is a huge range of values. The lowest is a light unballasted beamy boat, very stiff at an angle of heel of 45°, but dangerous if knocked down. The highest value belongs to a boat which has a heavy lead bulb on the bottom of its drop keel. In short, a high SRI is the result of ballast stability rather than form stability. Experience has shown that boats with an SRI greater than 1.0 will most probably self-right in 30 knot winds.

## §5.4 Hull Drag

In every chapter so far, fluid drag in one form or another has been discussed. Although the basic physics of drag are the same for both air flow and water flow, there is one aspect of hull drag which has no counterpart. This arises because a yacht is not a submarine; it is partly in the water and partly out and this produces surface waves. Waves are stored energy, their destructive power is known to all. Producing them requires work in the physical sense, which is measured by force multiplied by distance. The force we are talking about is just the drag force.

Hulls are subject to all the forms of drag that have already been discussed: surface friction drag, normal pressure drag, vortex drag and in addition wave drag.

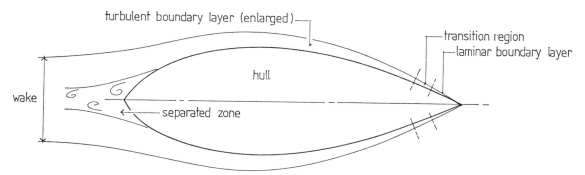

potential flow

turbulent boundary layer (enlarged)

transition region
laminar boundary layer

hull

wake

separated zone

*Fig. 5.14* The general pattern of flow around a yacht hull at zero leeway. The flow may be divided up into two regions: one, close to the surface (the boundary layer) where the effects of viscous forces predominate and the other, an outer region where the flow can be treated as though there were no viscosity, is the so-called potential flow discussed in chapter 4. Providing there is no turbulence present in the water a short region of laminar boundary layer flow can be expected near the bow, however transition to turbulent boundary layer flow soon occurs. In this region the boundary layer thickens and the drag per unit area increases. Finally, near the stern the flow may separate from the hull producing large eddies. This is analogous to the stall of an aerofoil.

Since the first three have been discussed at length, this section will concentrate mainly on wave drag.

First we will look at the overall flow pattern around a yacht hull. There are two physically different aspects to the flow: potential flow, in which viscous forces can be neglected, and boundary layer flow where the viscous forces play an important role. Viscous forces (chapter 2) give rise to a velocity gradient or change in velocity over a thin region close to the hull surface. The layer of water adjacent to the hull moves with it whereas water a few millimetres away is fixed with respect to the main body of the sea. If you like you can think of the boat as taking along with it a thin shell of water.

Fluid flow develops with time and distance in a characteristic way. Near the bow the boundary layer flow is very smooth or laminar. After travelling a certain distance along the hull surface it becomes unstable and random fluctuations in velocity or turbulence appear. The region where turbulence grows is called the transition region (fig. 5.14). In certain hull designs separation of the flow may occur near the stern. This region contains large scale eddies which move along with the yacht, and if present produce considerable drag.

The boundary layer gives rise to a resistance which has two components. The main one is due to tangential forces between the hull and the water brought about by viscosity. The other is the boundary layer normal pressure drag which is due to an imbalance in the pressure forces between the forebody and the afterbody. The growth of the boundary layer results in lower pressures near the stern which are unable to fully counteract the pressures in the bow region.

Although viscous resistance is dominant in the low

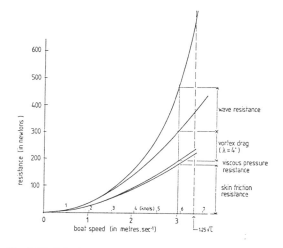

*Fig. 5.15* Calculated resistance components as a function of speed for a 5.5 Metre yacht. The vortex resistance is a result of the lift produced by the hull when making 4° of leeway. As can be seen the wave resistance component increases rapidly for speeds above $1.25\sqrt{L}$.

speed region, it is the resistance due to the generation of waves which determines the performance of a yacht at high speed. If the boat is making leeway there is additional vortex drag due to the lift being produced by the hull and keel. As discussed in §3.5.3 the hull side force or lift is proportional to the angle of heel, which in turn is proportional to the leeway angle. Since vortex drag is proportional to the square of the lift, we see that this component of the hull drag is proportional to the square of the leeway angle and the square of the angle of heel. Fig. 5.15 shows the relative importance of these drag components for a 5.5 Metre class yacht.

## §5.4.1 Wave Drag

The surface of the sea is the only part of the fluid flow around a yacht that can be observed and the waves produced by a moving yacht are familiar to all. Although the drag is roughly proportional to the square of this wave height, there is nothing the crew can do to reduce the wave size as unlike the sails the hull shape is fixed. Nevertheless, since there have probably been times when you have watched them for hours on end, you owe it to yourself to learn a little more about how they are formed and how they produce drag.

The most obvious characteristic of a bow wave is that it appears stationary when viewed from the boat. This means that the wave system travels at the same speed. It is as though the waves were attached to the boat which was trailing them astern. Yet if the boat suddenly stops the waves go on; there is no connection between them and the boat once they have been formed.

Fig. 2.20 shows how the waves are produced. At the bow water is pushed outwards as the boat advances whereas at the stern water is dragged in to fill the space vacated by the hull. Thus the boat sets up one process near the bow and a similar but reversed process near the stern.

Another obvious characteristic of bow waves is that they all look rather similar even though they are produced by different shaped hulls. There are two reasons for this. One is that all forebody shapes have one action in common; that is they displace water away from the bow as the hull moves forward. The other reason is that the overall pattern of the wake is a result more of the way in which the spreading of waves takes place rather than how they were produced.

To understand how the wave pattern in the wake is

*Fig. 5.16* The Fourier components of a lump in the sea. When the lump subsides waves of many lengths and speeds are propagated away.

formed we must first understand how water waves of different length are propagated. Imagine how the water surface would look just after a boulder is dropped into a lake. The water would be momentarily raised in a lump just like the lump in front of a bow. But here the boulder that caused the lump immediately disappears so its existence is only fleeting. It immediately starts to decompose into a whole set of waves of different wavelengths. As shown in fig. 5.16 the original lump can be described mathematically as the superposition of many waves if their amplitude, phase and wavelength are correctly chosen. These are called its Fourier components, named after Jean Baptiste Joseph Fourier (1768–1830) who is now remembered mainly for his mathematical decomposition of waves. Like many seemingly abstract ideas in mathematics the Fourier components of the lump really do exist, as we see as soon as the lump starts to subside. The energy of the surface of the water has been raised by the boulder so the lump cannot just simply subside down flat again, as that energy would be destroyed rather than conserved as it must be. (In a very viscous fluid the lump would simply subside, but its energy would be used up against the large frictional forces in the fluid.) Instead, as the lump subsides the many Fourier components which comprise it move away as individual waves. Their separate identity is very noticeable with water waves because their speed depends on their wavelength. The relationship is: speed $= \sqrt{g\lambda/2\pi}$. $\lambda$ (lambda) is the wavelength, the distance from crest to crest.

We see from this that the longer waves travel faster than the shorter ones, so some are overtaking others and where they overtake they build up in amplitude. It is a bit like cars racing around a circuit at different speeds. There will be a buildup in traffic density where the fast cars are making their way through the slower ones. Thus an aerial view would show a region of dense traffic moving around the circuit at a slower speed than that of the individual cars. This is precisely what is happening with water waves: there is a wave *group* with individual waves in the group moving through it at twice the group speed. It is a characteristic of water waves that the individual waves always move at twice the speed of the group in which they find themselves. Fig. 5.17 attempts to depict this situation where there are groups of various speeds. Since the individual crests always travel at twice the group velocity they must speed up as they pass through faster groups, and clearly the groups farthest away from the lumps must be the fastest, so we have the seemingly strange situation where the wave crests speed up with time. In fact it can be shown mathematically that the distance an individual crest has travelled from the original lump is proportional to the square of the time it has been travelling.

We are now in a position to understand the shape of the wake of a boat. Fig. 5.18 shows the positions of wave crests produced when the bow of the boat was at the numbered points: 0 corresponds to the position of

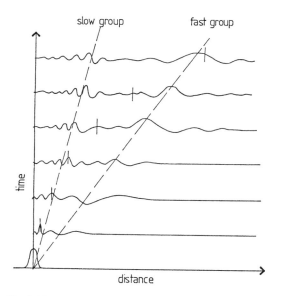

*Fig. 5.17* A simplified picture of how a lump in the sea propagates away with time. Waves of different lengths travel with different speeds. As they overtake each other they form groups; there are in practice many such groups. Those farthest away at a given time are going faster. A curious property of water waves is that individual crests travel twice as fast as the group through which they are moving. This means that the distance travelled by an individual crest is proportional to the square of the time. The progress of an individual crest is shown by the vertical dash on each wave train.

the bow now, 1 is where it was one second ago and so on. Individual waves propagate out from the lump caused by the bow in the same way as they do from that caused by the boulder in the lake. Since the distance they travel is proportional to the square of the time, the radius of the circle around 2 is proportional to 4, that around 3 is 9, that around 4 is 16 and so on. Of course these waves are emanating continuously from the bow so one should draw many more waves than just one per second. If all the rest are filled in it is easy to see that they form a wedge-shaped envelope like an arrowhead. But the V shape does not continue backwards indefinitely: from about position 9 on, the envelope turns back on itself, because after 9 or 10 seconds the wave is travelling faster than the boat. As can be seen by the crowding of the waves that originate from points 9, 10 and 11, we expect this part of the wave pattern to have greater amplitude than the rest. It is indeed found in practice that this transverse part of the wave pattern has the greatest amplitude and therefore contributes greatest to the drag.

What fig. 5.18 shows is, of course, incomplete since there are many wave crests in the disturbances that emanate from each point. The overall picture then looks like fig. 5.19. The transverse waves form a wave train moving at the same speed as the boat since they appear stationary in relation to it. As the speed of water waves depends on their wavelength this means that the distance between the crests of the transverse waves is defined by the boat speed. The formula quoted above says that wave speed = $\sqrt{g\lambda/2\pi}$ = boat speed in the present case.

Since the bow and the stern each produced a similar transverse wave pattern we expect these two patterns to interfere with each other (fig. 2.20). When the wavelength of the bow wave is such that it reinforces the stern wave, the transverse wave height rises and with it the wave resistance. This occurs when the wavelength is about equal to the waterline length of the boat. Since the wave is moving at the same speed as the boat, the boat speed is then just that of a wave equal to the waterline length. That is $V_B = \sqrt{gL/2\pi} = 1.25\sqrt{L}$, where L is the waterline length in metres and $V_B$ is the boat speed in metres per second. The numerical factor is different if other units are used: $V_B$ (knots) = $2.5\sqrt{L(\text{metres})}$
$$V_B \text{ (knots)} = 1.34\sqrt{L(\text{feet})}.$$
This is the so-called 'hull speed' and is shown in fig. 5.15 which are resistance curves for a 5.5 Metre yacht with a waterline length of 7.41 m. As can be seen this speed corresponds to a point where the wave

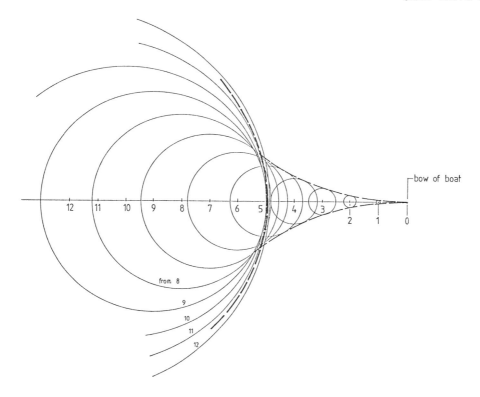

*Fig. 5.18* The bow of the boat is now at point 0. The other numbered points show where the bow was 1, 2, 3, 4 etc seconds in the past. The distance a wave crest has gone from these points is proportional to the square of the time that has elapsed, so the radii of the circles are proportional to 4, 9, 16, 36 etc. As can be seen, crests from points near 9, 10 and 11 tend to pile up forming a transverse wave of greater height than the Vee-shaped wave near the bow.

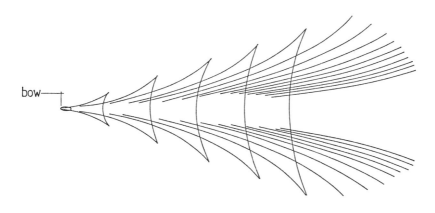

*Fig. 5.19* Since the bow is continually producing waves the pattern of fig. 5.18 is repeated. The distance between the transverse waves must be such that a wave train of this wavelength has a speed equal to the boat speed. This results from the requirement that the wave pattern moves with the boat.

resistance component of the total drag is increasing very rapidly.

It is obvious that the larger the bow wave the larger the drag. Can we make a connection between the two? It cannot be done precisely but the crude argument given in §10.2 can give us some idea. There it is shown that the drag is proportional to the square of the wave height. Furthermore for the kind of waves we are producing the wave height is proportional to the wavelength which in turn depends on the square of the speed. From all this we find that the drag due to wavemaking depends roughly on the fourth power of the speed. This accounts for the very rapid rise in the vicinity of the hull speed.

Since the surface friction drag is roughly proportional to $V^2$ (see §2.8) we expect that a rough expression for overall drag will be: Drag = $aV^2$ + $bV^4$. a and b are constants; a can be predicted theoretically but b must be measured. Thus in principle, a measurement of the resistance at one speed is all that is needed to determine b. Knowing a and b, the drag may then be predicted at all other speeds. But remember the expression is only a very crude approximation to the drag of a yacht when not heeled and not making leeway. If heel and leeway are included powers of $\varphi^2$ and $\lambda^2$ would also be involved. Of course it is possible to include enough terms and constants that the experimental drag curve can be reproduced to any degree of accuracy. Expressing experimental results in the form of such a *power series* is useful if a computer is to be used for subsequent performance calculations.

A curious thing happens to the speed of waves when the depth of water is less than the wavelength. The wave speed then no longer depends on the wavelength but is determined solely by the depth of water according to the equation $V = \sqrt{gD}$ where g is the acceleration of gravity and D is the water depth. If we put in the numerical value of g and put V in knots and D in metres, we get: $V(knots) = 6.2\sqrt{D}$. This means that all the waves move at the same speed regardless of whether their wavelength is short or long.

This changes completely the wake pattern from a moving boat. For example, assume we are moving at 4.4 knots in a depth of 2 m. This is half the wave speed for this depth. Thus waves produced at the bow will move ahead of the boat (fig. 5.20a). Waves originating from points 1, 2, 3 etc in the past have moved way out ahead of the boat. Now consider what happens when the boat speeds up to the wave speed

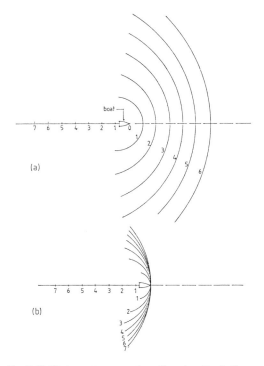

*Fig. 5.20* Wake patterns produced by a boat in shallow water. The wave speed in shallow water depends only on the depth and not on the wavelength as is the case in deep water. In (a) the boat speed is half the wave speed so the bow wave disturbance radiates out in front. In (b) the boat speed and the wave speed are the same. All disturbances from the past remain at the bow and build up a very large wave giving a limiting speed which depends on water depth rather than hull length.

of 8.8 knots. It is at the same speed as the waves so all those from past positions of the bow are still right at the bow. This results in the buildup of an enormous bow wave which the boat is trying to climb, with an associated rapid increase in drag. This is independent of the length of the boat and so might be called 'depth speed' as opposed to 'hull speed'.

This is not a situation that faces fixed-keel yachts very often as sailing at high speed in only 2 m of water is not good seamanship! Power boats do, but then they usually plane. This effect is only experienced therefore by shallow draft displacement power boats.

## §5.5 Scaling Laws and the Problems of Tank Testing

Tank testing concerns itself with predicting the lift and drag forces of a full size hull from measurements on a model. It sounds simple enough, but like many great ideas the more one looks at the details the more difficult it becomes. For instance, having measured the drag force on a model moving at 2 knots, how do we calculate the drag of the full size boat? And what will be the 'equivalent' speed of the full size boat anyway? To begin to answer these questions we must first think a little about the laws of scaling.

Back in §2.1 a general equation for drag was quoted: $D = f(R_n, F_n)\varrho V^2 S$. D is the drag force, $f(R_n, F_n)$ means some function of Reynolds' number and Froude's number, and S is the wetted surface area. This equation tells us which quantities drag depends on. For instance if both $R_n$ and $F_n$ could be kept the same for the full size boat and the model, then the drag would simply scale in proportion to the wetted area. Unfortunately this is not practical. Reynolds' number is given by $VL/v$, where V is velocity of flow, L is 'an appropriate length' which can be taken as the waterline length and $v$ is the kinematic viscosity. If tank tests are carried out in water, the viscosity is unchanged so Reynolds' number is kept constant by keeping the product $V \times L$ constant. Thus if the model is one-tenth the length of the real boat we must test it at ten times the speed to maintain a constant $R_n$. Now look at the Froude number which is given by $V/\sqrt{Lg}$ where g is the acceleration of gravity which cannot be changed unless one is thinking of tank testing on the moon. Now instead of keeping the product VL constant, we want to keep the ratio $V/\sqrt{L}$ constant if the Froude number is not to change. This requirement demands that our one-tenth scale model be tested at $1/\sqrt{10}$ of the speed of the full size boat, that is roughly one-third of the speed. Clearly it is impossible to keep both Froude's and Reynolds' numbers constant at the same time.

This is the dilemma of tank testing. It is basically because viscous and wavemaking forces scale differently. Although Reynolds' number can be adjusted by changing the kinematic viscosity of the fluid in the tank, this is no real solution. First, because of the small range of values of kinematic viscosity it differs only by a factor of 12 between air and water, and second, the inconvenience and expense of filling a large towing tank with anything other than water.

The key to escaping from this dilemma was supplied by William Froude. He proposed that only $F_n$ be kept constant in the drag equation and that the surface friction part of the drag which depends mainly on $R_n$ be calculated from data derived from towing thin planks, which don't make waves, in a towing tank. A summary of such data is given in fig. 2.10. It should be realised that the separation of resistance components in this way is done for engineering convenience rather than being dictated by the physics of fluid flow.

The procedure for determining the resistance of a full size boat from model measurements is then as follows. If the resistance of the full size boat is required at a speed $V^s$ ('ship speed'), then the model must be towed at a speed $V^m$ given by $V^m = V^s(L^m/L^s)^{1/2}$ or $V^s\sqrt{L^m/L^s}$. This is the scaling law for wave-making resistance (see §10.3). It means that at this speed the wake pattern of the model will be the same as that of the full size boat at speed $V^s$.

The measured total resistance of the model is then given by $D_T{}^m = D_F{}^m + D_R{}^m$. The subscript F refers to surface friction drag and R to *residual resistance*. $D_F{}^m$ can be calculated from the model surface area, Reynolds' number and the drag coefficient given in fig. 2.10. The difference between the measured total drag and the calculated surface friction drag is then the residual or wave-making drag $D_R{}^m$ of the model. The next step is to scale up the residual drag from the model to the full size boat. This is proportional to their volume displacements which in turn are proportional to the cube of their linear dimensions: $D_R{}^s = D_R{}^m(L^s/L^m)^3$.

A numerical example may help to clarify this. Assume the model is one-quarter the length of the full size boat and we wish to determine the full scale wave drag at 6 knots. The model must then be towed at a speed given by $6(1/4)^{1/2} = 3$ knots. Having determined the model residual drag by the above method it must then be multiplied by $(4)^3 = 64$, to get the wave drag of the full size boat. It can be seen from this how important accurate measurements are since errors are also multiplied by 64. This is also one of the several reasons why test models are made as large as possible.

Just as $D_F$ was calculated for the model, it can now also be calculated for the full size boat except that of course Reynolds' number is now different. Adding these two then gives the total full scale resistance. Although this procedure works reasonably well for large commercial vessels, there are a number of added difficulties in the case of the sailing yacht. The first is that because the underwater body of a yacht is cut away, the effective Reynolds' number, which depends on length, is less than the waterline length. It has been found empirically that a value of 0.7 of the LWL is better for determining Reynolds' number.

Another problem concerns the relative amounts of

laminar and boundary layer flow. It was pointed out earlier that laminar flow could extend for half a metre or so from the bow. Clearly this will represent a much greater fraction of the model hull surface than of the real boat. Thus to model the flow correctly in the testing tank, turbulence stimulators in the form of small studs or sand stuck on the hull near the bow are used to ensure turbulent boundary layer flow over the model analogous to that of the full scale boat. Unfortunately these devices themselves introduce additional drag and other unpredictable effects on measured model quantities which must be corrected for. These corrections will be larger the smaller the model.

Experiments have shown that regions of separated flow may appear on model yacht hulls. Although this has a considerable influence on the drag it is not known precisely how this effect scales.

The ultimate test of whether the towing tank procedure is correct is to check predictions against full scale measurements. This is not easy as towing tests on real water are notoriously inaccurate. There are some very large testing tanks which have enabled tests of identical hull shapes of several sizes up to full size. The trend that has been found is that smaller models predict too large a resistance at low Froude numbers and too small a resistance at high Froude numbers.

A peculiarity of the sailing vessel as opposed to a powered one is that most of the time the hull is both heeled and yawed. That is, it is making leeway, and since leeway is the result of sail side force which also produces heeling, it is clear that the two are inextricably intertwined. Thus if tank testing is to be of any use for points of sailing other than dead downwind, many more measurements are required at various heel and leeway angles. Since the flow around a yawed hull is no longer symmetrical the total force will now contain a lift component as well as a drag component. It is usual to consider lift forces as essentially free from scale effects, in spite of the fact that it has been generally recognised by sailors that yachts do not make as much leeway as their model predictions. There is evidence that model hull lift contributions are sensitive to Reynolds' number and roughness variations. Roughness is however required on both hull and keel of the model to induce turbulent boundary layer flow.

In view of the various sources of error it is clear that the test model should be as large as possible. The cost of construction of a model rises roughly in proportion to its length so one is faced with a tradeoff between

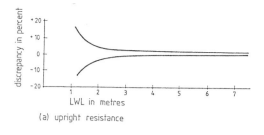

(a) upright resistance

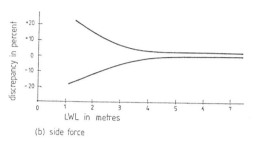

(b) side force

*Fig. 5.21* The average correlation error between model and full size boat as found for a large number of tests; (a) is for upright resistance measurements and (b) for lift force measurements. Like all statistical data, this represents the average range of variation for a large number of data but can make no predictions about a particular set of data. What it says, in effect, is that models should be at least 5 m long for 2–3% reliability. These graphs represent the state of the art in the mid-seventies.

cost and confidence in the test results. To get some idea of the magnitude of the errors fig. 5.21 was compiled. The correlation error is defined as the percentage difference between a particular model prediction and the measurement of the prototype or larger model. The uncertainties are the averages of a large number of cases and do not imply that a particular measurement will be in error by that amount or that future improvements in technique will not decrease the uncertainties.

Hopefully this brief synopsis of tank testing will have given you some idea of what is involved. Knowing the difficulties aids in appreciation of the results. Although tank testing has now emerged as a valuable input to yacht design, it was not always so. The history of the America's Cup provides a good insight into the highs and lows of tank testing over the years.

The first designer to use tank testing for a yacht did not fare well: G. L. Watson experimented for nine months testing eleven models in which both upright and heeled resistance measurements were made, resulting in the design of the J boat *Shamrock III*. She was beaten by Herreshoff's *Columbia*. Watson is reported to have said, 'I wish Herreshoff had had a towing tank.' It has subsequently been suggested that the one thing lacking in his tests was a model of *Columbia*.

When America's Cup competition resumed in 1958 with 12 Metre boats, the British challenger *Sceptre* was the best of a number of designs as decided by the interpretation of tank tests. Unsuccessful boats like the American Twelves *Valiant* and *Mariner* gave tank testing a bad name. It was only after the magnitude of scaling errors was better understood and larger models were tested (*Australia II* was tested with a 22 ft model) that the tank test finally emerged as a fruitful aspect of yacht design.

## §5.6 Keels

A good measure of the state of development of a device is to look at the range of variation in its design. Standard cloth sails which have reached an advanced stage of evolutionary development are all very similar in design. On the other hand, the hydrodynamic partner of the sail, the keel, does not seem yet to have evolved so fully if the great variety of sizes and shapes is a true indicator. One can only conclude that, until recently at least, designers have not understood how keels work. A possible reason why keel development has languished in the doldrums until recently is that all the answers appeared to be laid out on a platter from classical aerodynamics theory. Although a great deal can be learned from aerodynamics, the real breakthroughs have only come when it was realised where the classical theories break down when applied to a sailing boat. The following discussion draws on theory and experiment and ends up with a keel design different again from most of those of the past.

In answer to the question: why does a yacht have a keel? perhaps the following are in order of importance. A keel provides a hydrodynamic side force to oppose the wind side force when sailing to windward. Second, its configuration and placement help with balance and directional stability. Third, it is a repository for ballast providing a longer righting moment arm than would otherwise be possible. Finally the keel is a prop to stand on when on the hard. Though one must confess, putting a boat without a keel on the hard is easier.

As was pointed out in chapter 3, leeway is mainly determined by the size and shape of the keel; fig. 3.24 shows immediately the need to minimise $\lambda$ if $V_{mg}$ is to be maximised. It also shows that the mathematical connection between speed made good to windward and boat speed is: $V_{mg} = V_s\cos(\gamma' + \lambda)$. $\gamma'$ is the true wind angle measured to the boat's heading and $\lambda$ is the leeway angle. As the cosine decreases with increasing angle it is desirable to keep the leeway $\lambda$ as small as possible. Since boat speed is determined by resistance it is clear that any keel design that results in reducing leeway should be accompanied by the smallest possible increase in resistance. One can thus think of the hydrodynamic function of the keel as to minimise a certain combination of leeway and drag.

Off the wind, only the second of the four purposes of the keel is required. Apart from this its only function is to reduce speed through the frictional resistance over its wetted surface. Thus the particular combination of leeway and resistance which is to be minimised depends on the use to which the yacht is put. Racing around the buoys involves going to windward about 60% of the time. At the other extreme is a boat designed for tradewind cruising where its owner might only sail to windward when caught by an unexpected wind shift, but might have to do so efficiently enough to get out of trouble. In general a deeper, bigger keel will be required if a lot of windward sailing is anticipated. Choice of keel also depends upon wind speed. In stronger winds the true wind angle $\gamma'$ for optimum speed made good to windward tends to decrease, while leeway angle increases. Thus the relative importance of leeway in the above expression for $V_{mg}$ increases as the wind speed increases. On the other hand the relative importance of keel resistance decreases since at the higher boat speeds associated with strong winds the boat speed decreases less rapidly with an increase in drag. This is easily verified by looking at a typical resistance curve like fig. 5.15. In addition, a smaller fraction of this total resistance has its origins in the keel because of the rapid increase in hull wave-making resistance. From this it follows that a larger keel will give better windward performance in strong winds.

The conditions under which classical aerodynamics can be applied to keels are those where the sea surface is flat. The flow lines at the surface then have no vertical displacement, which is also the case for the flow on the axis of symmetry of an aeroplane wing/body assembly. This leads to the idea of an 'image' used already in the discussion of sails in fig. 4.4. The underwater version of this figure is given in

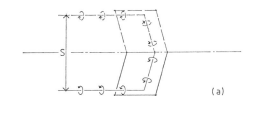

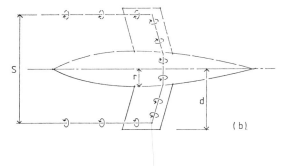

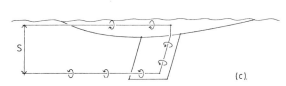

(a)

(b)

(c)

*Fig. 5.22* A keel which sheds vortices only at the tip is shown in (a). Such a keel has the same hydrodynamic behaviour as keel plus image (dashed), so its effective span s is approximately double the draft. This doubles the aspect ratio so improving windward efficiency. If the sea is smooth and there is no separation along the hull surface, a canoe body plus keel behaves hydrodynamically like a totally immersed body of revolution with two fins attached. Under these conditions the hull alone provides no lift but with fins attached the bound vortex continues into the hull thus increasing the span s. By increasing the aspect ratio in this way the hull is contributing to the lift by a process called *wing-body lift carryover*. Except in very light conditions the free water surface allows pressure relief from the high to low pressure sides of the canoe body, as shown in (c). The horseshoe vortex span is roughly halved, with the same reduction in aspect ratio and attendant reduction in windward efficiency.

fig. 5.22b: if the water surface is flat there is no cross-flow and hence no shed vortices in the water plane. The flow is the same as that over a symmetrical body like a submarine which is completely submerged, or an aeroplane which is completely immersed in air. The direct use of aerodynamic data is then only valid in very light wind conditions. Nevertheless we will look first at the results this

produces and then see how the presence of a real sea surface modifies the design.

From a purely hydrodynamic point of view there are two aspects of keel design that must be optimised, determination of the best plan form and section shape. As the latter is the most straightforward to deal with, I will start with that.

Fig. 5.23a shows the parameters which define the cross-section. As has been mentioned previously, a few numbers cannot specify an arbitrary curve so 'families' of aerofoil sections have been developed by the National Advisory Committee on Aeronautics (NACA). We are interested only in symmetric sections and the ones of most use in sailing are those whose maximum thickness occurs at the 30% chord position. (Complete data on sections of use for keel and rudder design are given in figs. 8.16 to 8.21.) Fig. 5.23b shows how the hydrodynamic properties depend on these parameters.

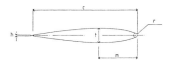

(a)

| parameter | increasing | decreasing |
|---|---|---|
| thickness ratio t/c | increases stall angle increases form drag increases volume for ballast | reduces stall angle reduces drag at zero angle of attack |
| position of maximum thickness | 0.5c reduces drag at zero angle of attack | 0.3c increases lift-drag ratio moves ballast forward |
| leading edge radius | increases lift-drag ratio | reduces drag at zero angle of attack decreases stall angle |
| trailing edge thickness | increases lift decreases lift-drag ratio h = 0.10 t absolute maximum | minimum ('knife edge') best |

*Fig. 5.23* (a) shows the parameters describing section shape: c is the chord length; h is the trailing edge thickness; m is the position of maximum thickness, usually specified as m/c; r is the leading edge radius; t is the section maximum thickness, usually specified as the ratio t/c. (b) shows the principal effects of varying the section shape. Except for special applications foils with a maximum thickness at 0.3c are best for keels and rudders. More details on the effect of trailing edge thickness are given in fig. 7.3.

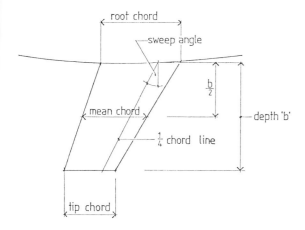

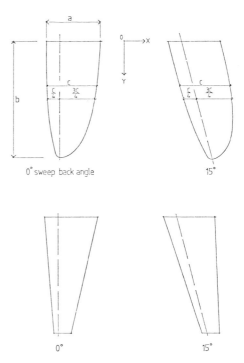

*Fig. 5.24* Parameters defining the planform of a keel or rudder. The hydrodynamic centre of pressure is located on the quarter chord line.

The third function of the keel, to provide stability, must be constantly kept in mind in keel design as the cross-section size and shape must always be sufficient to house the amount of ballast required.

It was explained in §4.1.1 how a loaded aerofoil, that is one producing lift, has an additional component of resistance called vortex drag. Vorticity is shed whenever the slower moving fluid in the boundary layer separates from the surface. As for sails, the downwash induced by the trailing vorticity reduces the effective angle of attack or leeway angle thereby reducing also the lift force. The effect is most pronounced at the tip of the keel and is therefore more important for low aspect ratios.

Keel planform thus determines the three-dimensional aspects of the keel hydrodynamic properties. Fig. 5.24 specifies the planform parameters.

Classical aerodynamics suggests that for windward sailing the best keel will have constant section shape with an elliptical planform, the highest possible aspect ratio and zero sweep angle. An elliptical keel is difficult to make and very impractical on a fixed-keel boat. Fortunately it is not necessary, as a keel with straight tapered sides and a flat bottom can give a very close approximation to the elliptic lift distribution. The theoretically optimum taper ratio is 0.33, shown in fig. 5.25 for sweepback angles of 0° and 15°. It is found in practice that the taper ratio is not at all critical. This is especially so for low aspect ratio keels where the lift distribution tends to elliptical regardless of taper. For this reason a shape

*Fig. 5.25* 'Elliptic' planforms are shapes that give rise to an elliptic lift distribution. To do this the planform does not necessarily have to be the shape of an ellipse (or semi-ellipse): a number of shapes could give an elliptic lift distribution if the section shape is varied appropriately. If one stays with a constant section shape, then an elliptic lift distribution results if the chord lengths c are related to their position y from the root by the equation of an ellipse given in the text. The dashed line is the quarter chord line. Since it is usually advantageous to make this straight and therefore with a constant sweep angle, the position of the leading edge is then fixed at c/4 ahead of the quarter chord position. A true semi-elliptic shape would have a curved quarter chord line. The two lower planforms have a 3:1 taper, are easier to make and are almost indistinguishable hydrodynamically from those above.

that allows proper placement of ballast is probably more important than optimum taper.

In fig. 5.25 the dashed line corresponds to the quarter chord point. For two-dimensional flow the net hydrodynamic force acts through this point. In more nautical language, the centre of lateral resistance CLR lies on this dashed line. This is also seen from the measurements given in figs. 8.17–8.19 which show the *moment coefficient* about the quarter chord point as being zero over a wide range of angles of attack. This means that even at an angle of attack there is no resultant twisting force about this point.

This can only happen if the resultant force acts through this point.

Note that the 'elliptical' planforms have been drawn so that their quarter chord lines are straight. This distorts the geometric shape from that of an ellipse, however the chord lengths are still 'elliptical'. All this means is that if the chord length is represented by x and the distance from the top by y (see fig. 5.25) then x and y are related by the equation for an ellipse which is: $x = a\sqrt{1 - (y/b)^2}$. From this you can determine the chord lengths at any depth. Draw in the quarter chord line at the required sweepback angle and mark in the chord lengths with one-quarter in front of the quarter chord line. Join the ends of the chords with a smooth line and you have an 'elliptical' planform.

Referring again to the aeronautical data in figs. 8.17–8.19, it will be seen that curves for lift and drag are also given when there is a trailing edge flap. In the case of a keel, a trim-tab. Both lift and drag are increased by its use, so they may have an advantage in strong wind conditions where drag is less of a penalty. On the other hand trim-tabs may be used to reduce lateral area without compromising leeway, an option that could be advantageous for light weather conditions. The sailing conditions under which trim-tabs are likely to be advantageous are difficult to ascertain by the crew without extensive instrumentation. For this reason and because of class rule penalties trim-tabs have not been used as extensively as their potential warrants.

If we continue with the assumption that there are no free surface effects, drag due to the keel can have only two origins: surface friction drag and vortex drag. Keel surface friction drag amounts to around 5% to 15% of the total hull drag; it depends on wetted area and the Reynolds' number which depends on speed and chord length. There is a slight dependence on other factors such as sweep angle and section shape but this is minor. As fig. 2.10 makes clear, drag is greatly reduced if laminar flow can be maintained over as large an area as possible. Laminar flow is possible at low Reynolds' numbers, which means near the leading edge and at low speeds. As one moves back along the chord Reynolds' number rises and a turbulent boundary layer becomes more likely. Thus a high aspect ratio keel will have a greater fraction of its area able to maintain laminar flow. It is also found that a small leading edge sweep is conducive to maintaining laminar flow. The large difference between the laminar and turbulent friction drag coefficients given in fig. 2.10 implies that a rough

rule of thumb for friction drag is that for every extra per cent of chord length having laminar flow there is a reduction of 1% in the keel friction drag. Surface smoothness is, of course, also an important parameter and was fully discussed in chapter 2.

Keel vortex drag has the same physical origins as sail vortex drag. Its magnitude is proportional to the square of the lift or heeling force and inversely proportional to the aspect ratio. It is usually of the order of 10% to 25% of the total drag when sailing upwind. Remember that the elliptic lift distribution minimises only the vortex drag component of the total drag and has little effect on the surface friction drag which depends purely on area.

A theory of fluid flow known as *slender body theory* has made it possible to calculate the interaction of a low aspect ratio keel with the canoe body. A schematic representation of this interaction is shown in fig. 5.22. A consequence of this theory is that for a fixed ratio of body 'radius' r to draft d, the lift force is given by the expression: $L = C(r/d)V^2 d^2 \pi \varrho \alpha$. Here $\alpha$ is the angle of attack (leeway) and $C(r/d)$ is a coefficient which depends on r/d. This therefore makes the surprising prediction that the lift and hence also the vortex drag do not depend on keel area or shape but only on the square of the draft!

How can this result be reconciled with all that has just been said about elliptic lift distributions and aspect ratio? It arises because fluid dynamics is so complicated that theories tend to be approximations valid only for a limited range of conditions. The slender body theory gives the interaction between hull and keel only for low aspect ratio keels but still does not take into account free surface effects. The results so far quoted for high aspect ratio keels stem from aeronautical data which ignore both the effect of the hull and the water surface. Unfortunately real boats are neither of these two extremes: they have keels of intermediate aspect ratio and pressure relief effects always occur to a greater or lesser degree at the sea surface. The picture is then like fig. 5.22c with very little vorticity being shed from the hull in light conditions but with the full amount being shed in strong winds.

From the foregoing it should be clear that only by careful measurement of the properties of keels attached to representative hulls can we hope to progress in keel design. It should be realised that such measurements are quite difficult since very small differences in large total forces must be measured. Nevertheless a number of useful studies have been made.

The first concerns the effect of sweep angle (defined in fig. 5.24). By tank testing a model with keels of different sweep angle but the same aspect ratio, it was found that when sailing upright but making leeway (small boat situation) zero sweep angle was best. This is shown in fig. 5.26.

The situation is different, however, for fixed ballast boats which must heel when making leeway. The upper curve in fig. 5.26 shows the vortex drag at 20° heel. It is seen to have a minimum for a sweep angle of about 30°. This result has its explanation in the origins of wave-making resistance. Wave drag in a yacht can be divided up into two parts, that due to displacement and that due to side force. The latter is a result of the asymmetric pressure distribution around keel and hull when making leeway, causing a distortion of the free sea surface. The contribution of the keel to the surface waves depends on its distance from the surface and this depends on angle of heel. Just as the bow wave can interfere with the stern wave, adding to it under the right conditions to increase drag, so the keel wave can interfere with the hull wave. Furthermore it has been found that the centre of effort moves down with increasing sweep angle. This gives us the ingredients for an explanation of an optimum sweep angle for a heeled boat. The shift of side force away from the root due to sweepback reduces the losses due to free surface

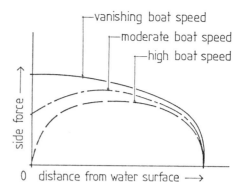

*Fig. 5.27* The variation of side force with depth from the root of a keel depends on boat speed. This is because in smooth water there is no vortex formation at the keel-hull interface and the keel behaves in the 'classical' manner of an aeroplane wing. At high boat speeds the keel contributes to wave-making and there is complete pressure equalisation on each side at the root and hence no lift there. This modifies the otherwise elliptic lift distribution.

effects; on the other hand the basic effect of sweepback is to increase vortex drag. Thus for a given angle of heel there is an optimum sweep angle where the combination of these two opposing effects gives minimum drag.

Following the same lines of argument, one might expect that deep high aspect ratio keels will suffer little from surface effects and therefore should have no sweep angle. This has indeed been found to be the case by one group of investigators.

Under heel the effective span s of fig. 5.22 is reduced as the keel tip comes nearer the surface. This is equivalent to a reduction of the aspect ratio with a concomitant increase in drag and leeway. This is evident in fig. 5.26 where even the minimum of the heeled vortex drag is more than twice that of the upright condition.

From the foregoing discussion it is clear that a common denominator in keel design is to put the hydrodynamic centre of force (CLR) as low down as possible. If, in addition, we accept that at high speed there is a full pressure relief effect at the surface, then there can be no side force developed at the root of the keel. For moderate boat speeds the loss at the root is not likely to be complete. Thus we expect a variation of keel lift with depth like that shown in fig. 5.27. At vanishingly small boat speed only is the lift distribution elliptic.

Consequently there is little point in trying to develop lift near the root of the keel. An optimum keel design is thus one in which the lift distribution is

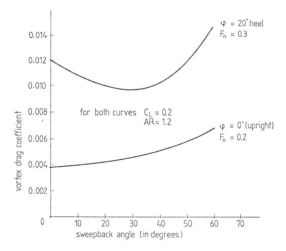

*Fig. 5.26* The effect of sweep angle on vortex drag when water surface effects are taken into account. The lift produced by leeway is the same in both cases. For small boats which are sailed upright, zero sweepback is best. When heeled some sweep is beneficial. The optimum amount of sweep decreases with increasing aspect ratio and draft.

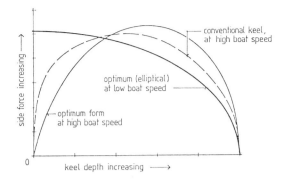

Fig. 5.28 Since at high boat speed it is impossible to expect keel lift near the root, it is better to concentrate it lower down. The exact shape of the optimum distribution shown here depends on the speed conditions under which the boat is expected to sail.

kept low but not so low as to give unacceptable drag at low speeds. The art of compromise is again the art of yacht design. The solution depends upon the use to which the boat is to be put, and any class or other restrictions on draft.

Fig. 5.28 shows a possible optimum lift distribution. If the vortex drag of these three distributions is calculated by classical aerodynamic methods they are in the ratio 4:5:10 for elliptic:conventional:optimum. This calculation ignores the effects of keel wave-making; when it is included the optimum distribution at high boat speed turns out to have about 10% less than the conventional keel at high boat speed.

How does one realise this optimum distribution of lift in practice? Since an elliptic lift distribution is no longer the aim, conventional tapers can be forgotten. In fact an obvious way to increase lift with depth is to increase the chord length with depth, making a sort of 'upside-down' keel. Because sweep angle tends to increase resistance when wave-making effects are reduced, an upside-down keel should have less sweep than a conventional one. Fig. 5.29 shows such keels with and without sweepback. An obvious practical disadvantage of keel (a) is that weed or fishing lines might easily remain caught on it. If a slight loss of performance can be tolerated then sweepback combined with inverse taper can give a more practical leading edge slope, as shown in (b).

Inverse taper first hit the headlines (though it was not unknown before) when it was used on *Australia II* in the 1983 America's Cup races. As is now well

known this keel also sported winglets as a means to reduce vortex drag when draft is limited.

If you think this was a giant step forward by yacht designers, then look up British patent no. 3608 taken out in 1897. In the patent specification an aeroplane was described as having at the extremities of the aerofoil two 'capping planes', intended to induce two-dimensional flow. In the words of their inventor, the capping planes were 'in order to minimise the lateral dissipation of the supporting wave'. These were the words of F. W. Lanchester, the man who first saw the connection between tip vortices and drag.

Winglets or end plates work by moving the tip vortex away from the tip of the keel to the tips of the winglets. This increases the effective span thus reducing vortex drag. Aerodynamic methods can be used to obtain the optimum spanwise distribution of loading along the main keel and winglets that minimises vortex drag. The keel parameters involved in producing this loading are chord length and section shape in both keel and winglets, and incidence and twist in the winglets.

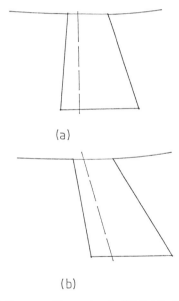

Fig. 5.29 One way of increasing the lift distribution with depth is to increase the chord length with depth. Another would be to increase the section thickness, and yet another would be to do both. The first option leads to the 'upside-down' keel, which will perform best at zero sweep angle as in (a). If leading edge sweep is thought necessary, the same lift distribution can be retained but with a slight loss in hydrodynamic efficiency by introducing sufficient sweepback as in (b).

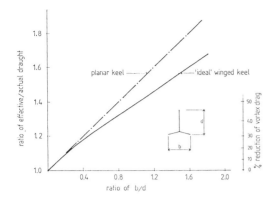

*Fig. 5.30* This compares a winged keel with a planar one whose draft is the same as that of the winged keel with its wings folded vertically down. As can be seen a planar keel is always better if draft is not a limitation. However if draft is limited to d then wings can improve windward performance.

The effect of winglets is summarised in fig. 5.30. The left hand ordinate is the ratio of the effective to the actual draft. For a planar keel the effective and actual draft are the same. Plotted horizontally is the ratio of winglet span to draft. The solid curve gives the effective draft of a winged keel and on the right hand ordinate the reduction of vortex drag. This is compared with a planar keel formed by folding the winglets down. This shows clearly that if there is no constraint on draft it is far better to use a planar keel. Just adding wings to an existing keel can therefore reduce leeway by increasing effective draft, but at the cost of more wetted surface. For the same reduction in vortex drag, the increase in surface friction drag is likely to be larger than by simply extending a planar keel to greater depth. Again it is a question of compromise and what is required of the boat.

It might be expected that winged keels would add resistance in a seaway by damping the pitching motion in a way analogous to increasing the weight in the ends of the boat (see chapter 6). Theoretical analyses indicate that this is indeed the case in following seas but that a net propulsive effect occurs in head seas.

Another clear advantage of the winged keel is that it provides excellent accommodation for the ballast at the lowest point in the boat. In this regard inverse taper is also an advantage.

To summarise:

(a) Boats which are sailed upright should have deep high aspect ratio keels with no sweepback, but tapered for an approximately elliptic lift distribution.

(b) Where the keel contributes to surface waves due to heeling, it should have an inverse taper and no sweepback. If this is regarded as inconvenient, only enough sweep angle should be used to give the leading edge some rearward slope.

(c) If draft is a restriction, correctly designed winglets might improve windward ability. The basic compromise is between reduced vortex drag and added surface friction drag.

(d) Downwind, the smallest possible wetted area is best regardless of shape.

(e) If inverse taper is not desirable then some sweepback should be used, more for low aspect ratio than for high.

## §5.7 Rudders

Much of what has been said about keels applies also to rudders. The purposes of a rudder are to steer the boat, keep it in balance and supply additional lift when on the wind. It is mainly the steering requirement that adds some design criteria not already discussed under keels.

The side force contribution of the rudder tends often to be forgotten. Compared with the keel it will be more or less in proportion to their areas. This means that even with the helm amidships that fraction of the total side force is taken by the rudder.

As with keels some aspects of rudder design can be settled by reference to classical aerodynamics theory. One of these is section shape. Unlike a keel, a rudder may be required to operate effectively at large angles of attack without stalling. This calls for a thick section (see fig. 5.23 and figs. 8.17–8.19 and 8.23). A NACA 0009 section stalls at about 12°, which means it is possible to turn the rudder fairly quickly through 12° without stalling. It does *not* mean that one can never put the tiller over more than 12°. This is because as soon as the boat begins to turn the new direction of flow onto the rudder reduces the angle of incidence, so the rudder can be turned continuously without stalling providing it is turned no faster than the boat itself is turning.

As explained in §5.6 the hydrodynamic centre of effort lies on the quarter chord line. A rudder with zero sweep angle pivoted on the quarter chord line will produce no torque whatever the angle of attack; this would provide no feedback to the helmsman and make steering extremely difficult. Either the rudder

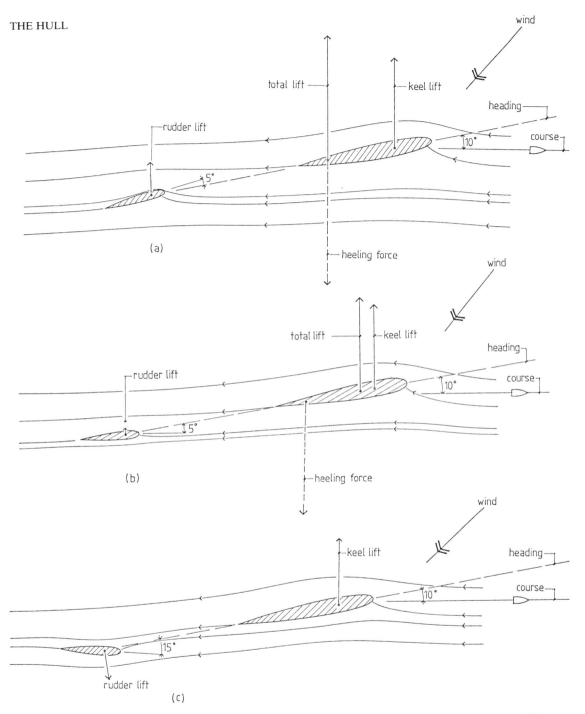

*Fig. 5.31* Potential flow measurements showing the interaction between keel and rudder but neglecting the effect of the hull. In (a) 5° of weather helm is necessary to produce a total lift force equal to and in line with the heeling force. This corresponds to steady balanced windward sailing. In (b) the helm is put 5° alee. The rudder lift is reduced almost to zero so the position of the total lift is shifted forward. This, combined with the heeling force, turns the boat into the wind. In (c) a greater rudder angle is used, reversing the direction of the lift and increasing the counterclockwise turning moment.

pivot must be forward of the quarter chord line or some degree of sweep used.

The action of the rudder in turning the boat is best seen by reference to fig. 5.31. The flow pattern was determined by the electric analogue plotter described in §8.2; it shows only two-dimensional flow and takes no account of the presence of the hull. Lift forces can be calculated but not drag. This is no great disadvantage when considering the action of the rudder which depends mainly on lift. In (a) the boat is sailing to windward with a leeway angle of 10° and a weather helm of 5°; it can be seen that both the keel and the rudder contribute to the total lift force. This resultant force acts at a point between the keel and rudder quarter chord points but nearer to the keel since its contribution is greater. Since this situation represents a boat sailing steadily and therefore perfectly in balance, the hydrodynamic lift force is not only equal and opposite to the heeling force but must act in the same fore-and-aft plane (see the further discussion on balance in chapter 6). When the rudder is turned 5° to leeward as in (b) the lift it generates is greatly reduced. Its contribution to the total lift force becomes very small and the resultant

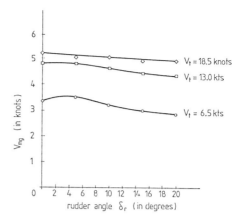

Fig. 5.33 Extensive tests on *Standfast* showed that a small amount of weather helm was advantageous when sailing to windward in winds up to 13 knots. There is good reason to believe this to be qualitatively true of most designs.

lift then moves its point of action forward since most of it now originates with the keel. The heeling force, on the other hand, remains where it was. These two opposing forces are now separated and will clearly produce a moment or turning force on the boat which acts in a counterclockwise direction. The boat is thus

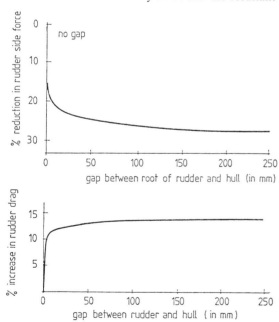

Fig. 5.32 Because of the pressure difference between the two sides of a rudder, water will flow through any gap between rudder and hull to equalise the pressures. This results in loss of lift and increased drag as shown here. The hull should be flat above the rudder so that this gap does not increase with rudder angle.

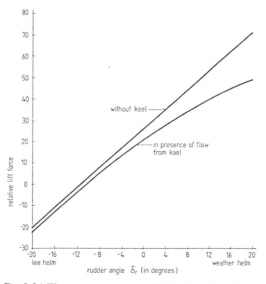

Fig. 5.34 These measurements with the electric analogue plotter show the interaction between keel and rudder assuming no flow separation and ignoring hull effects. The curves gives the rudder lift force as a function of rudder angle. The boat is assumed to be sailing with a leeway of 11°. Even this ideal flow from the keel reduces the effectiveness of the rudder. In real life the rudder is worse off due to turbulence and the distortion of the flow due to the hull.

*Fig. 5.35* It seems that wing keels are now so commonplace that the onlookers were more interested in the photographer than the boat. This is the wing keel on the first 12 Metre yacht to be constructed in glassfibre laminate. (*Photo: NZ Herald*)

turning head-to-wind as expected for this rudder position.

Turning the rudder even further to leeward can, of course, reverse the direction of the lift as shown in (c). Even in the absence of a heeling force there is a purely hydrodynamic turning moment set up by keel and rudder. The arm for this moment is the distance between quarter chord points of keel and rudder: the longer this is the more effective the rudder. It should be clear that the keel plays just as important a part in turning the boat as does the rudder. If you object that you have steered a boat without a keel, then I say that this was possible because all hulls can create a certain amount of lift on their own. You will also find that a boat without a keel is much less responsive to the helm.

At low speeds the hull acts as a good endplate for the keel, nearly doubling its aspect ratio. We would like to be able to say the same for a rudder hung under the hull. Unfortunately some clearance is needed for movement and even the slightest gap will allow flow from the high to the low pressure side, creating a vortex at the root of the rudder and reducing its effective span. This reduces the lift and since the turning action of the rudder is due to its lift, as shown in fig. 5.31, this gap drastically reduces efficiency. Its effect is summarised by fig. 5.32 which shows the loss in lift and the increase in drag as the gap is widened. As can be seen, a gap as small as 3 mm almost completely destroys the beneficial endplate effect of the hull. Even if the rudder is fitted to such a fine tolerance, it is in vain if the hull is shaped so that the gap widens with rudder angle.

Transom hung rudders must do without any end plate effect and those which pierce the surface, as most do, experience the added hazard of air

*Fig. 5.36* Despite the research into underwater appendages which specifies the best hydrodynamic design for rudders, their shapes and locations are possibly more varied than any other important yacht component. This probably reflects both a lack of appreciation of the hydrodynamics as well as the many varied requirements of different designs.

entrainment. The pressure in the water on the low pressure side of the rudder can easily be less than atmospheric. The air then pushes down into the water, completely destroying the local lift force. The standard way to reduce air entrainment is by attaching fences, as described for the hydrofoil in §8.11.

It is very common to see rudders hung from skegs, though difficult to justify this except as protection. The flow is always disturbed at the join, decreasing the stall angle and completely eliminating any possibility of laminar flow over the moving part of the rudder. Since surface smoothness is measured in fractions of a millimetre it follows that no joint, however carefully made, will be smooth. Another disadvantage of the skeg is that it inhibits manoeuvrability. A boat turns roughly around its centre of gravity so the skeg comes almost broadside-

on to the flow, acting as a brake and producing turbulence. An even worse form of skeg is the partial one in which the bottom part of the rudder goes through to the leading edge of the skeg. When the rudder is turned a large gap opens up completely upsetting the flow.

The statement is often made that a small amount of weather helm is advantageous when going to windward. This, indeed, seems to be the case as the measurements in fig. 5.33 indicate. They were from the *Standfast* trials and show that a weather helm of $3°–4°$ gives the best $V_{mg}$ up to a true wind of about 13 knots. In stronger winds rudder amidships was best. These conclusions must, of course, depend on the particular design; however there is good reason for thinking that some weather helm will generally be beneficial. The combination of keel and rudder produces most of the side force. When the tiller is pulled up to windward the rudder acts like a wing flap, increasing the lift of the underwater body at a given leeway angle. Or alternatively, one can think of it as producing the same side force for a smaller leeway. Of course every gain has its price, and a glance at fig. 8.16 will remind you of the increase in drag produced by a flap. So while leeway is decreased, drag is increased. Somewhere in between lies the winning optimum, probably not more than 5° weather helm for most boats.

The rudder's position at the aft end of the boat, though best for producing a large turning moment, is not without drawbacks. Not only is it situated in a region of downwash from the keel but it is also affected by the flow around the stern. If the hull lines are too full at the stern separation of the flow will occur and the resultant turbulent flow around the rudder will degrade the production of lift. Furthermore it has been found that the flow near the stern tends to follow the buttock lines more than the waterlines, so with sharply rising buttock lines it has a strong vertical component in the region of the rudder. This reduces its effectiveness, which is at maximum when the flow is perpendicular to the quarter chord line. A simple test of whether your boat is suffering from this problem is to see if it responds correctly to a very small rudder angle when sailing downwind. Cases have been known of yachts which actually turn the wrong way when given a small rudder angle of less than about 10°.

Some idea of the effect that the keel alone has on the rudder can be obtained by means of the electric analogue plotter. Fig. 5.34 shows the rudder lift force as a function of rudder angle for a boat with and without a keel, moving with a leeway of 11°. Even for conditions of ideal flow and without the disturbance of the hull, the rudder lift force is reduced by the presence of the keel especially at large angles of attack.

# Chapter Six

# DYNAMIC MOTION OF A YACHT

*Both men and ships live in an unstable element, are subject to subtle and powerful influences and want to have their merits understood rather than their faults found out.*

JOSEPH CONRAD, *The Mirror of the Sea*

## §6.1 Introduction

So far in this book we have talked only of sailing on smooth water with steady winds, or unaccelerated motion. If there is no acceleration there is no force, which is why we have always insisted that water forces and wind forces must be equal and opposite. Yet possibly the most characteristic thing about sailing is the constant feeling of being pushed. Forces are acting on the crew that constantly push them up and down and from side to side. Where there are forces there is acceleration and thus a change in velocity: the yacht is no longer in static equilibrium with its surroundings, moving forward at a steady pace. This chapter concerns itself with the much more complicated case of non-equilibrium sailing. Much is yet to be learned about the dynamics of yacht motion, so what will be said here is very far from the final word.

In a scientific description of the motion of any object in space a set of directions or coordinates must be defined to act as a framework by which to tag the various motions. As we have seen, forces or velocities which are mutually perpendicular can be regarded as independent of one another. Alternatively, one can think of the components of a force in different directions. The one at right angles is

zero. As it is possible to find three (but not more) directions in space which are all mutually perpendicular, motions along these three directions will be independent.

In principle any three mutually perpendicular directions will do, but by convention and for obvious practical reasons the three directions chosen are as shown in fig. 6.1. The origin of the coordinate system is the centre of gravity of the boat; x is parallel to the centreline and points toward the bow, y points out to starboard and z is down. Two independent kinds of motion are possible for each coordinate axis; translational movement along the axis or rotation about it. A variation in speed along the x axis is called *surge* and rotation about it is *roll*. Bodily motion of the boat sideways along the y axis is called *sway* and rotation about the y axis is *pitch*. Movement up and down the z axis is *heave* and rotation about it is *yaw*. The rotational motions are always depicted by the Greek letters shown and are taken to be positive when in a clockwise direction looking along the axis from the centre of gravity.

Although these motions are independent, the asymmetric shape of a yacht hull when heeled can give rise to coupling between them. For instance an analysis of measurements on the full scale tank tests of the 5.5 Metre yacht *Antiope* gives the static relationship between heel angle and trim angle shown in in fig. 6.2. This was calculated from the lines by determining the movement of the centre of buoyancy. It means that if a heeling force rotates the boat around the x axis by 30°, then it will also rotate about the y axis by about 2° in such a direction as to trim by the bow. It was also shown for the same boat that the yawing moment depended on the heel. These

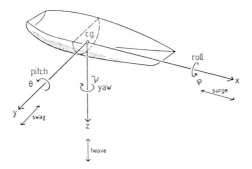

*Fig. 6.1* Any dynamic motion of a boat can be fully described in terms of the six movements shown here. Although the motions associated with each of the mutually perpendicular coordinates are independent of one another, there can be coupling between them under certain circumstances.

effects are likely to be more pronounced in a more modern hull form. What they mean is that if a boat is rocked back and forth (rolled) with an external periodic force, then after a while pitching and yawing motions will develop.

It is perhaps worth noting that because of the effects of buoyancy there is a fundamental difference between roll, heave and pitch on the one hand and the dissipative forces of surge, yaw and sway on the other. A sudden gust of wind may generate a surge, which will die out steadily due to resistance. The first three will generate restoring forces tending to bring the boat back to the initial position.

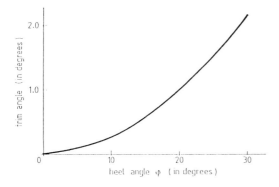

*Fig. 6.2* This curve gives an idea of the magnitude of the coupling between heel and pitch. For instance, as the boat heels to 20° it also trims 1° by the bow. More modern designs are likely to have greater coupling than this.

## §6.2 Anatomy of a Tack

Although the turning action and manoeuvrability of commercial ships have been studied in great detail, much of this work is not really relevant to small yachts. Unlike a powered vessel, the driving force of a yacht is variable during a turn giving the added complexity that always seems to be the case for sailing boats.

By applying the known laws of physics it is possible to make a qualitative description of a yacht's motion during a tack. When you put the tiller down the boat doesn't just turn the way your car does on a good road when the wheel is turned; it is more akin to turning at speed on gravel. The turning action of the boat can only be the result of applied external forces. These come in just two varieties: forces that produce a translation of the centre of mass, and *moments* that rotate the boat about its centre of mass. No matter what complex array of forces is acting, they can all be reduced to a resultant force and a resultant moment. To emphasise again: the resultant moment can *only* rotate the boat about its centre of mass, and the resultant force can *only* move the centre of mass of the boat in the direction of the force.

The procedure, then, is to look at the resultant force and moment acting on the boat at each stage of the turn. These, of course, vary continuously and the best we can do conveniently is to show six discrete steps (fig. 6.3).

In A the boat is sailing steadily to windward on the starboard tack, making 5° leeway and perfectly balanced with the rudder amidships. It is assumed, as in chapter 3, that this is a fin-keeled light displacement boat for which nearly all the hull lift force comes from the keel and rudder. Motion through the water at the leeway angle results in the keel and rudder forces shown; their resultant is a dashed arrow. For steady sailing this is exactly equal and opposite to the sail force, as can be seen. At B the skipper puts the helm down, the rudder angle of attack is switched to the other side and the rudder force reverses. The resultant hull force is reduced and a clockwise moment appears, indicated by a curved arrow. This turning moment results from the fact that the keel and rudder forces have components in opposite directions, which makes the hull rotate about its centre of gravity (marked with a cross). But this is not all, because now the resultant hull force is less than the sail force so there is a net force to port. Where there is a net force there is acceleration in the same direction, so the boat not only turns but moves

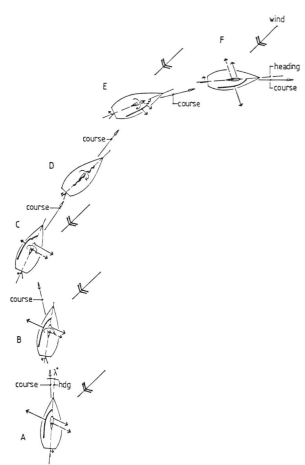

*Fig. 6.3* The net forces on a boat can always be decomposed into a turning moment and a resultant translational force. In A the boat is sailing steadily and has no net forces on it. At B the rudder is turned, upsetting the balance of forces. This gives rise to a new course and rotation of the boat. Then at C the keel force becomes larger than the sail force because of the excessive leeway and the boat moves to position D where there are almost no forces except a turning moment. The boat therefore rotates while its centre of gravity continues along the same line. Finally position F is reached where there is no moment and the forces are in equilibrium again.

off to port. But since it also has forward motion its course is something like that shown between B and C.

Now at C the boat has turned more, to the point where the sail is beginning to luff and the sail force is much reduced. Because of the increased leeway between B and C the keel force, on the other hand, has increased. The net hull force is now greater than the sail force and the boat moves to starboard as well

as forward, setting a new course to position D. The turning moment is still clearly present so by the time D is reached the boat is head-to-wind but assumed still to have way on. At D both the sail and the keel are luffing and produce no lift forces, just drag. Only the rudder is producing a force since it has an angle of attack and the boat is moving. Thus only a net turning moment is acting on the boat, so it turns. Since there is no net force, as Newton told us the centre of gravity 'will continue in its state of motion', i.e. it will continue to move in the same straight line.

Turning about the centre of gravity continues and at E the sail is beginning to fill on the port tack. Because the hull has turned, the keel develops an angle of attack giving a lift force to starboard. The net force is now to starboard further changing the course direction. Finally, at F all forces are in equilibrium and the turning moment has disappeared, the boat is sailing steadily on the port tack.

It is clear from this description that a boat doesn't simply rotate about its centre of gravity and then sail off on the new course. Hull and sail out-of-balance forces also contribute to the course change. In particular, the hydrodynamic efficiency of the keel is just as important as that of the rudder. (A further discussion of steering performance is given in §11.1.)

## §6.3 Balance

Of all the characteristics of a sailing vessel, balance is the one that has captured the attention and emotion of sailors the most. Perhaps this is a natural human reaction to the stress of hours at the helm constantly correcting any tendency to go off course in wind and waves. Since the hand on the helm is part of the feedback loop of the crew-boat system, it is not unnatural that the crew should endow the rest of the system of which they are a part with their own human characteristics. Many boats have been castigated as 'hard-mouthed' or 'ill-tempered': as they heel they constantly round up into the wind, requiring lots of weather helm to stay on course. The same craft when off the wind yaws all round the compass displaying an untameable 'wildness'. One which requires little helm adjustment on or off the wind is described as 'well-tempered'. The motion of a ship in a seaway has been described as '. . . one of the real wonders of this world, as something at once beautiful, inexplicable and often inspiring.' In the Book of the Proverbs of Solomon, Agur the son of Jakeh declared it to be beyond his comprehension.

It is curious that this attitude of sailors over the

centuries has been always directed toward the *hulls* of their ships and never toward the rigs or the whole of sails plus hull. Whatever the reason for this it led later generations of naval architects and others to formulate some of the most preposterous theories ever foisted on innocent sailors, the most notorious of which was the 'metacentric shelf' theory advocated by Rear-Admiral Alfred Turner. Although based on a fallacious physical argument it appeared to produce well balanced boats. Other ideas, also concerned entirely with hull shape, were that there should be a considerable degree of fore-and-aft symmetry, or that the section areas should follow some particular form of curve. It is undoubtedly true that if there is considerable coupling between heel and yaw, for instance, the hull alone may have a tendency to change course on heeling.

These effects are minor compared with the powerful influence of the rig. When a boat heels on the wind its driving force goes way outboard while its drag stays under the hull. These forces combine to twist it up to windward with a strength that no misshapen hull could ever muster. As we shall see, hull shape is only a small part of the balance problem so Admiral Turner's well-tempered boats probably all had some other characteristic in common which made them so. Balance is no problem for boats that don't heel very much; a very stiff boat or an undercanvassed one will fill this criterion and a well designed and efficient rudder can easily compensate for the inevitable balance variations on heeling. As you read on you will soon realise that it is physically impossible for a normal yacht to be even reasonably well balanced under all conditions of sailing. If you claim to be the proud owner of such a boat, it is probably because it is very heavy, very undercanvassed and very slow; or it is a multihull and very fast!

## §6.3.1 The Mechanics of Balance

When sailors speak of a well balanced boat they really mean one which can be steered with not too much rudder action under any normal conditions of sailing. We will therefore look now at the balance of forces and moments necessary to keep a boat on course under various conditions, keeping the hull and rudder forces separate.

When a boat is sailing steadily and unaccelerated, all forces and moments acting on it must be exactly balanced; as remarked a number of times, the total sail force and the total hull force must be equal and

opposite. A corollary of this is that the sail drive must be equal and opposite to the hull drag. This alone is not a sufficient condition for equilibrium since we must also ensure that there are no net turning moments acting, otherwise the boat will not stay on course. Balance is concerned with the conditions necessary for equilibrium of the yawing or turning moments.

Such turning moments are produced by the horizontal components of the various forces present, those we 'see' when we look vertically down on the yacht (fig. 6.4a). Although pressures act all over the surfaces of the sails they can be regarded as equivalent to a single force acting through the centre of pressure or CE. In the absence of a rudder force all the hull pressures can similarly be regarded as equivalent to a single force acting through the centre of lateral resistance or CLR. If a rudder side force is present then less hull lift $L_h$ is needed to balance the sail force component H. Since the rudder force is expected to be varied to keep the boat on course, the simplest way to show this in a diagram is to have the rudder force subtracting from the hull lift through the CLR. This representation has the further advantage that it is easy to see how the rudder develops a turning moment.

Remembering that moments are produced by equal and opposite forces that do not act along the same line, and that the magnitude of a moment is given by the product of *one* of the forces times the perpendicular distance between them, we can now see what the condition is for no net yawing moment. Fig. 6.4a shows that there are three turning moments present. Clearly the rudder force has the effect of rotating the boat in a counterclockwise direction and its moment is given by $N_r = r \times F_r\cos\varphi$. The second moment which rotates the boat the same way is formed by the equal and opposite forces H ($= F_H\cos\varphi$) and $L_h$ a distance d apart. This is the horizontal distance between the centre of effort and the centre of lateral resistance. The third moment turns the boat in the other direction and is formed by the pair of equal forces $F_R$ and D, the sail forward drive force and the hull drag. Fig. 6.4b shows their separation more clearly although the forces themselves are perpendicular to the paper. This moment is $F_R h\sin\varphi$. For the yacht to be balanced the clockwise moment must be equal to the sum of the counterclockwise moments, which gives rise to the equation:

$$r \times F_r\cos\varphi = h \times F_R\sin\varphi - d \times F_H\cos\varphi$$

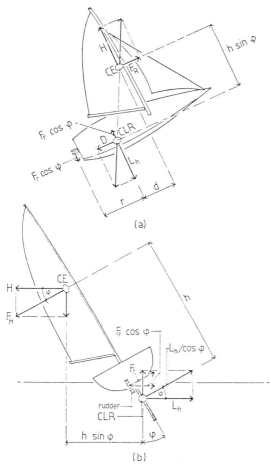

In light airs all three of the forces in this equation will be proportional to the square of the boat speed. Thus if a gust of wind speeds up the boat without heeling it, that is without changing $\varphi$, the expression will remain an equality and the boat will remain perfectly balanced. But of course it will heel, so $\varphi$ changes. If, for the moment, we regard all the other quantities as remaining fixed, then our balance condition is at the mercy of the factors $\tan\varphi$ and $1/\cos\varphi$. If the angle of heel is not too large and is measured in radians, it is possible to write the last equation in an approximate form: $r \times F_r = hD\varphi - dL_h$ which is accurate to about 5% for angles of heel up to about 17°. We see from this that as the boat heels, the rudder force $F_r$ must increase in order to maintain balance. In other words, to the extent that the assumptions made are correct, weather helm must increase with increasing heel angle.

Thus weather helm is predicted without the need to postulate a forward movement of the CLR due to trim by the bow when heeled, or some asymmetrical effect of the bow waves or any other true or imagined geometric property of a heeled hull. Such hydrodynamic effects will mainly make only small variations in the quantity d (the horizontal distance between CLR and CE), and these are completely swamped in the balance equation by the effect of heel. Thus balance is mainly a matter of geometry, common to all boats. This insensitivity to hull shape probably explains why the various hull balance theories of the past all seemed to work.

Although the above conclusions are drawn for light-air conditions where the two lift forces and the drag force were all assumed to be proportional to the square of the boat speed, they are no longer true when wave formation becomes important. Then the drag force D increases more rapidly with boat speed than the lift forces $L_h$ and $F_r$. This simply unbalances the equation even more, predicting that if the boat were to speed up while maintaining the same angle of heel it would develop more weather helm.

## §6.3.2 The Movement of the CE and the CLR

The balance equations or fig. 6.4 show that a quantity vital in determining how much rudder force is required to hold a boat on course is the magnitude of d, the horizontal distance between the CE and CLR. Unfortunately the positions of the CE and CLR vary somewhat with point of sailing and wind strength, which means that both d and h in the balance

*Fig. 6.4* (a) shows the forces in a horizontal plane acting on a yacht. There are three pairs of equal and opposite forces which do not act along the same straight line. Such pairs form torques or yawing moments tending to turn the yacht. These torques are: the rudder force with a moment arm r, the heeling force with a moment arm d (both tend to make the boat pay off from the wind), and the driving force with an arm $h\sin\varphi$. This latter torque tends to make the boat round up into the wind. It is because this torque increases greatly with angle of heel that all boats exhibit weather helm when heeled. This large aerodynamic effect swamps the smaller hydrodynamic effects on balance which are due to idiosyncrasies of hull shape.

Dividing through by $\cos\varphi$ gives finally:

$$rF_r = hF_R\tan\varphi - dF_H$$

Since $F_R = D$ and $F_H = L_h/\cos\varphi$ the moment balance can also be written using only water forces as:

$$rF_r = hD\tan\varphi - dL_h/\cos\varphi$$

These last two equations will be referred to as the *balance equations*.

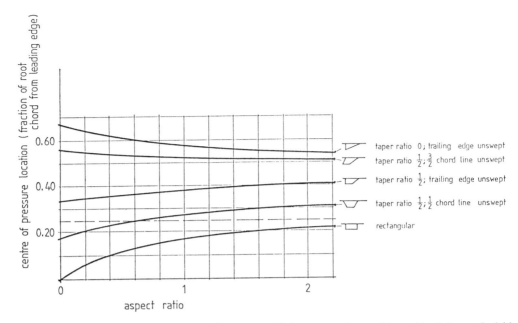

*Fig. 6.5* Although it is not possible to calculate easily the position of the CLR for the canoe body, it can be done for fin-like appendages such as keel and rudder. This graph shows the horizontal position of the CLR for a number of keel shapes. Since, in light displacement boats, most of the side force comes from rudder and keel, these calculable positions of the CLR carry the greatest weight in determining the overall location of the side force centre of action.

equation are not quite fixed quantities. First, let's see what can be said about the position of the CLR. The problem of calculating it for a hull in the presence of the free sea surface is extremely complex and the only sure way is by accurate tank testing. Nevertheless some useful ideas can be gleaned from theoretical considerations.

The first of these is the fore-and-aft position of the centre of pressure on some low aspect ratio keel forms (fig. 6.5). At all except very high sailing speeds the wave pattern on each side of a hull is fairly symmetrical and the pressures on each side of the hull due to these waves will be correspondingly symmetrical. Their contribution to the yawing moment is then likely to be small and any change due to different sailing conditions even smaller. It is certain, however, that an important contribution to the position of the centre of pressure on the canoe body will be the upwash in the flow induced by the keel and rudder (fig. 6.6).

Fig. 6.7 shows the profiles of three very different hull types. Tank tests made at the University of Osaka to determine the position of the CLR and calculational methods are here compared. Although

the lateral resistance of design A was considerably smaller than that of the two separate rudder types, the effect of rudder deflection to correct balance was similar for all three. Only 3° of rudder deflection moved the CLR as much as 10% of the LWL for all three models. The effect of heel on the CLR was small, e.g. 10° heel shifted the CLR forward by 6% of the LWL for design A and 2–3% for B and C. This can easily be cancelled by a very small rudder deflection, lending further weight to the contention that hydrodynamic forces on the heeled yacht hull are insufficient to affect balance materially. It was also

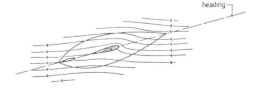

*Fig. 6.6* The upwash created by the keel and rudder lift produces an asymmetric flow across the hull and is an important factor in determining the CLR of the canoe body. Simplified theoretical analysis also suggests that the location of the CLR does not change much with variation in leeway angle.

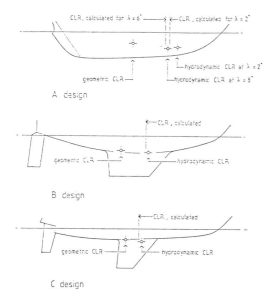

*Fig. 6.7* The position of the geometric and the true hydrodynamic centres of lateral resistance for three different types of hull. Traditionally the geometric CLR is determined by cutting out the underwater profile and balancing it on a knife edge. The hydrodynamic CLR is the result of tank testing. The calculated positions are based on firm hydrodynamic principles.

found that for the purposes of calculating moment the vertical position of the CLR could be taken as close to the bottom of the canoe body for all three models. This allows the heeling arm, the vertical distance between CE and CLR, to be determined.

Now refer again to fig. 6.7. The points labelled hydrodynamic CLR are of course what we have been referring to all along as CLR. It is the real physical centre of all the pressures on the hull. Unfortunately the time-honoured method of the naval architect for determining the position of the CLR has nothing whatever to do with fluid flow. The underwater profile is drawn on a piece of card and cut out; balancing it on a knife edge determines a position which, though traditionally referred to as the CLR, is called the 'geometric CLR' in fig. 6.7. It is actually no more than the centre of gravity of a piece of cardboard. As can be seen from the figure, it is located a long way from the true CLR, especially for designs A and B, and theoretical calculations of its horizontal position give a much better estimate of the true value. It is not appropriate to give the details of the calculation except to say that it is a lot more

complicated than cutting out a sheet of card. With a computer and the appropriate software the task becomes trivial, however.

A performance prediction calculation for the three hulls of fig. 6.7 makes it possible to sort out the amount of rudder angle needed to offset the three main causes of imbalance at each point of sailing, as shown in fig. 6.8. The curves marked $\delta_1$ are the

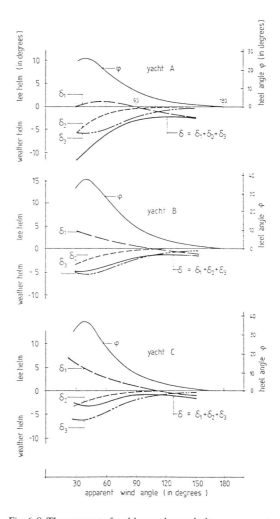

*Fig. 6.8* The amount of rudder angle needed to compensate for unbalance due to three separate causes for the three hull types shown in fig. 6.7. $\delta_1$ is the rudder angle needed to correct for unbalance of CE and CLR. $\delta_2$ is that needed because of the shift in CLR on heeling. $\delta_3$ is required to offset the turning moment due to leeward shift of the sail drive force when heeled. The solid curve labelled $\delta$ is the total rudder angle needed for balance. The solid curve labelled $\varphi$ is the angle of heel in a 16 knot wind.

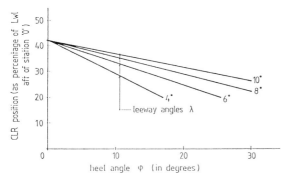

*Fig. 6.9* A simple argument given in §11.2 suggests that the position of the CLR and the heel angle are linearly related as shown by this graph. Since leeway angle and heel are connected these parameters are not independent. Small angles of heel correspond to small leeway angles.

rudder angles required due to the unbalance of CE and CLR. $\delta_2$ is that required due to the shift of the CLR due to heel, and $\delta_3$ that due to the leeward shift of the drive force of the sails with heel. The calculations have been made for a true wind speed of 16 knots and the graphs include also the heel angle under these conditions. The horizontal axis is the apparent wind direction or point of sailing. Again we see that the greatest single contribution to imbalance is the leeward shift of the sail driving force due to

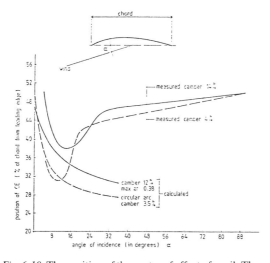

*Fig. 6.10* The position of the centre of effort of a sail. The approximate theoretical calculation shown here is valid only at small angles of incidence. Since soft sails can only be used at angles of incidence greater than about 20° we see that the CE moves linearly with angle of incidence, the movement being greater for a flatter sail.

heeling. The solid curve labelled $\delta$ gives the total required rudder angle due to all these causes. Note that the amount of rudder angle needed to keep the old style profile A on course is considerably greater than that for B or C, especially when hard on the wind.

Although the CLR is subject to variation due to heel and leeway, it turns out that the relationship between these variables is a rather simple one. It is shown in §11.2 that the position of the CLR is given by: $x' = x_o + \mu'\varphi/\lambda$ where $x'$ is the position of the CLR as a fraction of the waterline length behind station O (see fig. 5.2). $x_o$ is the position of the CLR for the hull when it is not heeled. $\varphi$ and $\lambda$ are the heel and leeway angles, $\mu'$ is a constant. Typical values for $x_o$ and $\mu'$ are 0.42 and $-0.052$, according to measurements made at Southampton University, so the CLR equation becomes: $x' = 0.42 - 0.052\,(\varphi/\lambda)$. Graphs of this equation are plotted in fig. 6.9.

To what extent can we determine the true position of the CE of the sails? It seems that less effort has been expended on this than on CLR determinations. As with the CLR the traditional method has been to assume that the CE lies at the centre of gravity of the sailplan. It can be done more easily for sails since one assumes they are exactly triangular in shape and the triangle has the geometric property that the three lines drawn from the corners to the mid-points of the opposite sides all meet at the same place. This is the centre of gravity of the triangular lamina, marked as M and G for main and genoa in fig. 6.11. The CE of the combination is then taken at a distance x along the line joining M and G, given by: $x = A_G \times l\,(A_M + A_G)$ where $A_G$ and $A_M$ are the areas of the genoa and main respectively and $l$ is the distance between points M and G. The question is, by how much does this geometrical point differ from the aerodynamic centre of effort?

There exists an approximate calculational procedure known as thin aerofoil theory by which it is possible to calculate the centre of pressure among other things. The nature of the calculation leads one to believe that the results are likely to be reliable only for small camber and small angles of incidence, and this is indeed the case as fig. 6.10 shows. Fortunately a few measurements have been made of the position of the CE using a wind tunnel. As can be seen the calculated values are only roughly correct up to angles of incidence of about 10°. Beyond that they are completely wrong. This is unfortunate because for soft sails the minimum angle of incidence without some luffing is of the order of 20°. We must therefore rely entirely upon the existing few measurements.

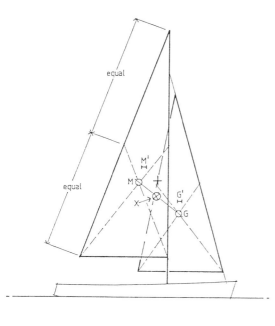

*Fig. 6.11* The traditional method for determining the centre of effort of sails is to draw straight lines (dashed) from the corners to the centre of the opposite edges. The CE is taken as the point where they cross, although this has no aerodynamic basis. The CE of the combination of sails is at a point x on the line between M and G given by $x = A_G \times l/(A_G + A_M)$. A better estimate of the CE is given by M' and G', based on aerodynamic considerations.

Since we are interested only in angles of incidence greater than about 25° we see from fig. 6.10 that the CE moves steadily aft with increasing angle of incidence from about 45% of the chord length behind the leading edge to 50%. The shift is greater with a flat sail than with a full one. The magnitude of this movement is indicated at M' and G' in fig. 6.11. The vertical position of M' and G' has been obtained from an approximate three-dimensional aerofoil theory and corresponds to 40% of the vertical height up from the foot.

Assuming both sails contribute equally according to area then the position of the CE is given by the symbol +. The geometric CE is located at the symbol $\otimes$ and it is interesting to note that there is very little horizontal difference between the two. Even the more or less aerodynamic determination above omits one important factor. We have assumed that each sail contributes in proportion to its area, but a reading of §4.2 on sail interaction should make it clear that such an assumption is not warranted. For instance a change of jib sheeting angle can have a major effect on the relative sail forces produced (fig. 4.16).

Fig. 6.10 shows that a full sail tends to bring the CE further aft when on the wind whereas a flat one shows a greater range of variation. In practice sail shapes do not remain constant on different points of sailing, however, as the crew will normally adjust them.

One further unexplained aerodynamic effect which was noticed during the extensive measurements on the 5.5 Metre yacht *Antiope* was that there was an aft travel of the CE amounting to about 6% of the waterline length with increasing speed.

To summarise:

(a)  The relative fore-and-aft positions of the CE and the CLR are one of the factors determining balance of a yacht.

(b)  Traditionally these positions have been determined by a simple geometric method which has no connection with the laws of fluid flow.

(c)  As a result, a 'fudge factor' called *lead* had to be introduced since boats were found to be in balance on the wind with the geometric CE about 10–20% forward of the geometric CLR. The amount of lead depended on rig and hull type and was determined empirically. In effect the existence of the need for lead was a measure of the inadequacy of the geometric method of determining the CE and the CLR.

(d)  The best way to determine the CLR is by tank testing. Failing this, fairly simple computing procedures give a reasonable result.

(e)  There is room for more scientific work on the determination of sail CE, especially the effect of sail interaction.

(f)  A procedure for determining the CE is given but it doesn't differ greatly from the traditional result. This suggests that the need for lead is mainly due to the inadequacy of the traditional CLR determination.

## §6.4  Added Resistance in Waves

No one who has sailed to windward in the kind of conditions which bring the bow of the boat crashing down into the next approaching wave will need

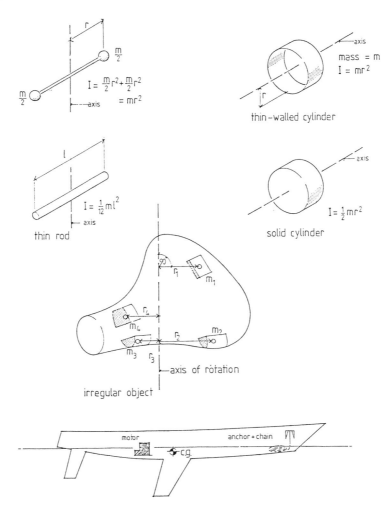

*Fig. 6.12* The motion of a boat in a seaway depends not so much on its mass as on the distribution of mass. This distribution is specified by the moment of inertia I. Here the moment of inertia of four simple shapes is given in terms of their total mass m, and the relevant dimensions of the object. The moment of inertia of an irregular object is obtained by taking all the elements of mass of which it is composed and summing up all the $mr^2$ products, where r is the corresponding perpendicular distance to the axis of rotation. Since added resistance in waves usually increases with moment of inertia, heavy objects like the engine and ground tackle should be moved as close as practical to the centre of gravity.

convincing, when the whole boat shudders and nearly stops, that waves have the effect of adding resistance. For a sailing boat the pitching motion produced introduces another nasty effect which slows it by disturbing the driving force as well as increasing resistance.

Consider a yacht driving to windward in a seaway and pitching 15° fore and aft about its normal equilibrium position. Let us assume that it is pitching rhythmically (sinusoidally) with a period of three seconds. It is then easy to calculate that the maximum fore-and-aft speed of the top of a 15 m mast is 8.2 m/sec. Putting it another way, the apparent wind has a forward vector added to it alternating between +16 and −16 knots. This rhythmical distortion of the effective wind will clearly have a devastating effect on the sail force if the mean apparent wind is also not much more than 16 knots. When pitching forward the sail will luff and when pitching back it will stall. Incidentally, under these conditions a crewman trying to clear a jammed halyard at the masthead would experience horizontal accelerations of about 1.7 times gravity. Although sail movement with respect to the air flow is extremely important in assessing windward ability in particular, it has so far received little scientific investigation. For this reason this section will deal only with the hydrodynamic effects of waves on a yacht.

For a yacht moving steadily on a smooth sea an important factor in determining its speed is the weight or displacement. When the boat's motion also involves oscillatory movement about its centre of gravity, then the *distribution* of weight becomes a vital factor. To pick up a 30 kg mast by its centre of gravity is no different from picking up a 30 kg ingot of lead, but to pick up the mast and rotate it through 90° is much more difficult than doing the same with the ingot. Although both objects have the same mass, its distribution is different. Since waves have the effect of rotating a hull about its centre of mass, this distribution must concern us.

How is mass distribution specified? Physicists use a quantity called the *moment of inertia*, shown in fig. 6.12 for various shapes. These are all based on the fact that the moment of inertia of a single point mass m, constrained to move at a distance r from a fixed axis, has a value of $mr^2$. If the object has a complicated shape the moment of inertia is calculated by adding up all the $mr^2$ products for each increment of mass.

The fact that the moment of inertia involves the square of a distance is worth noting. If, in fig. 6.12, the distance between the engine and the centre of gravity is halved, then its contribution to the moment of inertia is reduced by a factor of 4.

The moment of inertia depends upon the total mass and the distribution of that mass. The distribution is in turn determined by the size and shape of the object. For any shaped object it is always possible to find a radial distance from the axis at which the whole mass of the body could be concentrated that would give the same moment of inertia as the body itself. This radial distance is called the *radius of gyration* or the *gyradius* of the body. If the total mass is m, the radius of gyration k and the moment of inertia is I, then this may be written:

$$mk^2 = I \text{ or } k = (I/m)^{1/2} \text{ or } \sqrt{I/m}.$$

For a fixed-keel yacht, k typically has a value of about 0.25 of the overall length.

The added resistance of a vessel in a seaway is mainly caused by coupled heaving and pitching motions. Heaving is the vertical oscillatory motion of the centre of gravity. The natural periods of heave and pitch are very important for the behaviour in waves; if one of these periods is equal to the period of wave encounter, violent motion may ensue. In such resonant conditions the increase in resistance is found to be large, and furthermore damage may be caused.

At this point it is probably a good idea to inject into the discussion some basic ideas about oscillating systems. Heaving has an exact analogy with the suspension of a car (fig. 6.13). Buoyancy acts just like a spring: push down at the lateral centre of flotation and the boat sinks straight down only to bob up again when the force is removed. This applied force whose origin is in the waves does not move just the hull up and down but also a considerable mass of water entrained by it due to viscous drag. This entrained water adds to the mass of the hull giving the system a greater effective mass. This same viscous drag helps to damp out the motion in the same way as a shock absorber. But there is an additional effect in the case of the boat. A hull heaving up and down produces waves of its own which are radiated out in all directions. The energy of these waves is subtracted from the energy of the heaving motion and so constitutes further damping.

Such a combination of elements is known as a damped oscillatory system. If it is given a push it will oscillate at a fairly well defined *natural frequency*, but with an amplitude which steadily dies away. This natural frequency increases with a stiffer spring (fig. 6.13) and decreases with a greater mass.

If the boat is heading into uniform waves, then our oscillatory system is being forced up and down at the frequency of the wave encounter. This problem of forced oscillations is a classic one in physics; we are interested in how the amplitude of the oscillations varies as the frequency of wave encounter changes.

This relationship is shown in fig. 6.14. Look first at the upper part of the diagram and consider the heaving motion. The vertical axis gives the ratio of the maximum height of the upward motion of the boat compared to the wave height. The horizontal axis gives the ratio of the frequency of wave encounter $f_e$ to the natural resonant frequency of heave $f_n$. This is called a resonance curve. If there is little damping, the amplitude of heave when the frequency of wave encounter is close to the boat's natural frequency of heave can become very large,

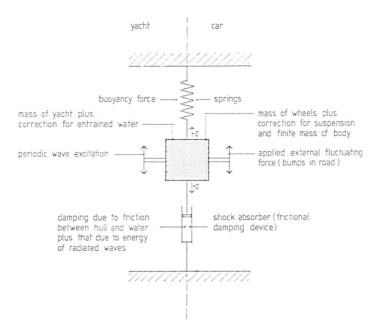

yacht | car

buoyancy force —— springs

mass of yacht plus
correction for entrained water

mass of wheels plus
correction for suspension
and finite mass of body

↑-z

periodic wave excitation

applied external fluctuating
force ( bumps in road )

↓-z

damping due to friction
between hull and water
plus that due to energy
of radiated waves

shock absorber ( frictional
damping device )

*Fig. 6.13* The heaving motion of a vessel in a seaway can be compared with the motion of a car suspension when driven over a bumpy road. The effective mass of the oscillating boat can be much greater than its actual mass, due to the amount of entrained water moving up and down with the hull. Damping is not only due to frictional effects but also to the energy in the waves created by the oscillating hull.

even greater than the wave amplitude itself. Fortunately for the sport of sailing most boats correspond to the curve labelled 'lots of damping', although there have been reports that very light catamarans can exhibit very large heave amplitudes.

There are several other important points to be noted about fig. 6.14. When the waves are very long, i.e. the frequency of encounter is small, then the amplitude of heave is the same as that of the waves. This is intuitively obvious: the boat simply rises and falls with the swell. The fact that the boat moves in synchronism with a long wave is shown in the lower graph of *phase relationship* as a function of the frequency ratio.

The meaning of phase relationship is seen in fig. 6.15. The solid line represents the rise and fall of approaching waves. The dashed curve which follows it closely represents the motion of the boat when the wave encounter frequency is much less than the natural frequency of heave. This corresponds to $f_e/f_n$ = 0. (The upper graph of fig. 6.14 shows that under these conditions the wave amplitude and the heave amplitude are the same, whereas the lower graph shows that there is no phase difference between the two and the boat and wave rise and fall together.)

Now when resonance is approached the boat actually heaves more than the wave; it overshoots both on the crest and in the trough, as represented by the dash-dot curve in fig. 6.15. Its motion is now out of step with the wave producing it, and is said to be 90° out of phase as shown by the lower graph of fig. 6.14. The heave of the boat reaches its maximum when the crest that produced it is already halfway back down to becoming a trough.

If we go now to the other extreme where the boat is encountering very short waves with a frequency much higher than its natural frequency of heave, we expect from fig. 6.14 that the amplitude of the motion will be small and the phase difference will tend toward 180° as shown by the dotted curve in fig. 6.15. The amplitude is smaller than the exciting waves and exactly out of phase: when the wave goes up, the boat goes down and vice versa.

Although this discussion has been in terms of heave, the conclusions are analogous for the pitching motion. In a very long swell the pitch angle of the

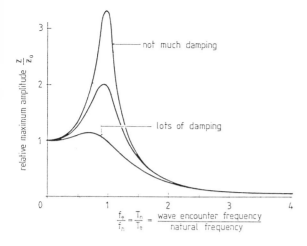

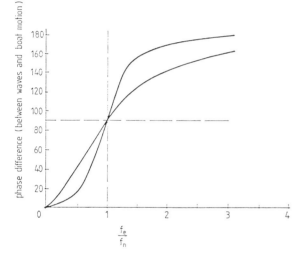

*Fig. 6.14* The upper graph is a classical resonance curve for a driven mechanical system with damping. Applied to the heaving motion of a boat, the driving force is the waves of amplitude $z_o$. The heave amplitude has a maximum when the wave encounter frequency is close to the natural frequency of heave. The same kind of resonance effect occurs for pitching. The lower graph shows the phase relationship between the boat's motion and that of the waves.

boat is the same as, and in phase with, the wave slope. Near resonance the 90° phase difference is what produces pounding when going to windward. The bow is still on its way down while the next wave crest is already rising. In very short waves the 180° phase shift is not a problem since the pitching amplitude is vanishingly small.

Before we can predict how a particular vessel will behave in waves we need to know its natural frequency of oscillation. Although this is difficult to ascertain for it is pitch that concerns us most. With the boat in calm water and preferably in a marina with the mooring lines slacked off, firmly push down on the bow in a rhythmic manner so as to build up a pitching rhythm. With a stopwatch determine the time for ten oscillations, say, and from this get the time or period for one. As the motion is heavily damped it is sometimes necessary to continue to apply rhythmic pressure to the bow while timing. Providing the pressure is applied in phase with the motion, this will not affect the natural frequency. The natural frequency is related to the natural period by $f_n = 1/T_n$. Thus if $T_n$ is 2.4 sec, $f_n$ is 0.42 oscillations per sec. It should be remarked, however, that this natural frequency is generally considerably less when a boat is moving.

The relationship between the moment of inertia of the boat for pitching and the natural frequency can be obtained from the following fundamental formula for oscillating systems:

$$f_n = \frac{1}{2\pi} \sqrt{\frac{\text{righting moment per unit displacement}}{\text{moment of inertia}}}$$

Writing this in the mathematical form appropriate to the pitching of a hull gives:

$$f_n = \frac{1}{2\pi} \sqrt{\frac{g(GM)_l}{k^2 + (a/\Delta)k_a^2}}$$

$(GM)_l$ is the longitudinal metacentric height and can be calculated from the hull lines. g is the acceleration of gravity. With metric units g can be taken as 10. a is the added weight of entrained water and $\Delta$ is the displacement. k is the radius of gyration of the boat 'in air', without the effect of the entrained water. $k_a$ is the radius of gyration of the added mass of water. Thus a measurement of the natural pitching frequency can allow the overall radius of gyration to be determined. Unfortunately k itself cannot be determined in this way unless the moment of inertia of the added mass of water is calculated. Values of k are typically about 0.25 × LOA. Because of their effect on the resonant frequency, k and $(GM)_l$ are the most important quantities in determining a boat's behaviour in a seaway. k is determined by the weight distribution and $(GM)_l$ is closely related to the prismatic coefficient, being larger for higher values of it.

To get a feel for typical figures, the moment of

169

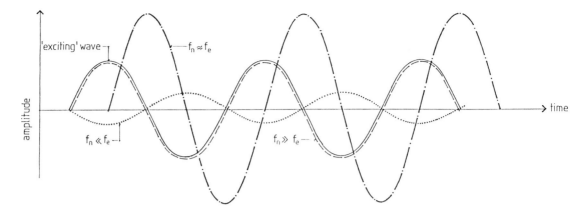

**Fig. 6.15** The pitching motion of a boat depends upon the frequency of wave encounter. Near resonance the pitching amplitude is large (dash-dot curve) and 90° out of phase with the exciting wave; i.e. the bow lifts after the wave has passed and comes down into the rising front of the next wave. With very long waves the boat simply follows them (dashed curve). In very short waves the amplitude of pitch is small but out of phase with the exciting waves.

inertia of the added mass of water was found to be 69% of the inertia of the boat alone for the IOR yacht *Standfast*. The added mass of the entrained water was 185% of the mass displacement. The natural frequency of pitch was 0.417/sec. The radius of gyration for boat alone was 0.25 of the LOA or 3.05 m in this case. The radius of gyration of the entrained water alone was 1.86 m. The longitudinal metacentric height $(GM)_l$ was 10.8 m on a waterline length of 10.03 m. An important factor contributing to the moment of inertia is the mast. Because of the great distance from the centre of gravity, and because the moment of inertia is proportional to $r^2$, weight near the top of the mast has a strong influence. For example if a weight at the top of a 15 m high mast is brought down to deck level its contribution to the overall moment of inertia is reduced by a factor of about 250!

A good example of the connection between resonance and added resistance in waves is seen in fig. 6.16. As expected, the added resistance reaches a peak for the same conditions when the pitch and heave amplitudes reach a peak. Clearly we would like to arrange things so that for the waves we mostly expect our boat is well off resonance. But should its natural frequency be above or below the frequency of encounter? Referring back to fig. 6.14, if we make the natural frequency low by having a large radius of gyration, we might hope that most of the wave encounters would be of higher frequency so that we are operating near the right hand side of the graph

where the pitch amplitude is small but out of phase with the waves. In other words, the boat is 'slicing through the waves'. If we try to operate mainly below the resonance by reducing the radius of gyration, we have a boat which follows the waves, pitching considerably in phase with them.

Which of these two alternatives is better? To answer this question calculations were carried out in the early 1970s at the Delft Shipbuilding Laboratory using a technique developed for merchant ships. When a yacht is sailing into waves its heaving and pitching generate damping waves which are superimposed on the incident wave system and carry energy away from the yacht, thereby adding resistance.

Since the sea contains a spectrum of waves of different length (see §11.3) the calculation must be carried out for a wide range of encounter frequencies and then weighted according to the spectrum of frequencies present. The results of one such calculation for three different values of the radius of gyration are shown in fig. 6.17. The vertical axis gives the added resistance due to waves as a function of the wave encounter frequency. Since one normally encounters waves over a wide range of frequencies, the total added resistance is then proportional to the area under these curves.

As expected, the largest value of the gyradius (largest moment of inertia) gives the lowest value of the added resistance at high frequencies of encounter. This is the situation of a heavy boat slicing

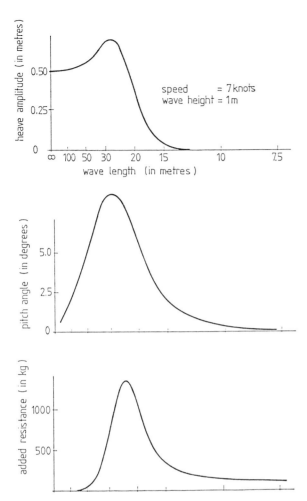

Fig. 6.16 Heave, pitch and added resistance of the 12 Metre *Valiant*. At a speed of 7 knots and waves of height 1 m clear resonance effects are seen for waves about 27 m long. Maximum added resistance occurs near to but not exactly at this resonance.

through the small waves. However the differences are much greater on the low frequency side of the peak where the lowest radius of gyration gives the lowest resistance. The reason the curves are separated more here is that the sea state assumed contains more long waves than short ones. Since the area under the curve with the smallest radius of gyration is the least, we conclude that it is best to minimise the moment of inertia by 'keeping the weight out of the ends' and not forgetting that at the top of the mast.

It must be borne in mind that this conclusion assumes a typical open sea state in which most of the waves have a length of about 100 m and the significant wave height is 2.9 m. It is possible that the differences would not be so marked for sailing in areas having very short steep seas.

To get an idea of the comparative effect of displacement and weight distribution the Delft calculations were made on three hulls, all of 10 m LWL (fig. 6.18). Their added resistance in waves give the results in fig. 6.19. The advantage of keeping the weight out of the ends is evident in all cases, but more so in large waves than in small. Notice that reducing the radius of gyration in the heavy displacement model gives a greater relative reduction in the added resistance than for the light displacement model.

In 1971 Gilbert Lamboley introduced an oscillation test to measure the weight distribution of Finn hulls. The boat is balanced horizontally by brackets having two pivot points, one above the other (fig. 6.20). Although the horizontal position of the centre of gravity can be determined from the balance point, its vertical position is not known. The period of such a pendulum is given by: $T = 2\pi \sqrt{(k^2 + a^2)/ga}$ where a is the distance from the pivot point to the CG. Unfortunately a is not known, so two measurements must be made at different pivot points. Corresponding to the two values of a, i.e. $a_1$ and $a_2$, we have two periods of swing, $T_1$ and $T_2$. This gives two equations in two unknowns so both k and the position of the CG can be determined; a fair amount of algebraic manipulation is required. Prior to the introduction of the Lamboley test in the 1970s, weight distribution in one-design dinghies could only be controlled by means of structural specifications.

To give an idea of the kind of values for the radius of gyration of small boats, the following table is taken from published figures for the Flying Dutchman and 470 classes.

| **470** | $k/LOA$ | **Flying Dutchman** | $k/LOA$ |
|---|---|---|---|
| hull only | 0.248 | hull only | 0.26 |
| complete boat | 0.327 | complete boat | 0.31 |
| | | complete boat + crew | 0.23 |

The large influence of the mast on the moment of inertia is seen in both boats. For the FD the mast was found to be 8.4% of the total weight and yet it contributed 30.8% of the total moment of inertia. Since the crew are generally heavier than the boat

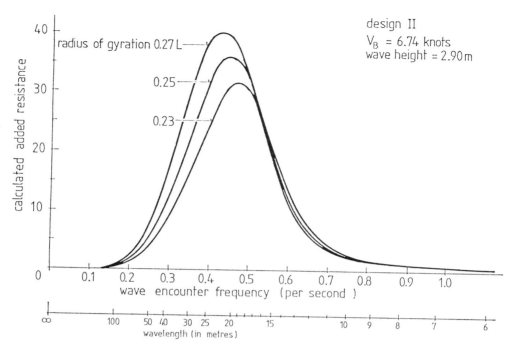

design II
$V_B$ = 6.74 knots
wave height = 2.90 m

*Fig. 6.17* The calculated added resistance as a function of the frequency of encounter for three different values of the radius of gyration. Since the total resistance is proportional to the area under the curves, clearly the smallest radius of gyration gives rise to the smallest resistance.

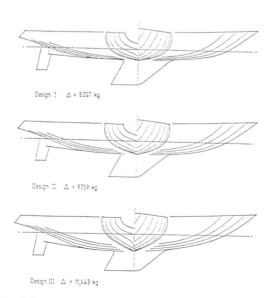

*Fig. 6.18* The lines of the systematic series used to compare added resistance in waves. They all have a waterline length of 10 m, beam of 3.66 m and draft of 2.15 m. Displacements are as shown.

and sit close together near the centre of gravity, they have the effect of strongly reducing the radius of gyration.

To put these considerations in perspective it is interesting to consider the results of moment of inertia measurements that were carried out on the Flying Dutchman fleet at the 1976 Olympics. The gold medal winning hull had the largest gyradius; the silver medal hull was lower than average; the bronze medal hull had the third largest and the hull with the lowest gyradius was fourth. Placings for the rest of the fleet showed the same scatter with no evidence of a trend. This might mean that wave conditions did not include those that give rise to resonant pitching of an FD. It does not rule out the possibility that if the winner had had a smaller gyradius he would have won with an even greater margin. This pinpoints the difficulty in drawing conclusions about any aspect of the rig or design of a boat just because it was used by the winning crew in a regatta. So many factors influence the outcome of a yacht race: fortunately for the sport, the crew is still the major one in most classes. Yet the way a crew sails, the techniques they have trained themselves to use, depend on their

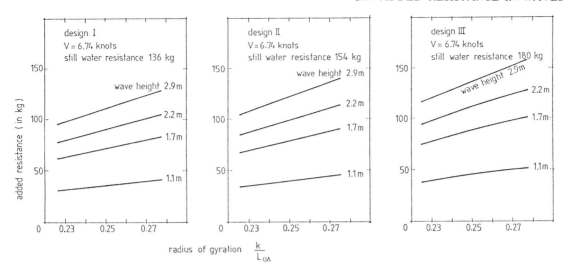

Fig. 6.19 The calculated added resistance in waves for the three designs in fig. 6.18. The wave heights quoted are *significant* wave heights. (This terminology is explained in §11.3.) The value of a small radius of gyration is clear for all cases.

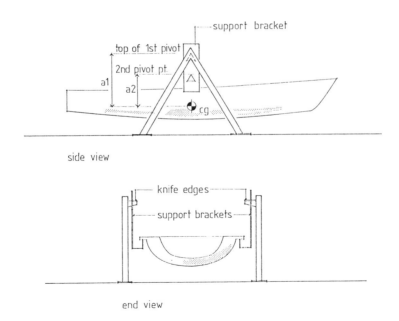

Fig. 6.20 The setup for determining the moment of inertia or radius of gyration of a small boat, known as the Lamboley test. Since the vertical location of the CG is not known two measurements must be made at different pivot points in order to extract the radius of gyration.

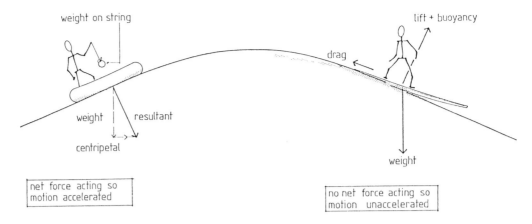

weight on string

lift + buoyancy

drag

weight | resultant

centripetal

net force acting so
motion accelerated

weight

no net force acting so
motion unaccelerated

*Fig. 6.21* The motion of a liferaft in large waves is nearly the same as that of the individual water particles, namely circular. The motion is therefore an accelerated one. This acceleration adds to that of gravity giving an apparent direction for 'down' which is always perpendicular to the wave slope. On the other hand a surfer who has caught a wave is moving with a constant unaccelerated speed. The force which keeps him going so fast is the component of his own weight parallel to the wave slope.

knowledge of the physics of sailing. Each individual factor, like radius of gyration, or fullness of sails, or design of the keel, or crew reactions is but one piece in the great and complicated jigsaw puzzle which is yacht performance. Remove one piece and the perfection of the picture is little disturbed, but if none of the pieces fitted properly the picture would be indistinct and blurred. It seems fundamental that humans are happy only if the prospect of continual improvement is present, and in the art of sailing this comes largely from perfecting each piece so that it fits perfectly into the jigsaw even though its effect on the overall picture seems difficult to discern.

To summarise: reducing the longitudinal radius of gyration is favourable for the windward performance of a yacht under most circumstances. This means keeping the weight out of the ends and off the mast. For a yacht with a radius of gyration equal to 0.25LOA, a movement of a weight equal to 1% of the boat's displacement from amidships to one of the ends will increase the total resistance by about 1%. Because of their large contribution to the total weight, the crew of a dinghy should sit close together and near the centre of gravity when sailing to windward.

### §6.4.1 Surfing in Waves

The last section was concerned with the *increase* in resistance when sailing against waves. We therefore ask: is there a *decrease* in effective resistance when we sail with waves? Most yachtsmen would now have little difficulty in answering this in the affirmative. Long ocean races in particular have confirmed this, perhaps the best example being the round-the-world races where boats deliberately head for the Roaring Forties to pick up the large waves which move incessantly around the bottom of the earth. At one time the possibility of a net gain in speed by a displacement boat sailing with waves was hotly disputed: understanding the dispute is the key to understanding the phenomenon of surfing itself.

As explained in §11.3 a drifting liferaft in waves much longer than itself actually moves along a path which is a circle in a vertical plane, as seen by a fixed observer. At the crest the water particles are moving in the direction of the waves, and in the trough against them. It was therefore argued that what a displacement boat gains in speed on the crest, it will lose in the trough. However, this ignores three other important effects. The first is surfing, in which gravity combined with buoyancy and lift produce a force acting down the forward slope of the wave which increases speed. Second, a boat sinks slightly farther into the crest of a wave because local gravity is slightly less there than in the trough (see §11.3). The wetted area in the trough is less and with it surface friction drag. Since local gravity is greater in the trough the boat's own wave disturbance travels faster (speed is proportional to the square root of gravity) increasing the 'hull speed'. The third effect is the change in apparent wind speed as the boat alternately catches a wave and then falls back into the trough. From all these effects a net increase in speed results, as we shall see.

A major difficulty in understanding surfing arises because we are considering the motion of a boat over a surface which is itself undergoing accelerated motion. A small boat which is not moving through the water feels an effective gravity which is always perpendicular to the wave slope, as pointed out in §11.3. One might well ask, then, how can the 'surfboard effect' work if there is no component of gravity down the slope? A strict application of Newton's laws of motion gives us the answer. The laws of Nature are like the rules of a game: you cannot disguise them so that they appear what they are not; you cannot ignore them even if you *think* they do not apply. The very different motion of a liferaft in waves and that of a surfer is a case in point. The liferaft, which is fixed with respect to the water, is undergoing *accelerated* motion. The surfer, once he catches a wave, moves at a uniform unaccelerated speed. In the former case net forces must be acting, in the latter there can be no net force. These two situations are shown in fig. 6.21. Viewed by a fixed observer unaware of the waves, the surfer would be moving in a straight line and the liferaft would be moving in a circle. The path of the surf rider and the path of the water through which she is moving are therefore different. She is moving with respect to the water and giving rise to a drag force. The surfboard also has dynamic lift and buoyancy. Both these forces act in the same direction which is perpendicular to the wave slope. Weight, as always, acts perpendicularly down. The combination of weight and lift plus buoyancy produces a powerful force tangential to the wave slope which accelerates the surfer until equalised by the drag. If the resultant equilibrium speed is the same as that of the wave, the surfer will remain fixed with respect to the forward moving wave shape but she will be moving quite fast with respect to the water particles under her. This is why one has to paddle furiously to catch the wave.

Waves in shelving water go slower and are more uniform than those in deep water. Catching a wave in deep water requires greater initial speed and the ride will not last, as deep water waves are continually disappearing and reappearing as they move at different speeds through other waves. Thus we have the characteristic motion where the boat catches a wave and stays with it only for a short time before losing it.

The driving force for the extra speed comes, of course, from the component of the weight of the boat acting parallel to the slope. If the wave slope is 30° then this component is Wsin30° or half the weight. This is an enormous force compared with available sail forces, so even a displacement boat can be pushed far beyond its normal speed. Remember that there is no absolute maximum speed for a displacement hull: given enough driving force the boat will always go faster. Of course if the hull has a tendency to plane, surfing will be more sustained.

The second effect that contributes to surfing is the variation in the volume displacement of the boat from crest to trough. Remember that the buoyancy force relies upon the increase in pressure with depth. This pressure depends upon the weight of water above and weight depends upon the gravitational force, so buoyancy is less on the crest where the local gravity is less, so the boat sinks down on the crest and rises up in the trough. This will certainly affect the resistance. Unfortunately the tank testing of yachts is usually carried out at only one displacement, unlike merchant ships where drag at different loadings is of major importance.

The third effect is the change in apparent wind that accompanies the varying downwind speed. In the trough, where the resistance is low, the apparent wind is greatest so the boat gets a good start for the acceleration back up to the crest.

Whether or not a net speed gain is made depends on how long the boat remains on the gravity-assisting front slope of the wave, and whether this speed gain more than offsets the losses in the trough. A calculation made by J. S. Letcher (see Bibliography) answers this question. It was based on the Cal/40, about 6670 kg and 9.8 m LWL. Reasonable assumptions were made of the resistance as a function of speed and displacement. A true wind speed of 25 knots was assumed with waves 60 m long and 3.7 m high. The speed of such waves in deep water is 18.9 knots and the period 6.26 seconds.

Fig. 6.22 shows how various quantities change during the passage of one wave, and the conditions experienced as a wave which is travelling nearly twice as fast as the boat approaches from astern and passes. The time axis starts at zero just as a crest is passing. The displacement reaches a maximum just ahead of the crest and falls by about 30% as the trough passes. Speed relative to the water is highest (11.2 knots) when the boat is on the forward slope of the wave, and lowest (6.6 knots) when the boat is falling down the back of the crest which has just passed. Apparent

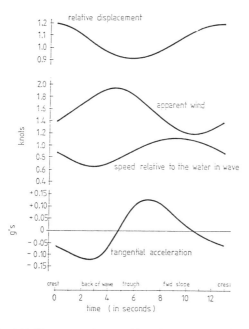

*Fig. 6.22* Here a wave is overtaking a boat travelling in the same direction. At time = 0 seconds the crest of the wave is under the boat; 3 seconds later the boat is on the back of the wave and sliding into the trough; 9 seconds later it is rushing down the forward slope of the next wave. Relative displacement refers to the heave amplitude of the boat as the wave passes. Apparent wind strength varies because the downwind boat speed is changing and because the wind flow is different in the troughs from that on the crests. The speed relative to the water is a maximum on the forward slope when the boat's weight contributes to the driving force. The bottom graph shows the acceleration along the wave in units of the acceleration of gravity. These calculations were carried out for a 6670 kg boat in waves of length 60 m and height 3.7 m. The wind was 25 knots.

wind is greatest (19.3 knots) in the trough and least (12.2 knots) just before the crest passes. The bottom curve in fig. 6.22 shows the variation in tangential acceleration. Under these conditions the average speed over a full cycle turns out to be 9.81 knots, a 13% gain over the steady speed that would be obtained in the same true wind speed in a smooth sea. This is certainly a substantial gain due to surfing although in this case the average boat speed over the ground is only 52% of the wave speed. In conditions where a yacht is only just catching the waves, very small differences in sail area, wind strength and wave height can make large differences in the average speed.

## §6.5  Dynamic Stability

Static stability was discussed in §5.3. The righting moment versus angle of heel can be measured fairly easily and is useful in determining sail carrying power. It is also useful in gauging righting ability after a complete knockdown. These are problems which crop up usually in racing when land or assistance is not very far away. For the cautious ocean sailor who usually reefs down long before he is knocked down, the biggest danger is not the wind but the waves.

Is a knowledge of the static righting moment a help in determining the stability of a yacht in waves? Answers to any question in sailing are rarely a simple yes or no, but in this case the answer is nearly all no. This will become clear as you read the following discussion which is based largely on published model experiments by Donald Jordan.

Waves which do not break pose little danger to the statically stable yacht. Their maximum slope is about 30°. Although breaking waves are a common occurrence, especially in shoal waters, it is uncommon for large deep-water waves more than 5 or 6 m high to break. Given the right conditions this can happen, and the most common cause is wave interference. Two wavetrains crossing each other at a small angle give rise to local high spots where the crests of the two waves add together (fig. 6.23). The contour lines are drawn at 1 m intervals and result from two wavetrains each consisting of 2.7 m high waves of length 32.4 m crossing at an angle of 20°. The dashed lines show the positions of the individual crests: where they cross the wave height is 5.4 m. A boat moving along the line B-B would encounter hardly any wave motion at all. On the other hand one moving along the line A-A just a short distance away would be experiencing 5 m waves. Of course the sea surface is never as uniform as this, but it is this kind of effect that gives rise to localised waves.

It has been established that if the ratio of the wave height to the wavelength is greater than 1:7 then the wave will become unstable and break. The development of such instability is shown in fig. 6.24. The combined effect of the reduced local gravity at the crest and the high wind drag causes a large part of the water in the wave to detach itself from the ordered circular rotation and come crashing down with a turbulent and random motion. The potential energy of the water which has been lifted up into the crest is now turned into kinetic energy. In breaking seas there is a considerable amount of water

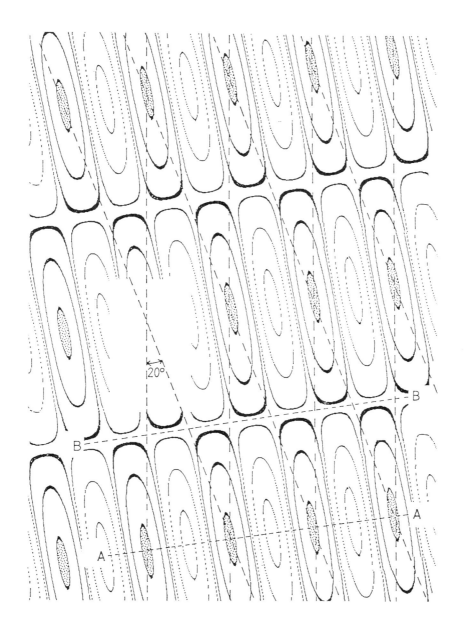

*Fig. 6.23* A computer generated contour diagram of wave heights resulting from the interference of two wave trains crossing at an angle of 20°. The dashed lines indicate the positions of the crests of the individual wave trains. The separate wave trains have a height/wavelength ratio of 1:12. In deep water, such waves whose height/length ratio is less than 1:7 are stable and do not have breaking crests. However when two such wave trains meet there are small regions in the resulting wave motion where this stability criterion is exceeded: breaking crests can be expected there and such areas are shown shaded. The dotted contours correspond to the troughs and the darker ones to the crests. Although real water waves are not so uniform, this is the basic origin of short-crested deep water waves.

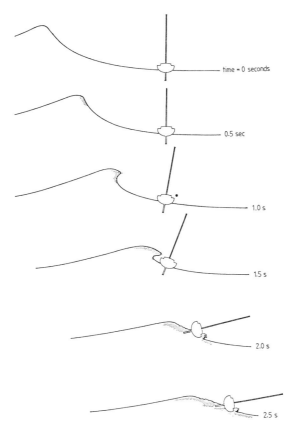

time = 0 seconds

0.5 sec

1.0 s

1.5 s

2.0 s

2.5 s

*Fig. 6.24* The development of an unstable wave in time and its effect on a boat lying beam-on to the seas. The crew's first awareness of danger is usually at 1.5 seconds with the roar of the breaking water. This is followed by a rapid roll and a simultaneous sideways accelerating movement.

movement downwind which does not happen in ordinary waves.

The individual waves which are combining in fig. 6.23 have a height/length ratio of 1/12 and are therefore quite stable. Where the peaks intersect and add up, the height/length ratio exceeds 1/7 and the wave breaks. The shaded regions on the contours of fig. 6.23 correspond to the parts of the combined wave motion where the height/length ratio is greater than 1/7. A wave formed in this manner will therefore break only over these limited regions.

To get a better idea of what this breaking wave is capable of doing it is worthwhile making some rough calculations. Fig. 6.24 indicates that about a third of the top of the wave breaks off, and for the waves shown the mean height of this upper third is about

4 m above the trough. In falling through 4 m there is a gain in speed of 9 m/s or about 18 knots. A 10 m boat caught broadside-on would be hit by about 10 tonnes of water moving at this speed.

When hit abeam by such a mass of water moving horizontally, the boat is first rolled over on its side and then 'picked up' by the water torrent, meaning that buoyancy lifts it, and at the same time it is carried rapidly sideways with the flowing water. It will be rapidly speeded up to 18 knots, going sideways, and then be halted with jarring suddenness when it reaches stationary water in the trough.

Fig. 6.24 shows the attitude of a yacht when hit by a breaking beam sea about 6 m high. What will a person below decks experience during such a knockdown? Nothing untoward will be noticed until the frame corresponding to 1.5 seconds, when the roaring noise of the breaking wave will be heard. The boat will then start to roll very rapidly and accelerate sideways so that a person on the leeward side will be flung into the bilges. About a second later the boat will be stopped dead by hitting still water. The person will now be flung even more violently into the deckhead. Imagine jumping off a 2 m high ladder and landing spread-eagled on the ground: the effect would be similar and serious injury is very probable. If you survive this you would likely find that the *leeward* cabin sides were stove in. Since the sideways motion is stopped more suddenly than it is started, the forces are greater. The effect would be similar to holding the boat with its mast horizontal 4 m above the surface of smooth water and dropping it.

That this scenario is basically correct has been verified by numerous reports and sailors have often spoken of a roaring noise just before capsize. K. Adlard Coles in his classic book *Heavy Weather Sailing* says: 'When a yacht suffers damage in a gale it is usually because it is struck by a sea and literally thrown down in the trough so the doghouse, or coachroof, is stove in on the lee side.'

It is clear from the above description that the form of the static stability curve will have little effect on the damage sustained. It turns out that about 95% of the energy given to the boat appears as translational kinetic energy and only about 5% goes into rolling it. Boats with a high centre of gravity have a certain degree of upside-down stability (see fig. 5.9b) and so may remain inverted for some minutes. This could be very serious if the coachroof has been damaged. The likelihood of damage is largely a matter of the strength/weight ratio. The impulsive force on the coachroof is directly proportional to the displacement of the boat, so a light boat built well of strong modern materials is likely to fare best. On the off chance that the coachroof is damaged, a low centre of gravity will provide a good backup for survival by righting the boat quickly.

Although the shape of the static stability curve has little effect, the moment of inertia for roll does. Like pitching which was discussed in §6.4, rolling is a dynamic phenomenon which depends upon the moment of inertia about the roll axis, i.e. the distribution of weight with respect to the x axis of fig. 6.1. A large moment of inertia or, what is the same thing, a large radius of gyration, will make the boat slow to roll after the application of an impulsive rolling force such as from a breaking wave. As with pitch, an important contributor to the radius of gyration for roll is the mast, one of the reasons why sailing boats do not roll as much as power boats. In the 1979 Fastnet Race the sloop *Ariadne* was dismasted and later, when struck again by a large wave, she rolled rapidly through 360°. As the mast supplies about half of the total moment of inertia about the roll axis it is clearly an important factor in reducing the vulnerability to capsize.

The investigation with models described by Donald Jordan was extended to study the effects of a wave striking at various angles from bow-on to stern-on, using the same size wave as would cause a knockdown if the boat were struck abeam. The boat can easily handle such a wave when struck bow-on. It will also safely handle a stern wave, but in this case high speed water pours into the cockpit and would go straight down the companionway if the cabin were open. The force of water on the helmsman would also be considerable. We have estimated that the breaking water reaches a maximum speed of about 9 m/sec. Water would not come aboard at this speed relative to the boat because it is simultaneously accelerated forward by the same water. Guessing that the water comes aboard at the relative speed of 4 m/sec, we can calculate roughly the drag force on the helmsman using the fundamental drag formula given in chapter 2: drag $= 1/2 C_D \varrho A v^2$. $C_D$ is the drag coefficient of a person. This is not known but for an order of magnitude calculation can be taken as 1. $\varrho$ is the density of water which is 1000 kg/m$^3$. A is the area of a helmsman, which assuming this to be a rectangle 1.8 by 0.3 m gives a drag force of 4320 newtons, a force equal to the weight of about 440 kg. In other words his safety harness would be subject to a force about six times his weight.

When the wave direction deviates from bow-on or stern-on by as little as 20° the boat is picked up and thrown violently on its side. Thus the technique of running before a storm has much to recommend it. A fast-moving boat will be less easily pooped, and if it is the relative water velocity will be reduced by the boat's own speed. The price for this safety is high: it requires that the helmsman's concentration never be broken and that he never makes a mistake that could result in a broach. Racing in the southern ocean is so fatiguing that the helmsman must be relieved about every 30 minutes.

Perhaps a better solution in most cases is to use a drogue or sea anchor to hold the bow or the stern toward the oncoming seas and slow the boat down to prevent surfing. Although the model tests showed that a sea anchor could perform these functions well, it was found that if the elasticity of a realistic warp were simulated the drogue lost much of its effectiveness.

Although hull shape has little influence on the vulnerability of capsize, the size of a boat certainly does. This comes as no surprise, but what may be surprising is the effectiveness of a size increase. It is shown in §10.3 that the righting moment of a boat scales in proportion to the fourth power of the length; if the length of a boat is doubled, it requires $2^4 = 16$ times as much wave energy to cause a capsize. Since wave energy is proportional to the square of the height, this means a wave four times as high would be required to knock down the bigger boat.

Higher freeboard would be expected to increase capsize vulnerability due to the increase in surface area being struck by the wave. A reduction in underwater area decreases capsize vulnerability since there is less drag below the centre of gravity when the boat is pushed sideways, with less tendency for the boat to 'trip over its own keel'. However lack of ballast low down will impair self-righting ability.

It is a sobering thought that the energy contained in even a small wave is, in principle, enough to destroy most boats. The energy per square metre of sea surface is given by: $E = \frac{1}{8}\varrho g H^2 = 1225 H^2$ Jm$^{-2}$ where H is the wave height. The international units used here are joules per square metre (Jm$^{-2}$). Another way of looking at this is shown in fig. 6.25. The energy per square metre of sea surface is the same as the energy obtained by dropping a 'slab' of water of area 1m$^2$ and thickness $\frac{1}{8}$ of the wave height, from the crest level vertically down to trough level. Such a slab of water weighs about 750 kg for a 6 m high wave, so the energy available per square metre of sea surface is the same as that available from

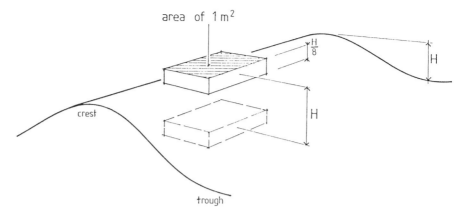

area of 1 m²

crest

trough

$\frac{H}{8}$

H

H

*Fig. 6.25* The energy per square metre in a wave of height H is the same as that obtained from dropping a slab of water 1 m² in area and in thickness ⅛ of the wave height from crest to trough. Such a slab of water would weigh 750 kg for a 6 m high wave.

dropping a 750 kg slab of say concrete from a height of 6 m. From fig. 5.11 it can be estimated that it takes about $10^5$ joules of energy to roll a 10 m boat through 90°. Such a boat occupies about 30 m² of sea surface and the energy contained in 6 m high waves in this area of sea is about $1.3 \times 10^6$ joules, so the energy available is of the order of 10 times that needed to roll the boat to 90°. It is fortunate therefore that waves treat us so kindly, rarely unleashing all the energy they are capable of. Good seamanship, however, demands that we recognise the magnitude of the energy contained in a wave even though we know that it will very rarely vent all its wrath upon us.

## §6.6   Dynamic Effects on Sails and Hull

When sailing to windward in a seaway it is easy to believe that the boat's pitching motion will result in an increase in hydrodynamic drag (see §6.4). However, it is easy to forget what effect the motion is having on the apparent wind on the sails.

To get a feel for this effect, have a look at fig. 6.26, which assumes a steady true wind of 10 knots, boat speed through the water a steady 5 knots, but that there is a sea running that makes it pitch with the fairly large amplitude of 10° above and below the horizontal. If we assume that the period of the pitching motion is 3 sec, it is a simple matter to calculate the effective forward velocity of any part of the sail. As an example let's take a point on the sail 10 m up from the CG around which the boat pitches. When the bow pitches down this point on the sail moves forward, adding to the boat speed. When the bow pitches up the sail moves back with a velocity which must be subtracted from the boat speed. The maximum fore-and-aft velocity of our point of interest works out at 7.1 knots for the conditions given. The 'boat speed' of this part of the sail is therefore oscillating between 12.1 knots forward and 2.1 knots backward, which effects the strength and direction of the apparent wind as shown. In smooth seas the apparent wind angle and strength would be 30° and 14 knots. With a pitch amplitude of 10° that part of the sail 10 m up the mast will experience an apparent wind varying in strength from 8.7 to 20.5 knots, and in direction between 20° and 55°.

If one could assume that under these varying conditions the sail force could be determined from a static polar curve like that of fig. 3.26, then the average sail driving force over many pitching cycles could be determined and matched against the drag in waves to get the boat speed. Unfortunately this simple assumption is not justified, basically because there is a time lag in the buildup of circulation and hence lift around an aerofoil. This is to be expected if you look back at fig. 3.20, which implies that it takes a time after the onset of motion equal to the chord length divided by the flow velocity to fully establish the trailing vortex and hence the circulation.

This is shown rather beautifully by fig. 6.27, a photograph showing the flow on the leeward side of an aerofoil at an angle of attack of 30°. The aerofoil itself is not visible, only the outline of the low pressure surface. The picture was taken after the air speed had been accelerated from rest to about 2 m/sec in 0.85 sec. By this time the starting vortex

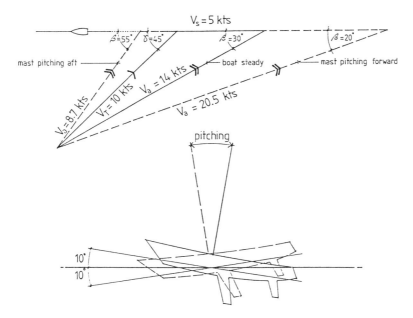

*Fig. 6.26* When a boat pitches the sail acquires a fore-and-aft velocity which is added to and subtracted from the boat speed. For the conditions assumed here, a point 10 m up the mast moves fore-and-aft at 7.1 knots. Combining this with the boat speed gives a resultant speed of 12.1 knots forward and 2.1 knots backward. The vector diagram shows the effect this has on the apparent wind speed and direction.

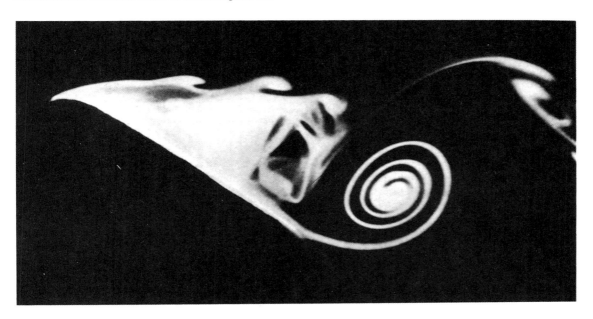

*Fig. 6.27* The development of the trailing vortex and conditions for circulation are shown here by means of a smoke generating liquid placed on the upper surface of an airfoil. The lower airfoil surface is not visible. The picture was taken 0.85 sec after the air flow started accelerating, indicating that the flow conditions which produce lift do not appear instantaneously but require a certain characteristic time to develop. *(Reproduced from* American Scientist, *vol. 72, p. 242 by kind permission of Prof. Peter Freymuth)*

from the trailing edge is well formed and lift is being generated. In this example the angle of incidence was fixed. If the angle of incidence is suddenly changed there will similarly be a delay before the new static value of the lift is generated. A change in angle of attack means a change in lift and hence a change in the vortices discharged from the trailing edge (compare fig. 4.1). These vortices induce velocities in the vicinity of the airfoil which cannot appear or disappear instantaneously. Thus the flow at any instant depends upon the acceleration and the immediate previous history of the flow as well as the flow speed and angle of attack as in the static case.

From the point of view of the sail, we have a situation in which the angle of attack is varying with the pitching motion, the wind speed is varying, and both these effects vary with height up the sail. To complicate matters further, a sail, being soft, will take up a curvature dependent on the pressure distribution. As far as I can ascertain nobody has ever attempted to measure or calculate sail forces under such dynamic conditions. The problem of measurement is a difficult one and yet for the science of sailing it is just as important as understanding added resistance in waves. The hydrodynamic case was investigated because the equipment and procedures had already been developed for commercial vessels. This is never likely to be the case for oscillating sails so we must wait for some non-profit research institute to take up the challenge for no other reason than the wish to know and understand.

Scientists have, of course, addressed themselves to similar problems. Experimental studies of oscillating aerofoils have historically been connected with problems of aircraft instability such as the flutter of wings and control surfaces. Aerodynamicists have been mainly interested in the conditions under which the amplitudes of oscillations start to grow from very small values, and consequently most experimental measurements have been for small amplitudes of oscillation of the angle of attack in a steady air flow.

In the 1920s and '30s various workers developed mathematical theories for what would happen to the lift when the angle of incidence of the airfoil was made to oscillate. The most useful of these theories was by Theodor von Karman. If the motion is non-uniform the total circulation around the aerofoil varies so that vortices are continually shed off the trailing edge, forming a wake which in turn modifies the flow over the aerofoil. The instantaneous unsteady lift is found to consist of three parts: $L(t) = L_0(t) + L_1(t) + L_2(t)$.

$L_0$, known as the quasi-steady lift, represents the force which would be produced if the instantaneous velocity and angle of attack of the aerofoil were permanently maintained. $L_0$ can be obtained by the usual stationary aerofoil methods.

$L_1$, called the apparent mass contribution, is a force the aerofoil would experience in a flow without circulation, due to the reaction of the accelerated fluid masses. It can be found by knowing the instantaneous acceleration of the aerofoil and its added mass. This contribution is only important if the airfoil is 'moving sideways' in the airflow, that is if the boat is heaving.

$L_2$ represents the wake influence. Because wake vorticity depends on the variation in circulation around the aerofoil, in turn caused by the unsteady motion, one can see that $L_2$ is a function of the time history of the motion of the aerofoil. The t in brackets indicates that these quantities are functions of time.

Although published measurements of lift for an oscillating angle of attack are not directly applicable to the sailing situation, a number of the qualitative characteristics are interesting. First, the variation in lift is different from that which would occur if the oscillation of the angle of attack were very slow. Second, the phase of the lift variation is different from that of the angle of attack (fig. 6.28). The static lift distribution follows the angle of attack and is represented by the dashed curve. After several steady oscillations the time-dependent lift variations are different in amplitude and are ahead in phase, $\varphi$. That is, they reach a maximum before the angle of attack does in each cycle. Third, the 'history' of cyclical lift and drag variations differs from that when the angle of attack is changed very slowly. Fourth, it is found that the onset of stall occurs at greater angles of attack in the dynamic situation. In addition, all these effects depend on the rate of variation of the angle of attack since the buildup of flow distribution requires that the air move a distance of the order of the chord length, c. Thus the ratio c/VT is a good non-dimensional measure of the effect of the rate of angle of attack variation. V is the wind average speed and T the period of the oscillation. Usually a quantity called the *reduced frequency* k is used and defined as $\pi$ times this: $k = \pi c/VT$.

Going back to the situation of fig. 6.26: V = 7 m/sec which taking c as 2 m and T as 3 sec gives k = 0.3. An approximate idea of the resulting phase angle of the lift can then be obtained from fig. 6.29 which is a

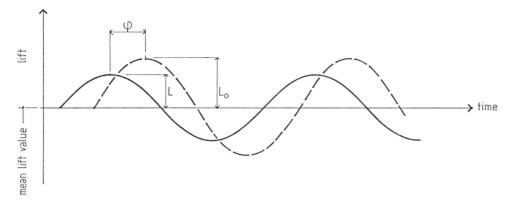

*Fig. 6.28* The dashed curve shows the variation in time of the angle of attack of an aerofoil. The solid line shows the variation of the lift that results. In this dynamic situation the lift does not follow the angle of attack but is out of phase by an amount $\varphi$.

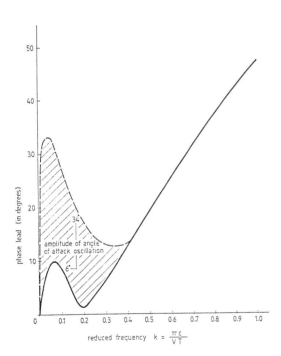

*Fig. 6.29* For an oscillating airfoil the phase difference between the angle of attack and the lift depends on a quantity called the reduced frequency. This is a measure of how far the air flows along the chord during one cycle of oscillation of the angle of attack. All curves for angles of attack oscillating between $\pm 6°$ and $\pm 34°$ lie in the shaded area and merge to the single curve for k greater than 0.4.

composite of measurements from several sources. The dashed curve labelled 34° refers to oscillations of the angle of attack by $\pm 34°$ about the mean value. All amplitudes of oscillation between 6° and 34° lie in the shaded area and join to the common curve for values of k greater than 0.4. For the pitching sail, T and the average value of V are more or less constant but the chord length c decreases with height, so that k becomes less near the top of the mast and the variation of angles of attack greater. This means that different parts of the sail experience maximum lift force at different times during the pitching cycle.

It is interesting to compare the static lift curve for an NACA 0012 section (fig. 6.30) with the lift amplitude obtained by oscillating the angle of incidence through $\pm 23°$ for instance. Under static conditions the lift coefficient reaches a maximum of 1.0 at an angle of attack of 10°. Fig. 6.31 is a plot of the maximum instantaneous lift generated during the oscillation cycle. At a reduced frequency of zero the maximum lift coefficient is just the maximum static value of 1.0. As the reduced frequency k is increased so is the maximum instantaneous lift. It is as though the linear part of the static curve were extrapolated beyond the normal stall limits as shown by the dashed line in fig. 6.1

This brings up the question of *dynamic stall*, which has several potential applications in sailing. If we measure the instantaneous lift coefficient and the corresponding instantaneous angle of attack while the latter is oscillating through 34° we get the curious looking diagram of fig. 6.32, sometimes known as a *hysteresis* curve. It is curious because as the angle of attack increases the lift coefficient takes up values

183

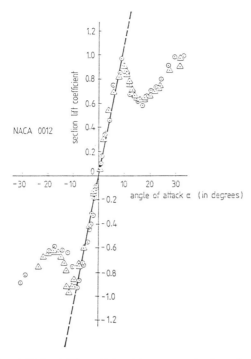

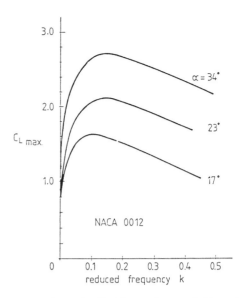

*Fig. 6.30* Static lift coefficient $C_L$ versus angle of attack $\alpha$. The lift rises linearly with angle of attack until stall is reached at 10°. Cyclical variation of $\alpha$ allows the lift to momentarily move to higher values along the dashed line.

*Fig. 6.31* If the angle of incidence of an aerofoil is made to oscillate through $\pm\alpha$ degrees the maximum lift coefficient can be pushed momentarily well above the maximum static value of 1.0. The speed of oscillation is measured by the reduced frequency.

which are different from those it has when the angle of attack is decreasing. For instance, for k = 0.036 the instantaneous value of $C_L$ is 2.0 when the angle of attack is 20° and *increasing*, but only 0.1 when the angle of attack is 20° and *decreasing*. The shape of the hysteresis loop depends somewhat on the rate at which the variation is carried out, as indicated by the reduced frequency. If the variation is carried out very slowly no open loop is produced: it is as though the fluid flow had a memory and doesn't want to change quickly from whatever state it is currently in. The stall delay implied by the hysteresis loop of fig. 6.32 is better shown by plotting the dynamic stall angle against the reduced frequency for various amplitudes of the angle of attack oscillation (fig. 6.33). The very large increase in dynamic stall angle compared with the static case is obvious.

Returning again to fig. 6.26 we see now that the large angle of attack on the sail which arises when the bow pitches up may not in fact give rise to stall as it would at this static angle. Perhaps a more important and certainly a more direct application of fig. 6.33 is

to the action of rudders. Their sections are commonly NACA 0012 and the skipper can easily oscillate the angle of incidence with the tiller. It is conceivable that going about could be speeded up by appropriate tiller action that allows greater lift without stall for long enough to turn the boat. In heavy weather helming rudder effectiveness might be improved by the right 'pumping' action. Large weather helm is usually associated with a large heel angle so that part of the rudder could be out of the water, with a danger of air entrainment which would only be worsened by high lift generation. Assuming this is not a problem, the question arises as to what rate of pumping is likely to give the best effect: fig. 6.33 indicates that the reduced frequency should not be less than 0.2. The reduced frequency, k, is given by $\pi c/VT$, where T is the time for one complete pumping cycle, c is the rudder chord and V is the boat speed. The pumping frequency should therefore be such that $\pi c/VT > 0.2$. As an example, take V = 2.5 m/sec, c = 0.75 m and we find $T < \pi c/0.2V = 4.7$ sec. This seems a surprisingly slow variation. The data of fig. 6.33 were taken at a Reynolds' number of about $10^5$ whereas our example corresponds to a Reynolds' number of about $10^6$. No data is available at this $R_n$ though there is some at the lower $R_n$ of $10^4$. This shows that the

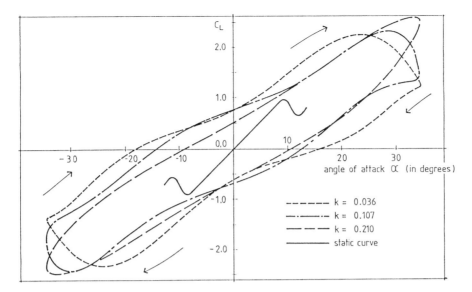

Fig. 6.32 The solid curve in the centre is the static lift versus angle of attack obtained by varying the angle of attack $\alpha$ very slowly. A cyclical variation of $\alpha$ at a reduced frequency k results in the hysteresis loops shown. Not only is the maximum lift greater than in the static case but its value depends on whether $\alpha$ is increasing or decreasing.

minimum pumping speed should be increased by a factor of 3 requiring a period of less than about 1.5 sec. Presumably at the higher $R_n$ of our example the pumping speed would certainly not need to be greater.

*To summarise*: enhanced rudder action is possible if, while increasing the mean rudder angle, the tiller is pumped quite slowly back and forth through about 20° to 30°.

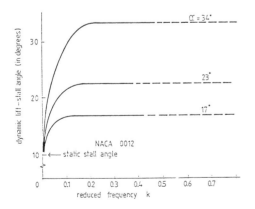

Fig. 6.33 Here the dynamic stall angle is plotted against the reduced frequency for various amplitudes of the angle of attack oscillation $\alpha$. Note the large increase over the static stall angle.

In order to get the maximum benefit out of pumping it should not be done too fast, as fig. 6.31 makes clear. Increased speed of pumping causes a drop in the maximum lift.

Along with the beneficial effects of momentarily delaying stall and increasing lift goes the penalty of increased drag, shown in fig. 6.34a which plots the increase in drag coefficient for an airfoil oscillating about zero angle of attack. The additional drag rises with the reduced frequency. In (b) the static drag curve is for comparison.

## §6.6.1 Downwind Rolling

Most sailors have experienced this or seen it happening to someone else. It can occur in large boats or small. It happens often when carrying a spinnaker, but is just as common for boats carrying main alone. Although it is influenced by wave motion it can also occur in perfectly calm seas. Its onset does not require particularly strong winds though it is not very likely in light conditions.

Characteristically the roll builds up to the mast-endangering point where spinnaker pole and boom are alternately dipping into the water. The experience is scary and seems impossible to control: a natural rudder action seems only to exacerbate the situation. The cause is by no means obvious and one

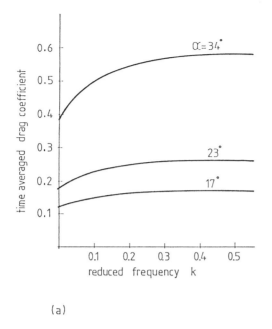

(a)

NACA 0012, $R_n = 0.37 \times 10^6$

(b)

*Fig. 6.34* (a) is a plot of the time averaged drag coefficient against reduced frequency for three values of the amplitude of the angle of attack oscillation. (b) shows the static drag curve versus angle of attack for the same NACA 0012 aerofoil, for comparison.

feels at the mercy of the boat. Though known politely as rhythmic rolling, you are more likely to refer to it as the death roll after experiencing it.

Because it can happen in smooth seas we therefore presume that rhythmic rolling has its origin in time-varying aerodynamic forces. Fig. 6.35 shows dramatically how one kind of time-varying flow can originate. In this series of pictures an aerofoil is fixed at an angle of attack of 60° with air coming in from the left. The aerofoil itself is not visible, only the outline of its leeward surface at the left of each picture. The 1.5 sec sequence starts at the top left and goes down. The air flow starts from rest in the first frame and accelerates at the rate of 2.4 m/sec² reaching a speed of 3.6 m/sec by the last frame. Although the Reynolds' number is much lower than for a sail, these pictures give a qualitative idea of what we might see during a gust of wind if air flow were visible. Vortices are produced alternately from the leading and trailing edges. Note that by the eleventh frame the simple spiral vortex of the first column develops into a complicated series of whirls within whirls reminiscent of Lewis F. Richardson's poem quoted in §2.7. The vortex from the trailing edge moves off

downstream first. Then, after the leading edge group of vortices has grown, it triggers off a second pattern of trailing edge vortices first seen in the sixth frame of the second column.

Let's assume first that an athwartships oscillatory aerodynamic force exists, and ask what effect this will have on the motion of the boat. The problem is analogous to pitching in waves as discussed in §6.4, except that now the oscillatory driving force is the wind. The important parameter is the natural frequency of roll, which depends on the lateral metacentric height (see §5.3.1) and the moment of inertia about the boat's fore-and-aft axis. The former depends on hull shape and the latter on weight distribution. It is a simple matter to measure this

*Fig. 6.35* The air flow from the leeward surface of an airfoil at an angle of attack of 60°. The airfoil itself is not visible, only the outline of the leeward surface on the left. The 1.5 sec sequence starts at the top left and goes down; the framing rate is 16/sec. The air flow is accelerating from rest in the first frame at 2.4 m/sec². Vortices alternately peel off the leading and trailing edges. *(Reproduced from* American Scientist, *vol. 72, p. 242 by kind permission of Professor Peter Freymuth)*

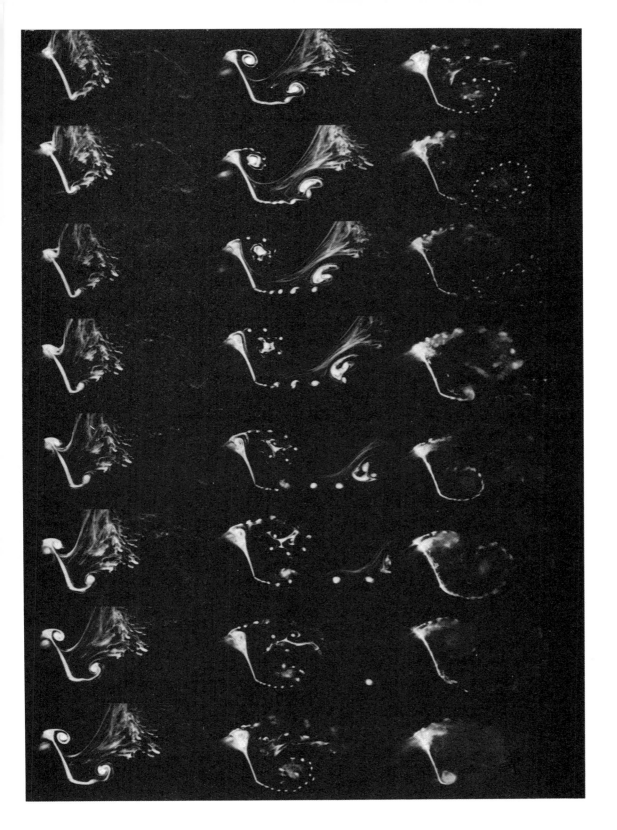

natural period of oscillation while at the same time finding how little force applied at the right time can get your boat into a rhythmic roll. With the boat beside a dock simply push down on the toerail about amidships in time with the roll: the amplitude of roll can become quite large, and is even more impressive when you realise how puny your weight is compared with the forces produced by the combined action of spinnaker and main in a 15 knot breeze.

An important factor in oscillatory motion is the damping. In the absence of a driving force the rolling motion will die down at a rate depending upon the amount of damping. For a hull this comes about first by the loss of energy in radiating waves. Second, there is a contribution from surface friction drag associated with the rolling motion, and third, there is the hydrodynamic drag of the rudder and keel moving sideways through the water. No measurements appear to have been made on the damping effectiveness of keels of various plan forms, though it has been suggested that modern cutaway keels and spade rudders might be less effective than the older style long keels, presumably because of the smaller area. However the considerations depicted in fig. 6.36 suggest that perhaps the reverse is the case.

Since the boat is moving forward as well as rolling, the flow velocity around the keel is the vector sum of the forward velocity and the sideways motion due to roll. As fig. 6.36 shows, the effective water flow direction is no longer from dead ahead, but, like making leeway, is such as to produce a lift force. Now it is clear from the diagram that when the direction of the movement is to the right the lift is to the left: the hydrodynamic keel force is always in opposition to the keel movement. This is a stabilising situation. As the boat rolls the angle of attack oscillates from one side of the keel to the other, so the enhanced dynamic lift variations discussed in the last section may add to the effectiveness of this form of damping. Since a high aspect ratio keel is a more effective generator of lift, it is expected that this contribution to the damping might well be greater for a modern cutaway underwater profile.

Fig. 6.35 shows how alternate vortices can be shed from a sail. If you try to move a plank of wood broadside-on through water you will see this same vortex formation and at the same time feel an alternate side-to-side force as the vortices peel off. Thus a sudden wind gust, even in the absence of waves, will initiate a rolling motion. Once it has started there is another aerodynamic mechanism which, given the right conditions, can produce a

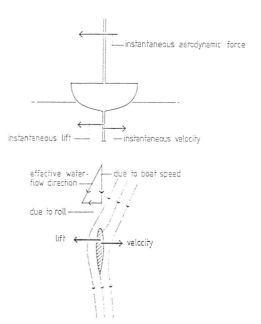

*Fig. 6.36* A keel which is effective in producing lift might also be expected to be effective in damping out roll. Providing the boat is moving forward, the rolling motion has the effect of giving the water flow an angle of attack which sweeps from side to side. As the diagram shows, the instantaneous lift direction will always oppose the roll. Although a high aspect ratio keel is effective under static conditions, it is possible that the instantaneous angle of attack near the tip of the keel will be high enough that stalling occurs. This will reduce the damping efficiency.

resonant rolling force. This has been thoroughly investigated in a series of wind tunnel experiments by C. A. Marchaj.

Qualitatively the origin of this resonant rolling force can be understood by reference to fig. 6.37 which depicts a boat sailing dead downwind just as it is also rolling to port. If we think about a slice of sail halfway up the mast, it will have some velocity v, along the negative y axis due to the roll. The sail therefore feels a component of air flow in the $+y$ direction of speed v. This must be added vectorially to the apparent wind speed $V_a$ to get the effective wind direction on the sail. This changes the angle of attack on the sail from 90° in the absence of roll to some smaller value labelled $\alpha$, which gives rise to a characteristic total sail force $F_T$. The direction of this force depends on the lift/drag ratio: a high L/D will angle the force to port, a low L/D will angle it forward or to starboard. If $F_T$ is angled to port it will have a component, labelled $-Y$ in the diagram, in the $-y$

direction. This constitutes a driving force in the same direction as the roll and will therefore amplify a roll which has already started. Furthermore, as the force has its origin in the rolling action it will always remain in synchronism with it no matter what natural frequency of roll the boat possesses. On the other hand, if L/D has a value such that $F_T$ is parallel to the +x axis then it has no y component and so cannot influence the roll. If L/D is smaller still, $F_T$ could have a component in the +y direction which would tend to damp out an already existing roll. Thus the occurrence or otherwise of rhythmic rolling depends on lift/drag ratios.

From the geometry of fig. 6.37 it can be shown that the transverse component of force is given by:

$$-Y = F_T \sin (\alpha - \varepsilon)$$

$\alpha$ is the angle of attack and $\varepsilon$ is the drag angle which is determined by the lift to drag ratio. If $\alpha = \varepsilon$, $Y = 0$, and there is no rolling force, the boat will be stable. Y points in the direction of the roll if $\alpha > \varepsilon$: this condition is unstable. Y opposes the roll if $\alpha < \varepsilon$. It is easy to see which of these is true for a single sail in the static case by referring to fig. 3.26. The angle between the vertical axis and the line from the origin to a point marking angle of attack is the drag angle. A measurement of these angles shows that they are always less than the angle of attack. This implies that the situation of fig. 6.37 is always unstable. Not too much should be read into this. Our model is an extremely crude one: as pointed out in the last section, the instantaneous dynamic forces can be very different from the static ones on which this argument is based. As the roll velocity varies with height the integrated effect is hard to predict.

Nevertheless it is interesting to draw some scale diagrams based on a static sail characteristic similar to fig. 3.26. This is done in fig. 6.38. The pair (a) and (b) are at the same wind angle of 180°, and (c) and (d) at the same angle of 160°. In both cases when the boom angle is reduced the y component of the total force is in the direction against the roll, so the boat will be stable. Due to the complexity of the problem there is little point in pursuing these arguments further. What we really want to know is: what are the conditions that give rise to rhythmic rolling and how do we prevent it?

Only experiment can answer these questions, which is why the wind tunnel experiments made by C. A. Marchaj are so valuable. Although he used a model of a Finn class dinghy the conclusions apply generally:

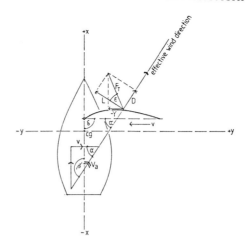

Fig. 6.37 The mechanism by which a synchronous aerodynamic rolling force may be produced. If the wind is dead aft a roll to port turns the effective wind as shown. Associated with the new angle of attack $\alpha$ will be a sail force $F_T$ which may have a component $-Y$ in the same direction as the roll. It thus tends to reinforce the already existing roll. A roll to starboard gives a similar result.

(a)  The apparent wind angle $\beta'$ has a considerable effect on rhythmic rolling. The instability is a maximum for $\beta' = 165°$ and decreases when $\beta'$ increases. This is unfortunate as most people when sailing downwind like to keep $\beta'$ less than 180° to avoid a gybe, but greater roll stability requires larger angles or even sailing by the lee.

(b)  The wind tunnel tests further showed that the boom angle $\delta$ had a strong effect on the rolling stability, as the argument of fig. 6.38 also implies. Large boom angles commonly used in boats with unstayed masts can be very unstable. Sheeting in the boom to about 65° can be very effective in damping out the roll.

(c)  Rolling instability depends on wind velocity. It seems to be determined by the reduced frequency, so that for a given boat there is a critical wind speed that is not necessarily very strong.

(d)  The wind tunnel experiments showed that a tall narrow sail in addition to the main was a very effective aerodynamic stabilising device. In practice this is usually a tall-boy or blooper.

## §6.7  Broaching

Ever since ships were first driven downwind the fear of broaching-to and its often deadly consequences

189

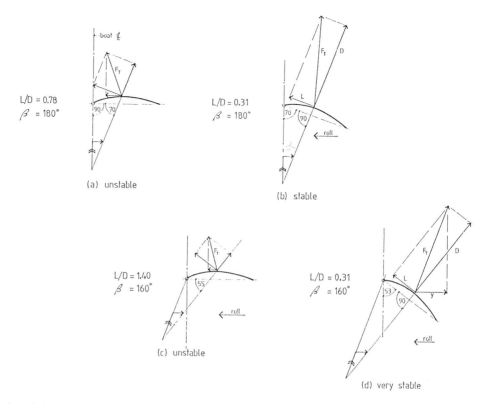

**Fig. 6.38** A crude idea of how sail forces can amplify or damp rolling is shown here. In (a) and (b) the static wind angle is 180°. Reducing the sheeting angle from 90° to 70° eliminates the transverse component of $F_T$: (c) and (d) correspond to a static wind angle of 160°: when the boom angle is reduced $F_T$ has a transverse component which opposes the roll. Different forces are produced when rolling back to starboard because the system is asymmetric. Ultimate stability or otherwise depends on overall effects averaged over a full cycle of roll. Furthermore, static L/D ratios have been used here instead of more realistic dynamic values.

has been a major concern of sailors. The following two quotes concerning ships of an earlier era remind us of the danger to and the helplessness of the crew once a broach has started. An early and vivid description appears in a book published in 1625 called *Hakluytus Posthumus*. In a chapter entitled 'A true repertory of the wracke, and redemption of Sir Thomas Gates Knight; upon and from the Islands of Bermuda' is found the following description of a broach.

'Once, so huge a sea brake upon the poope and quarter upon us, as it covered our ship from stearne to stemme, like a garment or a vast cloude, it filled her brim full for a while within.

'This source or confluence of water was so violent as it rushed and carried the Helme-man from the Helme and wrested the Whip-staffe [a sort of vertical tiller] out of his hand, which so flew from side to side that when he would have ceased the same again, it so tossed him from star-board to larboord as it was God's Mercy it had not split him.'

A later account from Alan Villiers in *Voyage of the Parma* is no less frightening:

'Then came the squall which almost robbed us of our morning, of all future mornings.

'With awful suddenness there came a mighty squall which made its predecessors seem like Doldrum airs. The helm would not go up, not for the combined striving of four boys and the mate: green water smashed along the length of her . . .

'Down, down she sank poop under!

'She pooped. Of all the sounds that a storm-lashed sea can make surely there is none more

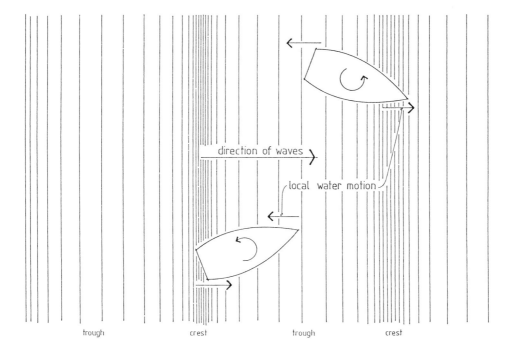

direction of waves

local water motion

trough          crest          trough          crest

*Fig. 6.39* One of the hydrodynamic causes of broaching has to do with the orbital motion of water in waves, which produces the local water motion given by the arrows. This in turn produces drag forces that tend to turn the boat beam-on to the waves when the stern is at the crest and stern-on when the bow is at the crest.

terrifying, more expressive of evil exultation, than the lapping of solid green water over a running ship's poop . . .

'The ship was now in a serious predicament. It was impossible to con the ship from the standard compass. She did not poop again. She avoided that. But a giant sea, before we were able to do anything about the compasses, came roaring at her while others of its fellows held her down; it hit her a mighty blow, carting her off before anything could be done to counteract it. She staggered badly as if she were a living thing that had been shot. . . . She fell on her side in the trough of the sea. . . . Broached-to!'

In these accounts broaching was mainly a result of wave action. It was also a once-in-a-lifetime experience. In yacht sailing broaching is far more common but hardly as dangerous to life and limb, though often fatal to one's chances of winning a race. Like downwind rolling, broaches can occur in flat water, so a purely aerodynamic cause is possible. Power boats can broach in a following sea, so a purely

hydrodynamic cause is possible. In general it will be a combination of both. Let's look first of all at how waves contribute to broaching.

Some characteristics of water waves are explained in §11.3 and fig. 11.10 contains facts relevant to this discussion. Since water particles in a wave motion rotate in fixed circles, water at the crest of a wave is moving forward and that in the trough is moving backward with respect to the direction of the wave motion.

The effect that this local water movement can have on the stability of a vessel moving downwind can be seen from fig. 6.39. The situation is worst if its waterline length is about half the wavelength as then the differential in water speed at bow and stern will produce turning moments as shown. If the boat is headed exactly in the direction of the waves these turning forces will have no moment arm. When the boat has its stern on the crest, a slight deviation from the wave direction results in a turning moment which increases the deviation: the situation is therefore unstable. On the other hand, when the stern is in the trough and the bow on the crest the turning moment

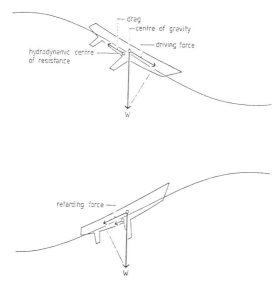

*Fig. 6.40* A second hydrodynamic effect which has an influence on broaching is connected with surfing, which all boats do to a greater or less extent. If the driving force due to weight is ahead of the hydrodynamic centre of drag there will be a 'weathercock' stability when on the front face of a wave, and instability on the back face.

tends to decrease the deviation from the wave direction. This condition is stable. It is clear that an important factor in wave-produced broaches is the relationship between wavelength and waterline length.

Like most things in sailing, the mechanics of wave-induced broaching are more involved than this simple description implies. Experience has taught us that boats can survive large waves if they are kept moving. In §6.4.1 it was pointed out that a boat moving through water in waves can receive an extra driving force from the component of its own weight tangential to the wave surface, as shown in fig. 6.40. This driving force, which can be much greater than the sail forces, acts through the centre of gravity of the boat. Opposing it is the equal and opposite hydrodynamic drag. The two are equal because at this point on the wave the acceleration is small (see fig. 6.22). This pair of forces constitutes a turning moment. It is clear that if the CG is in front of the centre of drag the hull will have 'weathercock' stability.

As the crest passes and the boat begins to sink back into the trough there is a large amount of deceleration, as fig. 6.22 shows, because the weight

force has a component retarding the boat. Since it will in general be different from the drag force, there is still a turning moment, albeit a weak one which in this case has a destabilising influence. If the CG is aft of the centre of drag, then of course the situation is reversed. We see therefore that this effect can add to or subtract from the effect of local water motion depending on the position of the CG.

Although waves are undoubtedly a contributing factor, most broaching of sailing yachts results from aerodynamic forces. Broaching is, of course, closely related to sail balance and the single main factor that changes that is the outboard movement of the sail centre of effort when heeled. This is also the main cause of broaching.

Fig. 6.41 shows a boat on a beam reach with a spinnaker. In (a) the boat is upright, the turning moment arm is small giving the boat a little weather helm. With the wind abeam the driving force is synonymous with the lift, so the heeling force is just the sail drag and that due to hull and rigging. As drag

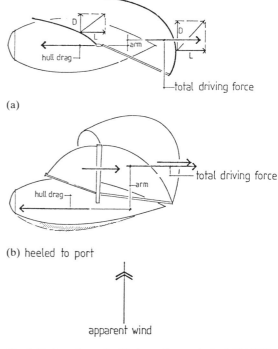

*Fig. 6.41* Aerodynamic effects are the most common cause of broaching. Excessive heel is the fundamental origin of the problem when on a reach. As the sail force moves outboard with heel the weather helm increases to the point where the rudder cannot cope and a broach turning up to windward becomes inevitable.

and lift are about the same under these circumstances, a gust of wind will quickly increase the heel angle to the situation shown in (b). The turning moment arm increases rapidly giving greater weather helm. If this can be offset by rudder action the boat will remain on course, otherwise sail adjustment is necessary to prevent a broach up to weather.

When sailing dead downwind one might at first think a broach was impossible since there are no heeling forces. Downwind broaching, which can occur either to weather or to leeward, is usually precipitated by downwind rolling. When the turning forces associated with the roll become more than the rudder can handle the boat broaches either way.

Note that sailing downwind with a main only is often more likely to cause a broach than with main and spinnaker since there is only a single sail force well out to one side, not balanced by a spinnaker on the other.

It is very important to avoid broaches, especially in racing where one broach can lose you more distance than you are likely to recover in a downwind leg, or cause damage. Knowing the cause, it is not too difficult to figure out how to avoid a broach.

Let's look first at reaching. When a gust hits and the boat heels, it is the mainsail which initially causes most of the problem. As fig. 6.41 shows its force will be farthest outboard and produce the most weather helm. If a broach looks imminent, ease the mainsheet first. To get a faster depowering of the main release the boom vang (kicker) also. This allows the boom to lift, lessening the chance that it will hit the water. If the boat is still out of control ease the spinnaker sheet. In an increasing wind, the main should first be reefed and then the spinnaker depowered by flattening it as much as possible. This can be done by keeping the pole end low to straighten the luff and move the draft to windward, also by oversquaring the pole so that it is about 20° back from the perpendicular to the apparent wind.

Since downwind broaches are usually precipitated by rhythmic rolling, the best cure is to stop the rolling using the procedures suggested in §6.6.1. In addition, it is often a help to ensure that the spinnaker doesn't oscillate with respect to the boat and to flatten it as much as possible. When sailing downwind with main alone, better balance will be achieved by flying a spinnaker or goose-winging the genoa.

# PART II

# Chapter Seven

# MORE ON DOWNWIND SAILING

## §7.1 Dimensional Analysis

The question: what quantities determine the hull drag of a sailing yacht and what is the relative importance of each quantity? is surely of fundamental importance.

Dimensional analysis cannot answer this completely – it cannot for instance tell us the best underwater shape – but it can tell us how drag depends on various quantities. All physical measurements depend in some way on only three basic quantities: mass M, length L, and time T. Various systems of units are used for these quantities; the scientific one, the simplest one and that favoured for use in most parts of the world uses kilogrammes for M, metres for L and seconds for T.

We start off by guessing what quantities might conceivably contribute to the hull drag of a boat in smooth water:

> waterline length $L_{wl}$
> viscosity of sea water $\mu$
> density of sea water $\varrho$
> speed through the water V
> acceleration of gravity g.

The reason gravity is included is that the boat will generate waves by displacing the water. The speed with which these waves move and the energy they contain depends on gravity.

Since drag is a force it has a known dependence on the quantities M, L and T. In fact Newton's Second Law tells us that force is mass times acceleration. The units of mass are kilogrammes and of acceleration, change in speed, metres per second per second, or metres per second squared, so the dimensions of force may be written: $ML/T^2$.

Similarly, each of the quantities $\mu$, $\varrho$, $L_{wl}$ etc which we think contributes to the drag have dimensions as follows:

| Quantity | Dimensions |
|----------|------------|
| L | L |
| $\mu$ | M/LT |
| $\varrho$ | M/L$^3$ |
| V | L/T |
| g | L/T$^2$ |

Now it should be clear that each of these quantities must contribute to the drag in such a way that their combined dimensions are that of a force, namely $ML/T^2$.

These requirements may be formulated more explicitly by writing:

$$\text{drag} = D = KL^a_{wl}\, \varrho^b\, \mu^c\, v^d\, g^e \qquad (1)$$

The exponents a, b, c, d and e are to be determined from the requirements that the dimensions of each side of the equation be equal. K is a dimensionless constant, i.e. just a number, which however cannot be determined from dimensional analysis.

If we write equation (1) in terms of the dimensions of the quantities involved we get:

$$\frac{ML}{T^2} = (L)^a \left(\frac{M}{L^3}\right)^b \left(\frac{M}{LT}\right)^c \left(\frac{L}{T}\right)^d \left(\frac{L}{T^2}\right)^e$$

Since M, L and T are independent quantities this equation is only true if the powers to which M, L and T appear on the left are the same as those on the right.

Equating the exponents of the dimensions on the left with those on the right gives:

for M    $1 = b + c$
for L    $1 = a - 3b - c + d + e$
for T    $-2 = -c - d - 2e$

so that $b = 1 - c$; $d = 2 - c - 2e$; $a = 2 + e - c$

Substituting these values back into equation (1) gives:

$$D = K \ L_{wl}^{2 + e - c} \varrho^{1 - c} \mu^c V^{2 - c - 2e} g^e$$

Now group all terms with the same exponent:

$$D = K \left( \frac{V^2}{Lg} \right)^{-e} \left( \frac{VL\varrho}{\mu} \right)^{-c} \varrho \ L^2 \ V^2 \qquad (2)$$

As well as the constant K, e and c cannot be determined by this procedure as we have only three equations corresponding to M, L and T and five unknowns a, b, c, d and e.

The quantities in brackets in equation (2) turn out to be dimensionless! The dimensions of $V^2$ are the same as those of $L_{wl}$ times g, and similarly the dimensions of $V L\varrho$ are the same as $\mu$. This means that these bracketed quantities are purely numbers whose value does not depend on the system of units used. This gives us the feeling that there is something fundamental about these quantities. The fact that each is named after a man eminent in fluid dynamics testifies to its usefulness.

The square root of $V^2/Lg$ is called the Froude number and is designated $F_n$. William Froude (1810–79) was an English engineer who first worked under I. K. Brunel on the Bristol & Exeter Railway and later conducted experiments on ship resistance at the Admiralty Establishment, Torquay. In 1877 he published a paper in the *Transactions* of the Institution of Naval Architects pointing out the connection between bow waves, waterline length and drag.

Froude's number can be thought of as the ratio of inertia forces to gravitational forces in the fluid flow. The inertia force is that associated with accelerating a mass M and is equal to Ma (a is the acceleration). Mass depends on density times the cube of a linear dimension, $\varrho L^3$. Acceleration is velocity divided by time, so the inertia force can be represented as:

$$\frac{\varrho L^3 V}{time}. \text{ But time} = \frac{distance}{velocity} \frac{L}{V}$$

so

$$\frac{\varrho L^3 V}{time} = \varrho L^2 V^2.$$

The gravity force is just the weight (mass) of the boat times the acceleration of gravity, namely $\varrho L^3 g$. Thus the ratio of inertia force to gravitational force is:

$$\frac{\varrho L^2 V^2}{\varrho L^3 g} = \frac{V^2}{Lg}$$

Since the energy contained in the bow wave depends among other things on gravity, we see that Froude's number characterises the part of the hull drag due to wave-making.

The dimensionless quantity $VL\varrho/\mu$ in equation (2) is called Reynolds' number after Osborne Reynolds. It is designated by the symbol $R_n$. The quantity $\mu/\varrho$, the viscosity of water divided by the density, is referred to as the kinematic viscosity and designated by the Greek letter $v$. As a result Reynolds' number is often written: $R_n = VL/v$.

Just as Froude's number could be regarded as a measure of the ratio of inertia forces to gravity forces, so Reynolds' number is seen to be the ratio of inertia force to viscous force in the fluid. As before, the inertia force is proportional to $\varrho L^2 V^2$. The viscous force is proportional to $\mu VL$. (See chapter 2 for discussion of viscous drag.)

$$\frac{\varrho L^2 V^2}{\mu VL} = \frac{\varrho LV}{\mu} = \frac{LV}{v} = R_n$$

Returning now to the drag equation (2) we see that it can be written:

$$D = Kf(R_n, F_n)\varrho L^2 V^2 \qquad (3)$$

$f(R_n, F_n)$ means 'some function of Reynolds' number and Froude's number', i.e. the drag depends on both $R_n$ and $F_n$ in some way but we don't know exactly how. This is because we cannot determine e and c in equation (2) by this method; in fact they depend upon such things as hull shape and smoothness.

We are now in a position to see the physical meaning of equation (3).

For a given hull the drag depends only on the Reynolds' number, the Froude's number and on the inertia force of the water $\varrho L^2 V^2$. This inertia force is analogous to the pressure of a water jet on a wall except that in the case of the boat the 'wall' is moving and the water is stationary.

$$\text{Drag} = \begin{bmatrix} \text{Dependence on} \\ R_n \text{ and } F_n \end{bmatrix} \times \begin{bmatrix} \text{Inertia force of} \\ \text{impinging water} \end{bmatrix}$$

Since the dependence on $R_n$ and $F_n$ cannot be readily calculated it must be measured in a wind tunnel or towing tank. It is then conventional to write the drag equation:

$$D = \tfrac{1}{2} C_D \varrho V^2 S \qquad (4)$$

$C_D$ is called the drag coefficient and embodies the dependence on $R_n$ and $F_n$. S (which has the dimensions of length squared) is the wetted area in the case of hull drag. $C_D$ has been extensively measured in wind tunnels for standard airfoil shapes and can be found in compilations such as that of Ira H. Abbott and Albert E. von Doenhoff *Theory of Wing Sections*. For the same conditions the drag coefficient is the same in air as in water. The drag force is of course greater in water, but this is taken care of in equation (4) by the density $\varrho$ which is about a thousand times greater for water than for air.

## §7.2 The Viscosity of Air and Water

The viscosity of all fluids changes with temperature. For air the kinematic viscosity goes down by about 2% for a temperature increase from 10° to 30°C; thus for all practical purposes it may be assumed constant and to have the value: $v_{air} = 1.5 \times 10^{-5}$ m$^2$s$^{-1}$.

Such is not the case for water, for which the viscosity changes by nearly a factor of 2 over the full range of water temperatures in which one might conceivably sail. This variation is shown in fig. 7.1, which also shows that on average sea water has a kinematic viscosity about 4.6% higher than that of fresh water.

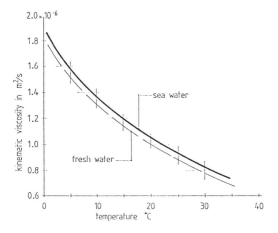

*Fig. 7.1* Variation of the kinematic viscosity of salt and fresh water with temperature. Since surface friction drag depends on the kinematic viscosity one might possibly steal an advantage over one's opponents by sailing in a region of warmer water.

Since surface friction drag depends directly on the viscosity, it is clear that it will be less in fresh water than in salt, and less at high temperatures than at low. The density of water does not change appreciably with temperature over the range of interest, but sea water is on average about 2.7% denser than fresh. It is therefore conceivable that when sailing downwind an advantage might be obtained by setting course through a river estuary where the water could well be warmer and less salty. Whether there would be an advantage when sailing to windward is not so clearcut as lift and hence leeway angle also depend on density.

## §7.3 Size of Turbulent Oscillations

Fig. 2.9 in Part I gives a crude idea of how turbulence might start in the laminar boundary layer. One might go further and ask what is the amplitude of the oscillations shown in Fig. 2.9d.

We can get some idea of the size of the turbulence from the following considerations. If the size of the turbulence is $\Delta y$ (fig. 7.2) which is a small increment of distance y measured in the vertical direction, then the velocity differences which contribute to the turbulent motion as in fig. 2.9 can be written $\Delta U$ (= a small change in velocity due to increase in distance $\Delta y$ from the surface). Clearly the time scale or period of these oscillations is just distance/velocity = $\Delta y/\Delta U$ = the inverse of the velocity gradient.

If we consider 1 square metre of surface area, the amount of mass involved in this motion is density $\times$ volume = $\varrho \Delta y(1)^2$. The acceleration of this mass is the change in velocity divided by time taken = $(\Delta U)^2/\Delta y$. The force which must have produced this acceleration is simply mass $\times$ acceleration (by Newton's Second Law) and is thus equal to $\varrho(\Delta U)^2$. But the physical origin of this force is the viscosity in the fluid, i.e. the shearing force between layers of fluid. As discussed in chapter 2 the viscous force per square metre is

$$\frac{\mu \, \Delta U}{\Delta y}.$$

So we have two expressions for the same force which when equated give:

$$\varrho \, (\Delta U)^2 = \mu \frac{\Delta U}{\Delta y} \quad \text{i.e. } \Delta y = \frac{\mu}{\varrho} \times \frac{1}{\Delta U}$$

And this is a measure of the size of the turbulence. To calculate this it is convenient to introduce the slope of the velocity gradient in the boundary layer. Call this slope S and note that it is just $\Delta U/\Delta y$. Turbulence size $\Delta y$ may then be written:

$$\Delta y = \left(\frac{\mu}{\varrho S}\right)^{1/2}$$

Thus the size of a turbulent cell increases with increasing velocity and decreases toward the surface where the velocity gradient is a maximum. At the outer boundary of the boundary layer $S \to 0$ and the size of the turbulence becomes much greater than the characteristic dimensions of the flow, i.e. no noticeable turbulence occurs in the absence of a velocity gradient.

For the turbulence to be significant it would need to be $\leqslant \delta$, the boundary layer thickness. We can check that this is indeed true for the case of fig. 2.5a where the slope at the surface is about $1600\,\text{s}^{-1}$, thus:

$$\Delta y = \left[\frac{\mu}{\varrho} \times \frac{1}{S}\right]^{1/2} = \left[10^{-6} \times \frac{1}{1600}\right]^{1/2} = 2.5 \times 10^{-2}\,\text{mm}$$

which is certainly less than $\delta$ which is 2.54 mm by fig. 2.6. Close to the outer edge of the boundary layer where the slope is about 1, the turbulence size becomes about 1 mm for these particular conditions. (The exponent ½ means 'the square root'.)

## §7.4 Increased Drag of Keel and Rudder Sections with Blunt Trailing Edges

A great deal of aeronautical data has been amassed over the years on the drag of various airfoil shapes. The data are of course valid for either air or water flow providing the Reynolds' number is the same. This ensures that the same flow conditions prevail. Most of the aeronautical data available is for Reynolds' numbers in the region of about $3 \times 10^6$. To see if this is applicable to the sailing situation we calculate the Reynolds' number for a fin keel of chord length 1.5 m travelling through water at 4 knots ($= 2$ m/sec):

$$R_n = \frac{Vl}{\nu} = \frac{2 \times 1.5}{10^{-6}} = 3 \times 10^{-6}$$

So the Reynolds' number for fin keels and rudders is about the same as that for the wings of slow-flying aircraft for which data are readily available. Even if the Reynolds' number differs considerably, it is of no great consequence since the characteristics of airfoils change only slowly with change in Reynolds' number. Care must be taken, however, in using aircraft data in this way. It is only applicable, for instance, to rudders and keels which are completely immersed and which are free of air entrainment and cavitation. Also the data available is always so called 'section data', applicable directly only for infinite aspect ratio. Although there are standard procedures for applying the data to lower aspect ratio situations, the very low aspect ratio of keels and rudders (from an aeronautical point of view) and the often strange plan forms used in yachts make it necessary to be very cautious about a too literal interpretation of the aeronautical data. The problem is really that sailing boats are more complicated than airplanes!

An important practical point in the design of keels and rudders, especially for small boats where they are made of wood, is to what extent it is possible to cut off the trailing edge without impairing the performance. Rudders and centreboards are generally made with a section shape corresponding to NACA 00xx; this is a symmetrical airfoil section (no camber) where xx is a number giving the maximum thickness as a percentage of the chord. The position of the maximum thickness is at 30% of the chord length from the leading edge. These sections have trailing edges which come slowly almost to a point. In some situations a sharp trailing edge may be structurally weak so it is important to know the effect on performance of a cut-off trailing edge. This has been measured and fig. 7.3 shows the increase in drag for NACA 0012 sections with standard roughness and perfectly smooth, as a function of the ratio of the thickness of the trailing edge to the maximum foil thickness. Note that this gives the effect on the section drag coefficient for the two-dimensional situation (infinite aspect ratio) and it is likely that the effect of trailing edge cut-off will be less in the real three-dimensional world in which rudders and keels work. Nevertheless this shows the importance of retaining as much of the trailing edge as possible especially if the drag coefficient is already low. It would certainly seem desirable to keep h/t down to less than 10% wherever possible. For the upper curve in fig. 7.3 this would correspond to an increase of about 20% in the drag.

## §7.5 Wetted Area of Hull Sections

Sometimes the question is asked: given a certain displacement, what hull form has the least wetted area? It can be answered by considering short sections of hull over which the section area does not change appreciably. This turns the difficult three-

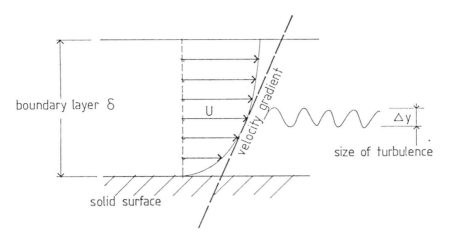

*Fig. 7.2* With the aid of this diagram and a simple mathematical argument it can be shown that the approximate size of a turbulent cell is proportional to the square root of the kinematic viscosity and inversely proportional to the square root of the velocity gradient in the boundary layer.

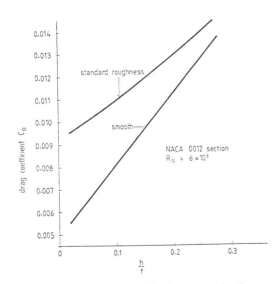

*Fig. 7.3* Increase in drag resulting from removing the trailing edge of a foil section; h is the thickness of the trailing edge and t the maximum foil thickness. The graph shows, for instance, that a 10% increase in drag can be expected for a hydrodynamically smooth section if the thickness of the trailing edge is 3.6% of the maximum thickness. Although these results are for the 'ideal' two-dimensional situation they give us a clue as to the importance of the trailing edge even though the effect is likely to be less marked in the real three-dimensional world.

dimensional problem into a simple two-dimensional one.

For comparison and to get a feel for what the results might be for other more realistic intermediate shapes, three extreme cases have been taken (see fig. 7.4). We would like to know the wetted area of each of the hull sections shown, with the constraint that they all have the same displacement. The two-dimensional problem boils down simply to determining the circumferences of geometrical figures having the same area. For a short section of hull the area of the section is proportional to its contribution to the total displacement.

In fig. 7.4 B is the beam of the semi-circular section which therefore has a draft of $B/2$. The beam and draft of the rectangular section (dotted) are $B_1$ and $D_1$ and those for the Vee section (dashed) are $B_2$ and $D_2$. The requirement that all sections have the same displacement means that the areas are the same, i.e. $\pi B^2/8 = B_1 D_1 = \frac{1}{2}(B_2 D_2)$. This puts a constraint on the relationship between beam and draft. The distance around the sections (excluding the horizontal waterline) is then a measure of the wetted area.

It will be left to the reader, if he is interested, to calculate the ratio of the wetted areas for the rectangular to circular, and Vee to circular, sections. If the ratio of the wetted area for the rectangular

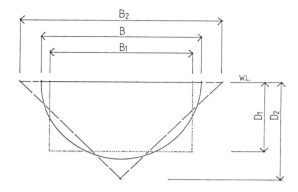

*Fig. 7.4* For a given displacement the hull section with the least wetted area is semi-circular in form. In this diagram two other extreme sections are shown; each has a beam:draft ratio such that its wetted area is a minimum consistent with both having the same displacement as the semi-circular section. Even this minimum wetted area is about 13% greater for both the rectangular and Vee sections than for the semi-circular. Any other beam: draft proportions having the same displacement will have even greater wetted area. No other shape can be found having the same displacement and less wetted area than the semi-circular form.

section to that for the semi-circular section is called $A_r/A_s$ and if the beam ratio is $B_1/B$, then it turns out that:

$$A_r/A_s = 2B_1/\pi B + B/2B_1$$

A slightly more complicated formula is found for the Vee section. It can further be shown mathematically that for both the rectangular and Vee sections the minimum wetted area, with the same displacement, is 13% more than for the circular section. These minimum values occur when $B_1 = 0.886B$ and $D_1 = 0.443B$ for the rectangular section, and $B_2 = 1.253B$ and $D_2 = 0.627B$ for the Vee section. Fig. 7.4 has been drawn to scale with these values.

## §7.6 Performance Prediction Downwind: an Approximate Method

If you remember some algebra it is a nice exercise to study analytically the buildup of downwind speed as discussed at the end of chapter 2. To do this one can make the approximation that the hull resistance is proportional to the square of the boat speed, i.e. $R = kV_s^2$ where k is a constant, R is the hull resistance and $V_s$ the boat speed. This is a reasonably good

approximation to the true facts for speeds well below hull speed in the displacement mode, but greatly underestimates the drag at and above hull speed.

The sail force can be represented by a similar equation $F = KV_a^2$ where F is the sail force, K is another constant different from the above, and $V_a$ is the apparent wind speed. The constants k and K have to be determined from experimental measurements; a reasonable value of k taken from measurements on a 5.5 Metre class boat is 26 $Ns^2m^{-2}$. K can be calculated from $1/2\varrho AC_D$, where $\varrho$ is the density of air which is 1.2 $kg/m^3$, A is the sail area in sq m and $C_D$ is the downwind sail drag coefficient for which a typical value is 0.7 (for main with spinnaker). Thus K has the value 0.42A.

When sailing directly downwind $V_a$ is related to the boat speed $V_s$ and the true wind speed $V_t$ by: $V_a = V_t - V_s$ so the sail force becomes $F = 0.42(V_t - V_s)^2A$. The net force which is producing acceleration is then just $F - R$ or $0.42A(V_t - V_s)^2 - kV_s^2$. Since $V_t$ is fixed this gives the variation of the resultant force with boat speed and is just what is plotted in fig. 2.32a. From this the acceleration can be determined by dividing by the mass of the boat. Then from the fact that $V = $ acceleration $\times$ time one can calculate the kind of result shown in fig. 2.32c.

Usually, however, one is not interested in the short acceleration period but only in what the final speed will be. This occurs when $F = R$ or $0.42A(V_t - V_m)^2 = kV_m^2$, where $V_m$ is the maximum boat speed. Using k = 26 as suggested above gives: $A(V_t - V_m)^2 = 61.9V_m^2$. The solution of this equation is:

$$V_m = \frac{(A/61.9)^{1/2} V_t}{(A/61.9)^{1/2} \pm 1}$$

The plus or minus sign ($\pm$) comes from the mathematical fact that this is a quadratic equation which must have two solutions. In this case only the plus sign is physically meaningful. The exponent $\frac{1}{2}$ means the square root.

We now have an equation that allows us to calculate the effect of true wind speed and sail area on downwind boat speed, assuming the hull remains in the displacement mode. Fig. 7.5 has been plotted from the above equation for $V_m$.

If you would like to get a rough idea of how your own boat's downwind speed will be affected by the true wind and its sail area you should find the following procedure interesting.

With spinnaker and main set in a 5 to 10 knot true wind measure your boat speed directly downwind in smooth water. The general expression giving the

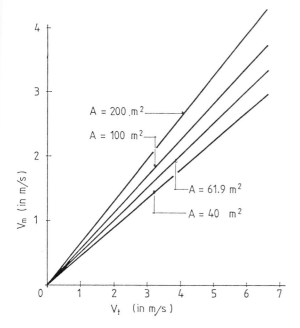

Fig. 7.5 The approximate way in which downwind speed $V_m$ varies with true wind speed $V_t$ and total sail area A for a particular boat. This result is only a reasonable approximation for speeds somewhat less than hull speed. A procedure is outlined in the text for establishing such curves for your own boat, which will allow you to extrapolate from a few measurements to a wider range of conditions within the limitations discussed.

maximum boat speed $V_m$ as a function of true wind speed $V_t$, sail area A and sail coefficient $C_D$ is:

$$V_m = \frac{V_t}{(k/\tfrac{1}{2}\varrho A C_D)^{1/2} + 1}$$

From measurements carried out at the Delft University of Technology a reasonable sail coefficient for main and spinnaker on a dead run is $C_D = 0.7$. Thus all the quantities in the above equation are known except k. One or two careful measurements on your own boat of $V_m$ for one or more values of $V_t$ will allow k to be determined from

$$k = 0.42\left(\frac{V_t}{V_m} - 1\right)^2 A$$

where A is the total sail area in sq m. Then to determine the approximate boat speed for other sail areas and true wind strengths one simply substitutes into the equation:

$$V_m = \frac{V_t}{(k/0.42A)^{1/2} + 1}$$

Since this procedure makes the assumption that the hydrodynamic resistance is proportional to $V_s^2$ it is only accurate for speeds somewhat less than 1.25 LWL, the 'hull speed'. Nevertheless it is a useful indicator of what to expect in light airs and smooth water.

# Chapter Eight

# MATHEMATICAL EXTRAS ON UPWIND SAILING

## §8.1 Albatross Flight

The fundamental physics of the flight of the albatross is closely related to the fundamentals of how a boat sails to windward. Thus the age-old affinity of the sailor for the albatross is strengthened by a common *modus operandi*. Many species of birds are noted for their soaring flight, but very few can do it in the absence of updraughts; yet the albatross can rise and fall and even glide to windward without ever flapping its wings.

At first sight it might be thought that providing there is a constant wind a bird could rise up like a kite. Considerations like those for the hot air balloon in chapter 1 show that unless the bird flaps its wings it will just be carried downwind and, since it is not buoyant like the balloon, will sink lower and lower. A kite is kept aloft by the fact that it is anchored to the ground: cut the string and it will move off downwind constantly losing altitude.

The kite has two parts, the string which is fixed to the ground and the lifting surface immersed in the air which is moving with respect to the ground. If the kite is to remain in a fixed position in the sky all the forces acting on it must be in equilibrium. Without the tension in the string this is not possible. In order for the kite to stay up against gravity it requires an intimate connection with two media moving with respect to each other.

As long ago as 1883 Rayleigh showed that for soaring flight to be possible one of two conditions had to exist. Either the air must have a horizontal velocity non-uniform in space or time, or it must have a local upward velocity. The first of these conditions is satisfied at sea because of the wind shear effect in the

first 30 or 40 m above the surface. This kind of soaring flight is known as dynamic soaring as opposed to the static soaring of birds like vultures which use thermals.

Rayleigh originally presented his argument in a very simplified form where, instead of assuming a continuously increasing wind speed with height, he assumed just two regions. The upper one with a wind speed u and the lower with a wind speed u − δu, where δu means some small change in u. To quote Rayleigh: 'In a uniform wind the available energy at the disposal of the bird depends on his velocity *relative* to the air about him. With only a moderate waste this energy can at any moment be applied to gain elevation, the gain of elevation being proportional to the loss of relative velocity squared.'

The law of conservation of energy tells us that the relation between the height h and the relative velocity V (fig. 8.2) is $2gh = V^2$ where g is the acceleration of gravity.

Similarly, when the bird comes back into the upper stratum heading upwind we have $2gh' = (V + 2\delta u)^2$. A little algebra then tells us how much higher point B is than point A: $h' - h = 2\delta u/g \times (\delta u + V)$.

If there is no velocity change between the two strata δu will be zero and the height gain will be zero. A gain in height means a gain in energy, and clearly with a slightly different flight path this could have been turned into a gain to windward at the same height.

The important point which comes out of the above mathematical expression is that the gain to windward depends on the velocity change δu with height. Unlike the kite or the sail boat which has two parts *simultaneously* associated with regions having

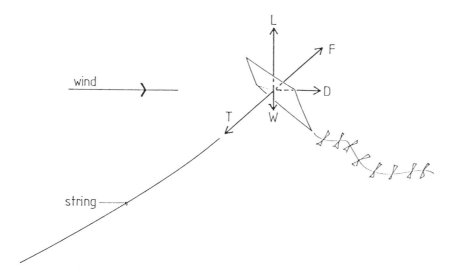

*Fig. 8.1* Although a kite is normally flown where wind shear exists (wind sheer is an increase in wind speed with height) it would fly just as well in a region of constant wind velocity. Nevertheless a kite will only fly if it is fixed to the ground by a string: it is thus intimately connected with two media moving with respect to one another. The kite remains stationary because the aerodynamic force F, the string tension T and the weight W are in equilibrium. There is no net force on the kite so it remains stationary. If the string breaks this balance of forces is destroyed and the kite falls to the ground.

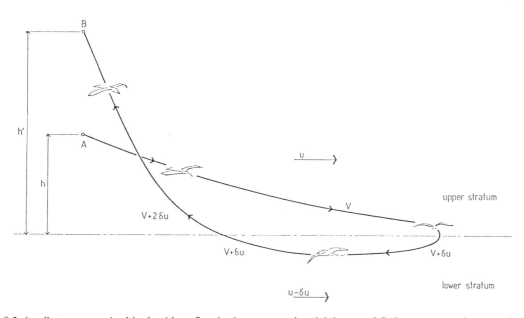

*Fig. 8.2* An albatross can gain altitude without flapping its wings. By losing altitude with the wind from A the bird increases its speed relative to the upper stratum of air which has a speed u with respect to the ground. When it moves across the boundary to the lower stratum its actual velocity does not change but its speed with respect to the slower moving air is increased. It then turns round, re-entering the upper stratum producing a further gain in relative velocity and hence in aerodynamic lift. This enables the bird to rise to B, higher than A, without any input of energy. V is the speed of the bird with respect to the air.

relative motion, the albatross moves *alternately* from one region to another and accomplishes the same objective.

Thus no matter by what means one extracts energy from fluid flow, the requirement for two or more regions moving relative to one another is always mandatory.

## §8.2 Determining the Potential or Ideal Flow by Means of an Electric Analogue

In physics it sometimes turns out that the same mathematical description fits more than one physical situation. The fact that mathematics is so suited to the description of physical phenomena is interesting in itself. Distinct physical systems which are described by the same mathematics are said to be analogues of one another. Automobile exhaust systems and early acoustic gramophones were designed using previously established electric circuit theory.

The analogy that interests us here comes about because the mathematical equations that describe electrical conductivity and two-dimensional non-viscous fluid flow are identical. This analogy was first formally exploited by G. I. Taylor and C. F. Sharman in a paper read to the Royal Society of London in 1928. The problem in those days was with the lift of aircraft wings. We will be concerned with the interaction and lift of sails, keels and rudders.

The electrical analogue of fluid flow is set up as follows. A uniform electric current is established between two parallel plates in a tank of water (fig. 8.3). If the plates are better electrical conductors than the water then the application of a voltage between them results in a uniform current flow as indicated by the solid lines. Assume that the bottom plate is at +10 volts with respect to the top plate. A voltmeter connected with one probe to the top plate and the other in the water will register a voltage somewhere between 0 and 10. It will be found, for instance, that all points in the water which are at 2.5 volts will lie on the straight dotted line which is parallel to the plates and located a quarter of the way down from the top. Similarly all the points at 5 volts will lie on a line going straight across the middle. These equipotential lines are the electric analogue of streamlines in fluid flow and thus we are representing here a uniform flow of wind from one side to the other of the diagram.

Because of electrode polarisation and other problems 'electrolytic tanks' are not used much these

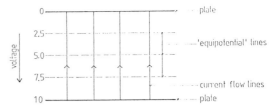

Fig. 8.3 The electrolytic tank, an electric analogue of fluid flow. If two flat conducting plates are placed parallel to one another in a tank of water and a voltage is established between them, an electric current will flow. This current flows uniformly in straight lines perpendicular to the plates. The dotted lines at right angles to the current flow lines are called equipotential lines. A high-quality voltmeter would register 7.5 volts only along the dotted line so labelled, assuming that 10 volts was supplied between the plates. Conducting strips on special conducting paper can also be used for plotting streamlines.

days, but an equivalent more practical method is to use a sheet of conducting paper instead of a tank of water. Electrodes of much higher conductivity than the paper are clamped to opposite edges to form the uniform equipotential lines just as in fig. 8.3.

If one now introduces a symmetrical conducting shape, the cross-section of a keel for instance, exactly in the middle then its potential will be 5 volts if there is 10 volts applied between the plates. This is shown in fig. 8.4 where it can be seen that the 5 volt equipotential line is unaltered except for the conducting region which is all at 5 volts, whereas the other equipotentials are distorted by the presence of the conducting shape between the plates. This distortion is exactly the same as would occur to the streamlines in irrotational fluid flow without viscous effects.

The streamline which strikes exactly at the centre

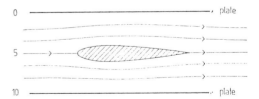

Fig. 8.4 Flow around a NACA 0009 section such as is commonly used for keels and rudders. The flow was plotted by means of the electric analogue device shown in fig. 8.9. The angle of incidence is zero so the flow is symmetrical and no lift is produced. Since this corresponds to potential flow it also predicts no drag, but this is not a disadvantage in many investigations.

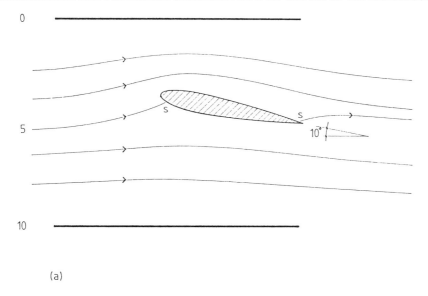

(a)

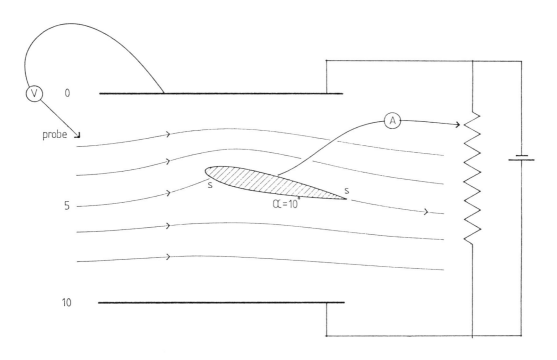

(b)

*Fig. 8.5* (a) Potential flow without circulation around a NACA 0009 foil at an angle of incidence of 10°. Because of the mass and viscosity of a real fluid this kind of flow does not occur. Instead, circulation is produced, distorting the flow to that shown in (b). It can be shown mathematically that the electric analogue of circulation is current flowing out of the model. This current is adjusted by varying the voltage of the model until the 5 volt equipotential line runs smoothly out from the trailing edge. The lift produced can then be calculated from this current.

207

of the leading edge divides the flow between the upper and lower surfaces and is known as the stagnation line. It is always continued on the downstream side and in this case comes exactly off the sharp trailing edge of the keel profile. This is a good approximation to the flow around a keel when sailing downwind.

The situation is more complicated when sailing on the wind. Now the boat makes leeway, so the angle between the undisturbed flow direction and the centreline of the keel section is perhaps 10°. A measurement with the analogue plotter gives the strange looking result shown in fig. 8.5a. The flow around the leading edge is correct but that around the trailing edge is all wrong. This is because we have not introduced into the electric analogue the effect of circulation as described in §3.3.3.

It turns out that there is also an electric analogue for circulation. It is directly proportional to the electric current fed into the model. Because of the mass and viscosity of a real fluid the flow off the trailing edge of a streamline body has the same direction as its centreline; this is the Kutta condition, fully explained in §3.3.3. In practice the correct circulation is obtained by tapping off a voltage somewhere between that applied to the plates and adjusting this until the Kutta condition is found, as shown in fig. 8.5b. The equipotential lines that are now plotted will be a true representation of the real flow outside the boundary layer region providing there is no separation of the flow.

One must bear in mind, however, that this is only a two-dimensional analogue of fluid flow which assumes that the model has a constant cross-section and infinite aspect ratio. It would not be possible to study, for instance, the effect of end plates or wings on a keel, or twist in a mainsail. Nevertheless important and valid conclusions can be drawn for a number of situations. One of the best examples is the contribution to the understanding of sail interaction as discussed in chapter 4.

## §8.3 Bernoulli's Law and Pressure Coefficients

As pointed out in §3.3.1, Bernoulli's equation is really just a statement of conservation of energy in fluid flow. The mathematical form appropriate to the sailing situation is: $p + \frac{1}{2}\varrho v^2 = $ constant. p is the static pressure in the flow, v is the speed of the flow and $\varrho$ is the density of the fluid. The quantity $\frac{1}{2}\varrho v^2$ is the kinetic energy of the fluid and is referred to as the dynamic pressure, whereas the static pressure p is a measure of the potential energy in the flow. So where the fluid is flowing fast the kinetic energy is high, forcing the pressure to be low so that the sum of the two always remains the same. Thus if one knows the velocity distribution around a sail, for instance, the pressure distribution can also be determined.

Velocity distributions over standard airfoil shapes have been measured in wind tunnels and published in a book by Ira H. Abbott and Albert E. von Doenhoff called *Theory of Wing Sections*. Some of these sections are useful in keel and rudder design, but before we can calculate the forces a little more mathematics is required.

First of all it is worth getting a rough idea of the magnitudes of the quantities in the Bernoulli equation. If we are considering a point on a keel at say 1 m below the surface of the sea, then the static pressure in the fluid will be the atmospheric pressure plus the pressure due to a head of 1 m of water above; the latter pressure is only about 1/10th of atmospheric. Thus to a rough approximation it is reasonable to ignore the effect of depth for boats of average draft and say that the static pressure near the keel is roughly equal to atmospheric pressure which is $100,000$ $Nm^{-2}$ (newtons per sq m) or about 15 lb/in². This is the pressure when the velocity v = 0. So the constant in the above formula is about $10^5$.

The contribution of the dynamic part of the pressure to the total is relatively small both for water flow and for air flow. For example if the water speed is 4 knots or 2.0 $ms^{-1}$ then the quantity $\frac{1}{2}\varrho v^2$ has the value $\frac{1}{2} \times 1000 \times (2.0)^2 = 2000$ $Nm^{-2}$ (Remember that the density of water is 1000 $kgm^{-3}$ also written kg/m³. $ms^{-1}$ is m/sec; $Nm^{-2}$ is N/m².)

Thus the static pressure decreases from 100,000 to 98,000 $Nm^{-2}$, a change of only 2%. A more useful way of writing Bernoulli's equation is as follows. Since the sum of the dynamic and static pressures is constant we can equate this sum in the undisturbed part of the flow ahead of the boat to that in the region of interest near the keel or sail, and write: $p_o + \frac{1}{2}\varrho v^2_o = p + \frac{1}{2}\varrho v^2$. In this equation $p_o$ and $v_o$ refer to the pressure and velocity in the undisturbed part of the flow. Since the dynamic pressure is such a small fraction of the total pressure it is clear that small changes in velocity have an even smaller effect on the static pressure. Therefore, instead of working in terms of actual pressures it is more convenient to use a quantity which depends on pressure differences. This quantity is called the pressure coefficient $C_p$ and is defined by: $C_p = (p - p_o)/(\frac{1}{2}\varrho v^2_o)$. p is the pressure at the point of interest and $p_o$ is the pressure in the undisturbed part of the flow. Dividing by $\frac{1}{2}\varrho v^2_o$ is done simply to make $C_p$ dimensionless and give it a value close to 1.

To see how the pressure coefficient is used let's look at the pressure at two points in the flow of water around the keel of a boat sailing dead downwind. There is no leeway so the flow around the keel is symmetrical. Fig. 8.6 shows the flow lines for a NACA 0009 section (see chapter 5), a standard symmetrical airfoil section of a type suitable for a rudder. From the method by Abbott and von Doenhoff the pressure distribution around such a shape can be determined. We will do this just for the simple case of downwind sailing and compare the pressure at the leading edge with that at a point 30% of the chord length from the leading edge.

A simple algebraic manipulation of Bernoulli's equation and the expression for the pressure coefficient tells us that $C_p$ can also be written: $C_p = 1 - (v/v_o)^2$. The tables in Abbott and von Doenhoff give for this situation $v/v_o = 1.119$ so $(v/v_o)^2$ is 1.252. Thus $C_p = 1 - 1.252 = -0.252$. Thus in the region

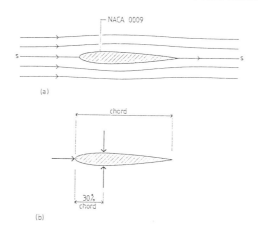

*Fig. 8.6* (a) The flow around a NACA 0009 section shows crowding of the streamlines near the leading edge, where we expect the pressure to be less than elsewhere. The arrows in (b) represent the magnitude and direction of the actual pressure at three points. Although the pressure at the leading edge is greater than the other two it differs by only about ½%. Even though this diagram shows the actual pressures it is not very informative; the next two show alternative ways of representing the pressure distribution.

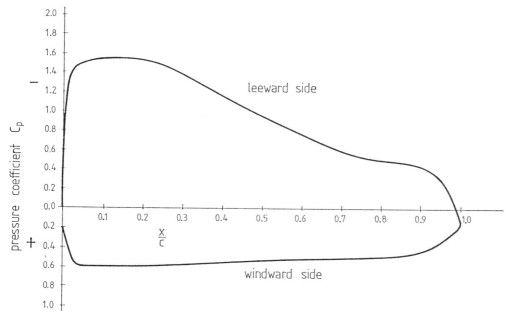

*Fig. 8.7* The standard method of plotting the pressure distribution over a keel or sail. $C_P$ is the pressure coefficient, a quantity which depends on the difference in pressure of the undisturbed flow and that of the point of interest. Those parts of the flow where the pressure is reduced have a negative coefficient, although the pressure itself can never be negative of course. It is conventional to plot $C_P$ with negative values upwards and positive downwards. This means that for a sail the upper curve applies mainly to the leeward side while the lower one corresponds to the windward side. This plot is for a sail with 10% camber and an angle of attack to the apparent wind of about 15°. c is the chord length so x/c is position as a fraction of the chord from the leading edge.

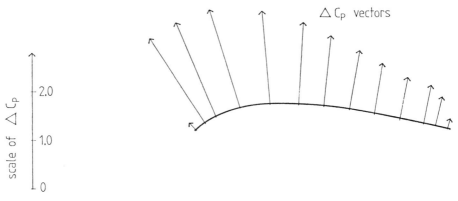

**Fig. 8.8** This method of displaying the pressure distribution over a sail is quantitatively accurate and conveys more information in a more easily understood way than the conventional aeronautical plot of fig. 8.7. The length of the arrows represents actual net sail forces calculated from the difference between the pressure coefficients on the windward and leeward sides in fig. 8.7. Since they are pressure forces their direction must always be perpendicular to the sail. This type of plot is only possible with a cloth sail where the forces on opposite sides are in line.

where the flow speeds up the pressure coefficient becomes negative but not the actual pressure as we can see from the following calculation:

From the original definition of $C_p$ we have: $p - p_o = -252 \times \tfrac{1}{2}\delta v^2_0$. For $v_o = 4$ knots $= 2$ ms$^{-1}$, $p - p_o = -.252 \times \tfrac{1}{2} \times 1000 \times 2^2 = -504$ Nm$^{-2}$. If $p_o = 10^5$ Nm$^{-2}$ then $p = p_o - 504 = 100,000 - 504 = 99496$ Nm$^{-2}$, i.e. a reduction in actual pressure of only about $\tfrac{1}{2}\%$.

The direction of this pressure is, as always, perpendicular and toward the surface as shown in fig. 8.6b. If the arrows in this figure genuinely represent the pressure vectors then they will have a length directly proportional to the magnitude of the pressure. However, since there is only $\tfrac{1}{2}\%$ difference between them the arrows will appear to be the same length, so that although they represent physical reality they hardly convey much useful information in a diagram of this kind.

This is the reason for using pressure coefficients. An example of the standard method of displaying the pressure changes is given in fig. 8.7. They are usually plotted with negative values of $C_p$ above the zero line and positive values below.

For the special case of a cloth sail which has effectively zero thickness the actual pressure forces on each side of the sail are in line so a more effective way of displaying the total pressure distribution across a sail is to simply take the arithmetic difference of the pressures at opposite sides.

Formally this can be done by defining a differential pressure coefficient called $\Delta C_p$ where:

$$\Delta C_p = C^w_p - C^L_p = (p^W - p^L)/(\tfrac{1}{2}\varrho v^2)$$

In this expression the superscript W refers to the windward side of the sail and L to the leeward side. Simply multiplying $\Delta C_p$ by $\tfrac{1}{2}\varrho v^2_0$ gives the actual physical net force per unit area on the sail for the wind speed $v_o$. This is a net force pushing on the sail from the windward side. Talk of 'suction' or the greater effectiveness of the leeward side is irrelevant and confusing.

## §8.4 The Momentum Change Theory

Advances in science are made by observing how things behave and then attempting to describe that behaviour in the simplest possible way. Irrelevancies and red herrings must be purged from our thinking: once this is done the logic of science leads quickly to understanding. The trick is in determining what matters and what is irrelevant.

One way of doing this is to develop a *model*. This is not a small replica of the real thing but a simplified version of it – a version which contains only the things that matter. The model should be easier to understand than the real thing but ideally should behave in the same way. This ideal is rarely achieved; usually a model will simulate only some aspects of the

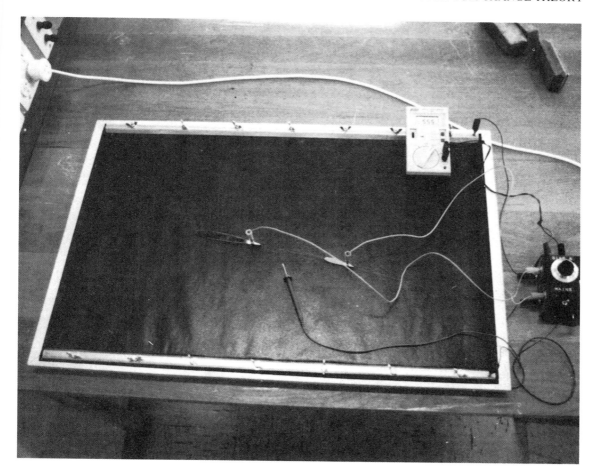

*Fig. 8.9* The analogue plotter used for determining most of the fluid flow and pressure distributions quoted in this book. An electric current is established through the black conducting paper between conducting plates at top and bottom. A conducting model of the sail or keel is placed in the centre. Current flowing into the model simulates circulation and a voltage sensitive probe is used to plot equipotential lines which are an exact analogue of two-dimensional fluid flow in situations where no separation occurs.

real thing so that a different model or picture is required to explain other aspects.

The momentum change theory of sail force as given in §3.3.2 is just such a model, though it takes no account of the details of the air or water flow. Basically it assumes that the air can be regarded as a stream of particles which are elastically deflected by the sail. With this simple model it is straightforward to calculate the total force and the lift produced by the sail. The relevant equations are given below and fig. 8.10 shows the predicted lift as a function of the deflection angle. This erroneously predicts that when

sailing on a beam reach the maximum drive is obtained when the sail deflects the air through 90°. This would require the sail to be sheeted in much harder than good practice dictates. Nevertheless despite its deficiencies this view of sail force is very helpful especially during the stress of sailing.

If the apparent wind speed is $v_a$, then the mass of air hitting the sail per second is $\varrho A v_a^2$ where $\varrho$ is the density of air, A is the sail area and $v_a$ the apparent wind speed. This mass of air is deflected through an angle $\theta$ by the sail (see fig. 8.10) resulting in a total force F at an angle $\beta$ to the incident wind direction.

211

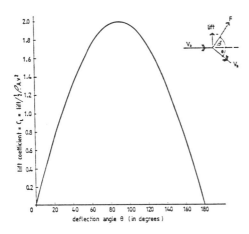

*Fig. 8.10* The variation of lift with deflection angle as predicted by a very crude model of the origin of sail force. Qualitatively this momentum change model is useful, but because it ignores the details of air flow its quantitative predictions are of little use.

The total force is then given by:

$$F = \delta A v_a^2 [2(1 - \cos\theta)]^{1/2}$$

and

$$\beta = \tan^{-1}[\sin\theta/(1 - \cos\theta)]$$

The lift is then $L = F\sin\beta$. The lift coefficient $C_L$ which is given by $L/\tfrac{1}{2}\varrho A v_a^2$ is the quantity plotted in the graph of fig. 8.10.

## §8.5  Flettner's Rotor Ship

Mention was made in §3.3.3 of a 'sailing ship' propelled by means of the Magnus effect. A rotating cylinder in a wind will develop a lift force perpendicular to the wind direction, as explained in chapter 3. This idea was researched in detail in the early 1920s by the German engineer Anton Flettner. Presented here are some of his experimental results, which make the Flettner rotor seem very attractive as a way of using wind power to economise on fuel.

Flettner, using the resources of the fluid mechanics laboratory at the University of Göttingen, made measurements of the lift and drag produced by a model cylinder rotating at various speeds in an air stream. Fig. 8.11 shows his results for a plot of lift coefficient $C_L$ versus drag coefficient $C_D$ for a rotating cylinder whose dimensions are given in the inset. The surface of the cylinder is smooth but it has end plates which effectively increase its aspect ratio.

The same graph gives typical characteristics of a modern design of cloth sail. Remember that plotting coefficients in this way removes the effects of different sail areas and wind speeds since $C_L$ = lift divided by $1/2\varrho A v^2$ and similarly for the drag coefficient. $\varrho$ is the air density, A is the sail area and in the case of the cylinder the superficial area, and v is the apparent wind speed. Thus a sail with an area equal to the surface area of the cylinder has a maximum lift only about one-sixth of that of the rotor. On the other hand the drag of the rotor is greater so that it has a smaller lift/drag ratio when sailing to windward.

Although the rotor looks impressively superior as to potential driving force on a broad reach, for instance, it can be deceptive to use such results in isolation. They must be combined with the hull characteristics to give a performance prediction as in §3.4.2 on several points of sailing. There is another point about fig. 8.11 which clouds comparison a little. For the cloth sail the different values of $C_D$ and $C_L$ are obtained by varying the angle of incidence on the

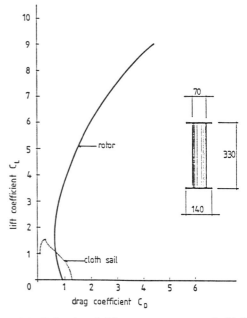

*Fig. 8.11* Polar plot of a Flettner rotor compared with that of a cloth sail. Although the maximum lift/drag ratio is not as high as for the conventional sail, the maximum lift coefficient is many times greater. However many other factors are also involved in determining the feasibility of wind-powered vessels. The inset shows the geometry of the rotor model used for the measurements; dimensions are in millimetres.

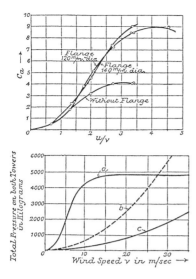

*Fig. 8.12* A reproduction of two figures from one of Flettner's original papers. The upper graph shows the effectiveness of end flanges which perform a similar function to wings on keels. Lift coefficient is plotted on the vertical axis and the ratio of rotor speed to apparent wind speed on the horizontal scale. In the lower graph curve (a) shows the self-regulating effect of a Flettner rotor wherein the force does not rise with the square of the wind speed as one might expect but saturates at very high wind speeds. This lift force is to be compared with the drag force on conventional rigging, curve (b), and the drag on the rotor when it is not turning (c).

sail, whereas for the rotor the points on the graph are obtained by changing the speed of rotation.

Fig. 8.12 shows two of Flettner's original graphs. The upper one is a plot of lift coefficient on the vertical axis against the ratio of peripheral rotation speed u of the cylinder to the apparent wind speed v. The effectiveness of the end flanges is very apparent. As can be seen the lift is a maximum when the ratio u/v is about 4. The fact that the lift force goes to zero when u/v tends to zero gives the Flettner rotor a very important and useful characteristic not possessed to the same extent by a normal sail. This feature is shown explicitly in the lower graph. Curve (a) shows how the total force varies with wind speed. Above about 12.5 m/sec or 25 knots the force on the rotor remains constant even when the wind speed is more than doubled. Normally a doubling of wind speed would be associated with a fourfold increase in forces. This lower graph applies to the three-masted topsail schooner *Buckau* and curve (b) shows the wind resistance of its normal rigging without sails set.

This is to be compared with curve (c) which is the resistance of a rotor when not turning.

From this it would seem that the Flettner rotor is a very seaworthy device indeed. Sudden wind gusts would not create excessive forces on the structure and the rig can be 'reefed' by simply slowing down the rotors. When they are stopped the drag is less than that on conventional rigging.

As fossil fuel costs wax and wane so does interest in the economics of wind power for merchant ships. A number of designs have been drawn up and some ships built, but the Flettner rotor has not been favoured. Most designs use some form of rigid but controllable airfoil.

## §8.6 Velocity Triangle Formulae

The geometrical relationships between the six quantities, apparent wind speed $V_a$, apparent wind angle $\beta$, boat speed $V_s$, true wind speed $V_t$, true wind angle $\gamma$, and speed made good to windward $V_{mg}$ are fundamental and applicable to all boats on any point of sailing.

Even without reference to the details of hull and sail characteristics these relationships impose severe restrictions on the range of possible performance of a sail boat. The following useful formulae are simply algebraic statements of the geometry of fig. 8.13.

$$V_{mg} = V_s\cos\gamma$$
$$V_t\sin\gamma = V_a\sin\beta$$
$$V_s\sin\gamma = V_a\sin(\gamma - \beta)$$

These are the simplest and most fundamental relationships that follow from the figure. Another relationship that follows from the diagram (remembering the theorem of Pythagoras) is:

$$(V_t + V_{mg})^2 + (V_{mg}\tan\gamma)^2 = V_a^2$$

i.e. $\dfrac{V_t}{V_a} = -\dfrac{V_{mg}}{V_a} + \sqrt{1 - \left(\dfrac{V_{mg}}{V_a} \times \tan\gamma\right)^2}$

When tacking, the change in course is unlikely to be outside the range 70° to 100° so that $\gamma$ is half this value as shown in fig. 8.14. Furthermore since maximum values of $V_{mg}/V_a$ are not likely to be much greater than 0.4, the square root term in the above equation never differs much from 1, so that a plot of $V_t/V_a$ against $V_{mg}/V_a$ exhibits only a very narrow region of physically possible values. This is illustrated in fig. 8.15. All possible values of the ratios $V_t/V_a$ and $V_{mg}/V_a$ are contained in the shaded area. The upper limit of the area corresponds to tacking through an

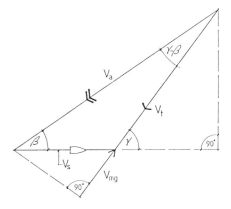

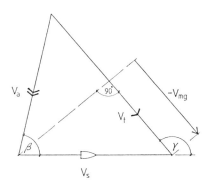

*Fig. 8.13* Velocity triangles for a boat sailing on the wind and on a reach. These diagrams show the relationship between the on-board observables of boat speed, apparent wind speed and direction on the one hand, and true wind speed and angle and speed made good to windward on the other. Note that when the true wind angle $\gamma$ (gamma) exceeds 90°, $V_{mg}$ is regarded as negative since it now becomes speed made good downwind rather than against the wind.

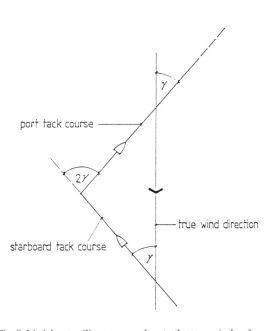

*Fig. 8.14* A boat sailing at an angle $\gamma$ to the true wind tacks through an angle $2\gamma$ as this shows. Those who were made to learn Euclidean geometry in their youth may remember the teacher intoning 'the exterior angle is equal to the sum of the two interior opposite angles'.

angle of 70°, the lower limit to 100°. Most boats will fall somewhere between these limits. Maximum values of $V_{mg}/V_a$ are not likely to exceed 0.45 so the shaded area comes to a fuzzy end about there.

The existence of these fairly stringent geometric limitations on performance is useful when checking yacht or model performance measurements for errors. Any point outside the shaded region is unlikely to be valid. The above discussion applies, of course, only to optimum windward sailing.

Apparent wind strength and direction are usually measured by instruments at the top of the mast, which means that when the boat heels an error is introduced by the fact that the measurements are being made in a plane perpendicular to the mast rather than in a horizontal plane. The following formulae convert the instrument measurements to those which would be obtained in the horizontal plane. The primed quantities refer to the raw instrument measurements and $\varphi$ is the angle of heel.

$$V_a = V'_a/\sqrt{\cos^2\beta + \sin^2\beta\cos^2\varphi}$$
$$\beta = \tan^{-1}(\tan\beta'/\cos\varphi)$$

The error introduced by this effect is only small for heel angles less than 20° as can be seen by the following example:

if $V'_a = 15$ knots, $\beta' = 30°$ and $\varphi = 20°$, then $V_a = 15.25$ knots, an increase of about 1.6%. Similarly $\beta = 31.6°$, an increase of 5%.

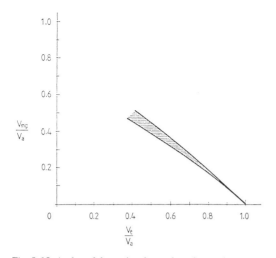

*Fig. 8.15* A plot of the ratio of speed made good to windward to apparent wind speed against the ratio of true to apparent wind speed. It shows the stringent limitations on these quantities imposed by the geometry of windward sailing. The upper limit of the shaded area corresponds to tacking through an angle of 70° and the lower boundary to a tacking angle of 100°. Plots of this kind are useful when making performance measurements of windward ability since valid measurements must fall within the shaded zone.

## §8.7 Application of Aerodynamic Data to Keels and Rudders

Because of their economic importance a great deal of effort has been expended on research into airplanes; basic yacht research is by comparison infinitesimal. Many types of lifting sections have been studied by aerodynamicists. None of them bears any resemblance to soft sails, but they do have a close affinity with the hull's two important underwater appendages the keel and the rudder. We can therefore benefit from the work of the airplane developers, and this section provides some aerodynamic data which are useful in the design and construction of keels and rudders.

There are certain limitations on the extent to which aerodynamic data can be used for boats and this must be made clear at the beginning. The diagrams given in this section are labelled with the Reynolds' number R, and strictly are only applicable to that value. Curves for three Reynolds' numbers are given. You will notice that lift is hardly affected by a threefold change in Reynolds' number, and drag only slightly so.

One's first reaction might be to wonder how data measured in air at 150 knots could be applicable to

keels in water at 6 knots. As pointed out in chapter 2, the shape of the flow is the same if the Reynolds' number is the same. Reynolds' number was defined in chapter 2 as the product of velocity times an 'appropriate' length divided by the kinematic viscosity: $R_n = Vl/v$. For airplane wings and keels the appropriate length is the chord length, the distance from the leading edge to the trailing edge. The Reynolds' number for an aircraft wing of chord length 2 m travelling at 100 knots (= 50 m/sec) is $(50 \times 2)/(1.46 \times 10^{-5}) = 7 \times 10^6$. Note that the divisor is the kinematic viscosity of air.

Now we calculate Reynolds' number for a keel of chord length 1.5 m in water with a kinematic viscosity of $1.11 \times 10^{-6}$ m$^2$/sec. If we assume a speed of about 3 m/sec or 6 knots, $R_n$ turns out to be $4.1 \times 10^6$. The difference in viscosity brings the $R_n$ for these seemingly very different situations sufficiently close together that lift coefficients are scarcely affected, although some change in drag coefficient is expected. Although it would be desirable to have data at lower Reynolds' numbers, the aeronautical data for $R_n$ between 3 and $9 \times 10^6$ is certainly useful.

Another problem in using wind tunnel data for keels and rudders arises because of the presence of the sea surface. At this interface the density of the fluid medium suddenly changes by a factor of about 1000. Air entrainment, which is common with transom-hung surface-piercing rudders, can radically change the lift and drag characteristics. Distortion of the flow by the presence of the hull can also have a considerable effect, especially for rudders. Despite these constraints aeronautical data are invaluable in keel and rudder design.

Figs. 8.16, 8.17 and 8.18 give wind tunnel measurements at three values of the Reynolds' number (here called R). These are 'section' data and refer to an infinite aspect ratio. For an elliptic lift distribution they may be converted to finite aspect ratios by means of the following formulae:

$$C_L = C_l(AR)/(AR + 2) \qquad C_D = C_d + C_L^2/\pi(AR)$$

where $C_l$ and $C_d$ are the section lift and drag coefficients given in figs. 8.16–8.18. AR is the aspect ratio defined as $AR = 2 \times (\text{depth})^2/\text{area}$. The 2 in this formula will be reasonable for keels and rudders under the hull and always totally immersed. For surface-piercing rudders this number could lie anywhere between 1 and 2, and of course calculations will be meaningless if any air entrainment is present.

The NACA number which specifies the wing section has the following meaning. NACA stands for

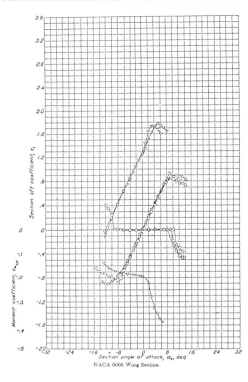

NACA 0006 Wing Section

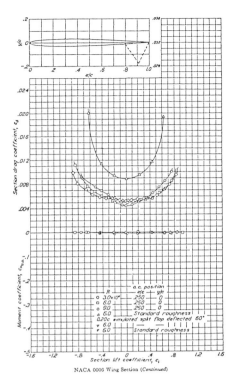

NACA 0006 Wing Section (Continued)

*Fig. 8.16* The 'section' lift, drag and moment coefficients for a NACA 0006 airfoil. 'Section' values refer to infinite aspect ratio; the flow is two-dimensional and devoid of end effects. The method of converting these results to finite aspect ratio is described in the text. The open circles correspond to a Reynolds' number of $3 \times 10^6$ which is the nearest value to conditions experienced by keels and rudders. The moment coefficient $C_M$ is a measure of the torque about the quarter chord point. The actual moment is related to the moment coefficient by the expression: $M = \frac{1}{2}\varrho V^2 A c C_M$. A is the keel area and c is the chord length. Note that for symmetrical sections the moment about the quarter chord point is zero until stall is reached. This means that a rudder pivoted at the quarter chord point would be perfectly balanced. (Diagram from Abbott and von Doenhoff, *Theory of Wing Sections.* Dover, New York 1959)

National Advisory Committee on Aeronautics. The first integer of the four-digit number is a measure of the camber, the second indicates the distance from the leading edge where the camber is a maximum. The last two digits indicate the section maximum thickness in % of the chord. If the mean line or centreline through the section is straight the camber is zero. Generally the only sections of interest in sailing are symmetrical sections with no camber: hence the first two digits are zero in figs. 8.16–8.18.

The aerodynamic centre (a.c. position) is equivalent to centre of lateral resistance and for these airfoils is at a quarter of the chord length back from the leading edge. The moment coefficient is a measure of the torque experienced about the quarter chord position. For smooth surfaces this is zero until the stall angle is reached, when the sections have a tendency to develop a torque about the quarter chord point which is in the direction of decreasing the angle of attack. This means that if a rudder were perfectly balanced by having the rudderpost at the quarter chord point, the onset of stall would be accompanied by a force on the tiller tending to decrease the angle of attack and lessening the stall.

Although there exist many families of airfoil sections the ones shown are possibly the most useful in yacht design. Their main characteristic is that they have a maximum thickness at 30% of the chord length back from the leading edge. So-called 'low drag sections' with the maximum thickness farther back are probably not so useful since they exhibit a low drag only for small angles of incidence, the drag

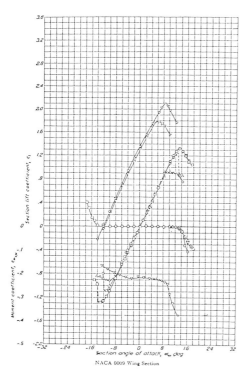

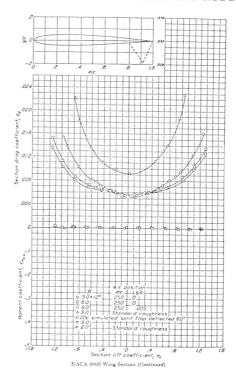

**Fig. 8.17** Lift and drag for NACA 0009 wing section. Note that roughness decreases the maximum obtainable lift and increases the drag. (From Abbott and von Doenhoff, *Theory of Wing Sections*)

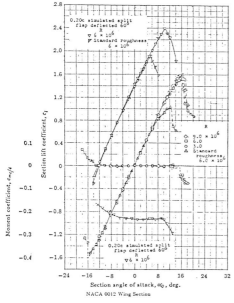

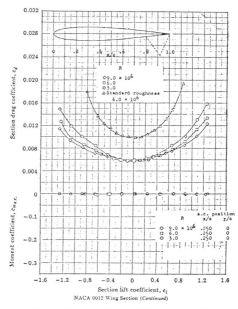

**Fig. 8.18** Lift, drag and moment characteristics for a 12% symmetrical section. Although these data have been measured in air using a wind tunnel they are equally valid for water flow at the same Reynolds' number. (From Abbott and von Doenhoff, *Theory of Wing Sections*)

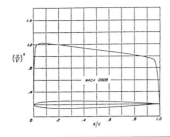

| x (per cent c) | y (per cent c) | $(v/V)^2$ | $v/V$ | $\Delta v_a/V$ |
|---|---|---|---|---|
| 0 | 0 | 0 | 0 | 3.992 |
| 0.5 | ..... | 0.880 | 0.938 | 2.015 |
| 1.25 | 0.947 | 1.117 | 1.057 | 1.364 |
| 2.5 | 1.307 | 1.186 | 1.089 | 0.984 |
| 5.0 | 1.777 | 1.217 | 1.103 | 0.696 |
| 7.5 | 2.100 | 1.225 | 1.107 | 0.562 |
| 10 | 2.341 | 1.212 | 1.101 | 0.478 |
| 15 | 2.673 | 1.206 | 1.098 | 0.378 |
| 20 | 2.869 | 1.190 | 1.091 | 0.316 |
| 25 | 2.971 | 1.179 | 1.086 | 0.272 |
| 30 | 3.001 | 1.162 | 1.078 | 0.239 |
| 40 | 2.902 | 1.136 | 1.066 | 0.189 |
| 50 | 2.647 | 1.109 | 1.053 | 0.152 |
| 60 | 2.282 | 1.086 | 1.042 | 0.123 |
| 70 | 1.832 | 1.057 | 1.028 | 0.097 |
| 80 | 1.312 | 1.026 | 1.013 | 0.073 |
| 90 | 0.724 | 0.980 | 0.990 | 0.047 |
| 95 | 0.403 | 0.949 | 0.974 | 0.032 |
| 100 | 0.063 | 0 | 0 | 0 |

L. E. radius: 0.40 per cent c

NACA 0006 Basic Thickness Form

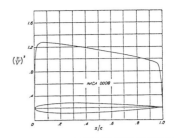

| x (per cent c) | y (per cent c) | $(v/V)^2$ | $v/V$ | $\Delta v_a/V$ |
|---|---|---|---|---|
| 0 | 0 | 0 | 0 | 2.900 |
| 0.5 | ..... | 0.792 | 0.890 | 1.795 |
| 1.25 | 1.263 | 1.103 | 1.050 | 1.310 |
| 2.5 | 1.743 | 1.221 | 1.105 | 0.971 |
| 5.0 | 2.369 | 1.272 | 1.128 | 0.694 |
| 7.5 | 2.800 | 1.284 | 1.133 | 0.561 |
| 10 | 3.121 | 1.277 | 1.130 | 0.479 |
| 15 | 3.564 | 1.272 | 1.128 | 0.379 |
| 20 | 3.825 | 1.259 | 1.122 | 0.318 |
| 25 | 3.961 | 1.241 | 1.114 | 0.273 |
| 30 | 4.001 | 1.223 | 1.106 | 0.239 |
| 40 | 3.869 | 1.186 | 1.089 | 0.188 |
| 50 | 3.529 | 1.149 | 1.072 | 0.152 |
| 60 | 3.043 | 1.111 | 1.054 | 0.121 |
| 70 | 2.443 | 1.080 | 1.039 | 0.096 |
| 80 | 1.749 | 1.034 | 1.017 | 0.071 |
| 90 | 0.965 | 0.968 | 0.984 | 0.047 |
| 95 | 0.537 | 0.939 | 0.969 | 0.031 |
| 100 | 0.084 | ..... | ..... | 0 |

L. E. radius: 0.70 per cent c

NACA 0008 Basic Thickness Form

Fig. 8.19 In order to make sections with the characteristics of the last three diagrams one needs to know the coordinates of the points defining the shape. This diagram supplies that information for the NACA 0006 and 0008 sections. y is the half-thickness (% of c) at a distance x (% of c) measured back from the leading edge. (From Abbott and von Doenhoff, *Theory of Wing Sections*)

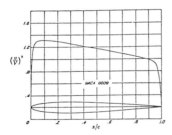

| x (per cent c) | y (per cent c) | $(v/V)^2$ | $v/V$ | $\Delta v_a/V$ |
|---|---|---|---|---|
| 0 | 0 | 0 | 0 | 0.595 |
| 0.5 | ..... | 0.750 | 0.866 | 1.700 |
| 1.25 | 1.420 | 1.083 | 1.041 | 1.283 |
| 2.5 | 1.961 | 1.229 | 1.109 | 0.963 |
| 5.0 | 2.666 | 1.299 | 1.140 | 0.692 |
| 7.5 | 3.150 | 1.310 | 1.145 | 0.560 |
| 10 | 3.512 | 1.309 | 1.144 | 0.479 |
| 15 | 4.009 | 1.304 | 1.142 | 0.380 |
| 20 | 4.303 | 1.293 | 1.137 | 0.318 |
| 25 | 4.456 | 1.275 | 1.129 | 0.273 |
| 30 | 4.501 | 1.252 | 1.119 | 0.239 |
| 40 | 4.352 | 1.209 | 1.100 | 0.188 |
| 50 | 3.971 | 1.170 | 1.082 | 0.151 |
| 60 | 3.423 | 1.126 | 1.061 | 0.120 |
| 70 | 2.748 | 1.087 | 1.043 | 0.095 |
| 80 | 1.967 | 1.037 | 1.018 | 0.070 |
| 90 | 1.086 | 0.984 | 0.982 | 0.046 |
| 95 | 0.605 | 0.933 | 0.966 | 0.030 |
| 100 | 0.095 | 0 | 0 | 0 |

L. E. radius: 0.89 per cent c

NACA 0009 Basic Thickness Form

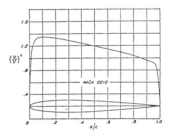

| x (per cent c) | y (per cent c) | $(v/V)^2$ | $v/V$ | $\Delta v_a/V$ |
|---|---|---|---|---|
| 0 | 0 | 0 | 0 | 2.372 |
| 0.5 | ..... | 0.712 | 0.844 | 1.618 |
| 1.25 | 1.578 | 1.061 | 1.030 | 1.255 |
| 2.5 | 2.178 | 1.237 | 1.112 | 0.955 |
| 5.0 | 2.962 | 1.325 | 1.151 | 0.690 |
| 7.5 | 3.500 | 1.341 | 1.158 | 0.559 |
| 10 | 3.902 | 1.341 | 1.158 | 0.479 |
| 15 | 4.455 | 1.341 | 1.158 | 0.380 |
| 20 | 4.782 | 1.329 | 1.153 | 0.318 |
| 25 | 4.952 | 1.309 | 1.144 | 0.273 |
| 30 | 5.002 | 1.284 | 1.133 | 0.239 |
| 40 | 4.837 | 1.237 | 1.112 | 0.188 |
| 50 | 4.412 | 1.190 | 1.091 | 0.150 |
| 60 | 3.803 | 1.138 | 1.067 | 0.119 |
| 70 | 3.053 | 1.094 | 1.046 | 0.094 |
| 80 | 2.187 | 1.040 | 1.020 | 0.069 |
| 90 | 1.207 | 0.960 | 0.980 | 0.045 |
| 95 | 0.672 | 0.925 | 0.962 | 0.030 |
| 100 | 0.105 | ..... | ..... | 0 |

L. E. radius: 1.10 per cent c

NACA 0010 Basic Thickness Form

Fig. 8.20 Coordinates for 9 and 10% sections. v/V is the ratio of the flow velocity at point x to that in the undisturbed flow. The pressure distribution is determined by $(v/V)^2$ through the equation for the pressure coefficient: $C_P = 1 - (v/V)^2$. (Data from Abbott and von Doenhoff, *Theory of Wing Sections*)

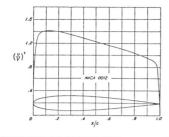

| $x$ (per cent $c$) | $y$ (per cent $c$) | $(v/V)^2$ | $v/V$ | $\Delta v_a/V$ |
|---|---|---|---|---|
| 0 | 0 | 0 | 0 | 1.988 |
| 0.5 | ..... | 0.640 | 0.800 | 1.475 |
| 1.25 | 1.894 | 1.010 | 1.005 | 1.199 |
| 2.5 | 2.615 | 1.241 | 1.114 | 0.934 |
| 5.0 | 3.555 | 1.378 | 1.174 | 0.685 |
| 7.5 | 4.200 | 1.402 | 1.184 | 0.558 |
| 10 | 4.683 | 1.411 | 1.188 | 0.479 |
| 15 | 5.345 | 1.411 | 1.188 | 0.381 |
| 20 | 5.737 | 1.399 | 1.183 | 0.319 |
| 25 | 5.941 | 1.378 | 1.174 | 0.273 |
| 30 | 6.002 | 1.350 | 1.162 | 0.239 |
| 40 | 5.803 | 1.288 | 1.135 | 0.187 |
| 50 | 5.294 | 1.228 | 1.108 | 0.149 |
| 60 | 4.563 | 1.166 | 1.080 | 0.118 |
| 70 | 3.664 | 1.109 | 1.053 | 0.092 |
| 80 | 2.623 | 1.044 | 1.022 | 0.068 |
| 90 | 1.448 | 0.956 | 0.978 | 0.044 |
| 95 | 0.807 | 0.906 | 0.952 | 0.029 |
| 100 | 0.126 | 0 | 0 | 0 |
| L.E. radius: 1.58 per cent $c$ | | | | |

NACA 0012 Basic Thickness Form

*Fig. 8.21* Coordinates for a 12% symmetrical section. The last column on the right gives the change in relative velocity for a change in angle of attack such that the lift coefficient changes by 1.0. It can be used for calculating the pressure distribution at an angle of attack other than zero. (For details see Abbott and von Doenhoff, *Theory of Wing Sections*)

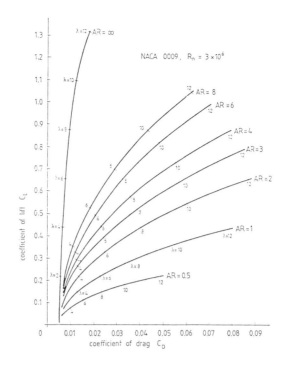

*Fig. 8.22* Polar curves for the symmetric wing section NACA 0009. These have been calculated from the data of fig. 8.17 and corrected for finite aspect ratio as described in §8.7. The numbers on the curves are the angle of attack or leeway angle in degrees.

usually being greater than other sections at larger incidence angles.

To construct sections, the coordinates of a number of points along the outline are required; this is the purpose of figs. 8.19–8.21. y is the half-thickness (as a % of the chord length) at various distances x (as a % of the chord length) from the leading edge. The leading edge radius of curvature is given at the bottom of each table.

v/V is the ratio of the velocity of flow at position x to that of the free stream flow well away from the foil: thus $(v/V)^2$ determines the pressure distribution (see the definition of the pressure coefficient $C_p$ in §8.3). $\Delta v_a/V$ is the relative change in v for a change of angle of attack such that it produces a change in lift coefficient of 1.0.

From the above results it is possible to calculate the lift and drag of a keel of any aspect ratio under ideal flow conditions for various leeway angles. This has been done for the NACA 0009 section (fig. 8.22); $\lambda$ is the leeway angle in degrees.

The actual lift and drag forces are related to the lift and drag coefficients by the following formulae:

$$\text{Lift} = L = 1/2\varrho SV^2 C_L \qquad \text{Drag} = L = 1/2\varrho SV^2 C_D$$

$\varrho$ is the density of water, S is the area, V the speed, and $C_L$ and $C_D$ the coefficients calculated by the formulae given earlier from the section coefficients in figs. 8.16–8.18. Note that the area to use in these formulae is the geometric area and not the wetted surface area which is a little more than twice as great since it includes both sides.

In International Units, L(newtons) = $500S(m^2)V^2(m/s)^2 C_L$. In non-metric units, L(pounds) = $0.995S(ft^2)V^2(ft/s)^2 C_L$ with similar expressions for drag.

Fig 8.23 gives the 'lift slope' as a function of aspect ratio for all three sections. The approximate stall limits are shown and these should be borne in mind especially when designing rudders. Notice that thinner sections stall at smaller angles of attack.

## §8.8 Estimation of Hull Polars for a Light Displacement Boat

The assumptions made in this calculation and their justification are fully covered in §3.4.1. Briefly, they are that the hull produces only drag and no lift, and this drag is independent of leeway. Hull lift is produced entirely by the keel which also gives a contribution to the overall drag due to a combination of boundary layer drag and vortex drag. The aeronautical data of §8.7 was used for the keel contribution assuming it to have a NACA 0009 section.

The drag of the hull alone (which is taken as independent of leeway) is calculated from a formula consisting of two terms. The first term accounts for the surface friction drag and the second the residual drag. This is a very simple parametrisation of hull resistance, but it has the necessary characteristics to typify a light-displacement centreboard boat.

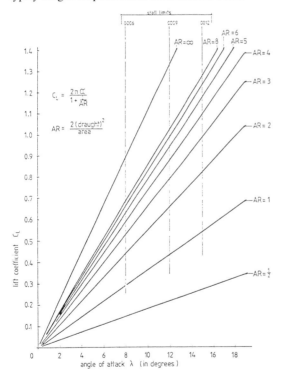

*Fig. 8.23* Rigid lifting sections generally have a lift coefficient which is proportional to angle of attack. The constant of proportionality or the lift slope depends on the aspect ratio: it is a maximum for infinite aspect ratio, reducing to about one-third for an aspect ratio of 1.0. The vertical dashed lines give the approximate stalling points of the three symmetrical sections indicated.

The wetted area was taken as 3.375 m² and the LWL as 3.7 m. This means that the Reynolds' number for flow along the full length of the hull at 4 knots is about $7 \times 10^6$. Reference to fig. 2.10 shows that the flow over the whole hull will be hydrodynamically smooth at this speed if $k/L < 2 \times 10^{-5}$. That is, surface irregularities must be no greater than about 0.1 mm. Since this is certainly feasible, we can assume boundary layer flow over the hull and use the surface friction drag coefficients given by fig. 2.10. For various speeds the coefficient $C_F$ may be read off fig. 2.10 and the friction drag $D_F$ calculated from $D_F = \frac{1}{2}\varrho SV^2 C_F$. We also know from §2.8 that drag due to turbulent boundary layer flow is proportional to $V^{9/5}$, or $D = kV^{9/5}$. Comparing this expression with the drag calculated using a wetted area $S = 3.375$ m² gives a value for k. If this is done for a range of speeds k is found to be constant (as it should be) and has the value 6.26.

The second term in the drag formula is assumed to depend on $V^3$ (see fig. 3.28). So the drag formula has the form $D = 6.26V^{9/5} + bV^3$ where b is a constant. b was determined by making the assumption that the residual resistance and the surface friction resistance are equal at $V/L^{1/2} = 1.27$ m$^{1/2}$/sec. This gives a value for b of 2.17. Thus the final hull drag formula is: $D = 6.26V^{9/5} + 2.17V^3$. This is shown in fig. 3.28.

Using a centreboard aspect ratio of 6 the lift and drag coefficients for the keel may be read off fig. 8.22 and the actual lift and drag calculated for various speeds. The centreboard drag was then added to the hull drag giving a set of values for total hull drag and lift as a function of leeway angle.

## §8.9 Determination of Maximum Heeling Force for Crew Ballasted Boat

Fig. 8.24 shows the heeling and righting forces on a small boat with two crew, one of whom is on a trapeze. For a boat which is sailed upright, hull buoyancy acts upward through the hull centre of gravity and so the pair of forces consisting of hull weight and buoyancy contributes nothing to the righting moment. The righting arm is therefore given by the perpendicular distance from the centre of buoyancy to the centre of gravity of the crew. The righting moment is therefore $W \times d$ where W is the crew weight. This must be equal to the heeling moment which from fig. 8.24 is $F_H \times h$.

If each crew member 'weighs' 70 kg then W = 1400 newtons. Reasonable values for d and h are 1.7 m and 3 m respectively, so the maximum value of $F_H$ which

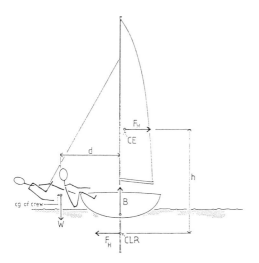

*Fig. 8.24* Maximum usable total sail force is determined by the heeling force component and the available righting moment. Small boats are best sailed upright so the righting moment comes entirely from the disposition of the crew. Under these circumstances movement of the centre of buoyancy and hence hull shape makes no contribution to the righting moment. At equilibrium $Wd = F_H h$. W is the crew weight and $F_H$ the heeling force.

can be sustained is $F_H = 1400 \times 1.7/3 = 793N$. Using a slight extension of fig. 3.31 you should be able to show that $F_H = F_T \cos(\varepsilon_h - \lambda)$. Since best $V_{mg}$ corresponds to values of $\varepsilon_h$ and $\lambda$ of about 12° and 3.5° respectively, we finally get for $F_T$ the value 802 N. An arc of this radius is plotted in fig. 3.29.

## §8.10 What the Skipper Controls and What the Boat Decides for Itself

In chapter 3 it was pointed out that, apart from sail shape controls, all that the sailor controls directly is the angle between the apparent wind and the boat's heading, and the angle of incidence of wind on sails. When the crew change these variables via the rudder and sheets a whole host of other quantities vary in a manner not determined by the crew but by the design of the boat. One might think of the boat as being sailed by the crew plus the ghost of the designer. Even the two quantities that the skipper does have control of are not independent since angle of attack is determined both by sheeting angle and course steered, as is clear from fig. 3.33.

Apart from the small effect of leeway, the skipper can, however, maximise driving force since this is a

function only of $\alpha$ and $\beta$ as is clear from fig. 3.35. Of course this does not necessarily maximise $V_{mg}$ unless $\beta$ is chosen correctly. Fig. 8.25 shows this chain of control in schematic form. Theoretical analyses which plot the relationship between one variable and another are of little use in practice unless one of those variables is under the sailor's direct control.

## §8.11 Sailing Hydrofoils

The maximum speed of a yacht is determined by its hull resistance which consists mainly of two parts, surface friction drag and wavemaking drag. Planing hulls effectively reduce the latter and so do long narrow hulls like those of racing catamarans or rowing shells. A much more radical way of reducing hull drag is to use hydrofoils, but like many great ideas they introduce a panorama of completely different problems.

The lift force produced by a hydrofoil, like that of an airplane wing or yacht's keel, depends on the product of four quantities: density of the fluid (air or water), the square of the speed, area of the foil and the lift coefficient. The last depends sensitively on the shape of the foil and its angle of attack to the flow, but hardly at all on the speed.

Since the density of water is about 1000 times that of air, a hydrofoil will generate about the same lift force as an airplane wing at only one-thirtieth of the speed. If one does the calculation it turns out that the lift generated by a hydrofoil moving at 5 knots is approximately 320 kg/m² (65 lb/ft²) of hydrofoil area. If we double the speed the lift goes up by a factor of 4. Clearly we don't need much foil area to hold up a light catamaran, for instance, so a set of horizontal hydrofoils will certainly lift a boat. But what is to stop it rising higher and higher? As soon as the foils come out of the water they find themselves in another fluid medium, the air, wherein the density is a thousandfold less. The lift suddenly drops by a factor of 1000 and the boat crashes back into the sea. Such unstable rabbit hops can be prevented in a number of ways. In principle, the depth of submergence could be controlled by an elevator just as an aircraft pilot does. So in addition to the tiller in one hand, the helmsman could have a joystick in the other and the main sheet . . . ? Clearly some form of automatic height control is desirable.

Some commercial powered hydrofoils use sonic sensing to keep their foils at a fixed depth below the average sea surface. A cruder method, known as the Hook hydrofoil, uses a kind of water ski on an arm

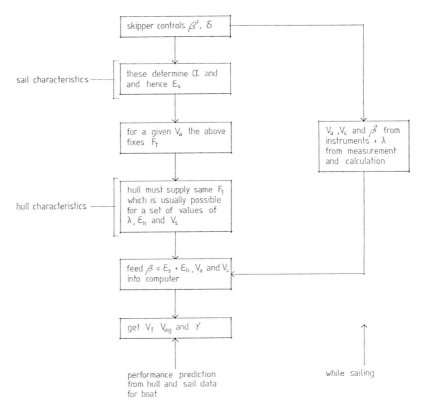

*Fig. 8.25* Apart from sail shape controls, all that the crew can normally adjust is the angle between the apparent wind and the boat's heading $\beta'$ and the sheeting angle of the sail $\delta$. Once these two adjustments are made the sail characteristics determine the sail drag angle and total sail force. The hull must then move in such a way as to generate this same total force. The boat does this automatically, producing the necessary force at a leeway angle and boat speed that are determined by the shape of its underwater body. The resultant leeway and drag angle are not under the direct control of the crew. If sail and hull polars are known the resultant windward performance can be calculated. Instrument measurements can also be used while sailing to compute $V_{mg}$.

projecting in front of the boat to sense the water surface. This arm adjusts the foil's angle of incidence by means of a mechanical linkage. A much simpler way of providing automatic height stability is to use angled surface-piercing foils (fig. 8.26a). The foils each produce lift forces which combine to give a resultant total lift L opposing the boat's weight W. If L is greater than W the boat rises, but as it does so the submerged foil area decreases so the lift decreases and the boat comes to equilibrium when L = W. For a given speed it is a simple matter to design the foils so that the hull will rise any prescribed distance above the water surface.

The same result can be obtained using ladder foils (fig. 8.26b). This was the method used on one of the earliest patents by Forlanini in 1898 and subsequently developed by Alexander Graham Bell, of telephone fame. In early designs the foils were horizontal giving a discontinuous change in lift as a foil came out of the water. Angling them as in (b) causes the lift to change more smoothly. Although there are other methods of getting height stability, sailing hydrofoils have used the above two methods almost exclusively.

The hydrofoil can counteract heeling in a unique way. To understand how it works it is a good idea to look back at fig. 8.24 which shows how an ordinary small boat remains upright. A heeling moment is produced by the equal hull and sail forces $F_H$ separated by the distance h. This overturning moment has the magnitude h × $F_H$. The righting moment is supplied by the weight of the crew acting through a CG located at a distance d from the

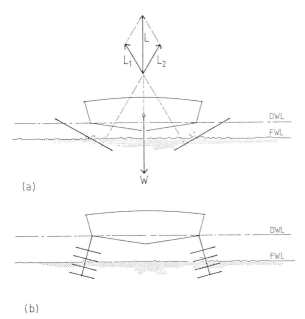

(a)

(b)

Fig. 8.26 Two systems of surface-piercing foils which give height stability. The lift forces act perpendicularly to the plane of the foils and their lines of action are shown by the dashed lines in (a). If the resultant lift L is too great the boat rises, reducing the area of foil in the water and thereby limiting the lift. The hydrofoil therefore adjusts itself automatically to the equilibrium point where L = W.

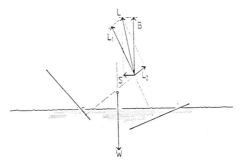

Fig. 8.27 On most points of sailing an unwanted side force is produced by the sails. This must be opposed by the hull force generated by water flow. Unlike a displacement boat with a keel, a hydrofoil generates more side force the more it heels. The resultant hydrofoil lift is now slewed to windward giving a force with both a vertical component to balance the weight and a horizontal component to offset the heeling force.

buoyancy force. The magnitude of the righting moment is $W \times d$ and is equal to $h \times F_H$ when the boat is sailing upright. Clearly these heeling and righting moments arise from the fact that the lines of action of the pairs of forces are separated: the greater the distance between them the greater the moment. If it could be arranged so that the lines of action of these forces all passed through one point there would be no moments at all and the boat would not heel no matter how hard the wind blew.

This apparent paradox may be accomplished with an appropriately designed hydrofoil. Fig. 8.26a could be thought of as a power boat fitted with hydrofoils. The resultant lift force L is exactly equal to the weight W and acts along the same line, so there are no heeling or righting moments. A sailing hydrofoil has, of necessity, a heeling moment. Let us assume that the crew does not move so that the centre of gravity remains in the middle of the boat. The hydrofoil boat will now tend to heel and in so doing the leeward foil will immerse a greater area in the water, so its contribution to the total lift will be greater than that

from the windward foil. This distortion of the lift forces is shown in fig. 8.27. The hydrofoil lift forces $L_1$ and $L_2$ combine to produce a resultant L which is now tilted to windward. L may thus be decomposed into two components: the vertical one B must be exactly equal and opposite to the weight W; the horizontal component S must be equal to the heeling force produced by the sail. W and B form a righting moment because they are parallel forces which do not act along the same line. The heeling moment is produced by the 'hull' side force S acting with the sail heeling force $F_H$ (not shown in the diagram). If $F_H$, which acts through the centre of effort of the sail, is above S, then there will be a clockwise heeling moment which must be balanced by the righting moment just as in an ordinary boat. But here there is a fundamental difference. In an ordinary boat the height at which the side force S acts is of necessity below the water level. In a hydrofoil boat this height is determined by the intersection of the line of action of the lift forces and can be changed simply by changing the foil dihedral angle. If S is adjusted to be at the same height as the CE of the sail, then $F_H$ and S lie along the same line and there is no heeling moment whatsoever! Even stranger still is the situation where S is above the sail's CE: the boat would then heel to windward.

Although this lack of heeling sounds idyllic, a small amount of heel is desirable in that it gives the helmsman 'feel' and also reduces leeway. A hydrofoil boat sailing upright can generate no side force unless it makes leeway. When it does so the flow direction with respect to the foil differs substantially between the leeward and windward foils because of their dihedral angle. $L_1$ and $L_2$ are no longer equal as in fig. 8.26 so their resultant $L$ is slewed to windward, thereby creating a component of side force to counteract the sail side force $F_H$. A better way of generating side force is to let the boat heel as in fig. 8.27 where the different immersion lengths of the foils create the necessary asymmetry to produce a side force. Unlike a normal boat, a hydrofoil can generate 'hull' side force simply by heeling. In general the total side force will be a combination of that produced by heeling and by leeway. To just what extent the design should be adjusted to reduce leeway or reduce heel is one of those questions that can only be answered by competition between many similar boats resulting in a sort of Darwinian evolution of the fittest.

For surface-piercing foils, pitch stability is provided in the same way as is roll stability. Any trim down by the bow will result in increased foil immersion area and hence increased lift which will tend to bring the boat back to an even keel. So effective are surface-piercing foils in smooth water that one experiences an unaccustomed feeling of stability when sailing.

As hinted at the beginning of this section, surface-piercing foils are simple, practical and the most often used system on sailing hydrofoils, but there are problems. The most severe is that of *ventilation* or *air entrainment* due to air being sucked down the low pressure side of the foil (or rudder). This gives rise to a sudden loss of lift which can result in a 'sea crash'. Even when ventilation does not occur, the low pressure on the upper side of the foil tends to produce a depression in the sea where it cuts the surface. This

*Fig. 8.28* A successful sailing hydrofoil built by David Knaggs of Auckland, New Zealand. Although the design is simple, surface-piercing foils suffer from ventilation in which air is sucked down the low pressure side producing a sudden loss of lift. The problem can be reduced by the use of 'fences' which may be just visible on the leeward main foil. Though necessary, they increase drag. (*Photo:* Sea Spray *magazine*)

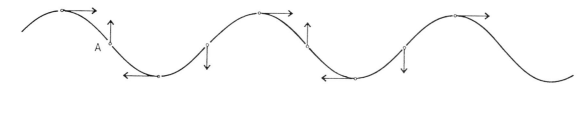

direction of waves

*Fig. 8.29* The forces on a hydrofoil in waves are quite different from those on a buoyant boat. This is because the hydrofoil lift results from the *motion* of the water: as the foil moves forward the angle of attack is a result of its own forward motion plus the local motion of the water. Thus at a point A the angle of attack is increased and the lift therefore increased for a hydrofoil going in the direction of the waves, whereas the reverse is true if it is going against the waves. Totally submerged foils are not so greatly affected by wave action, but then height control becomes a more complex problem.

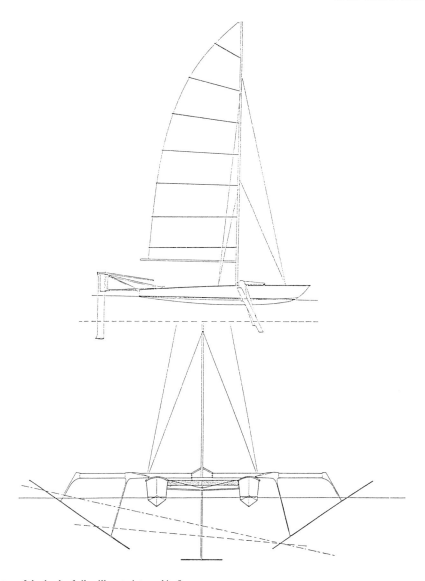

*Fig. 8.30* Geometry of the hydrofoil sailboat pictured in fig. 8.28. The main lifting foils are Wortman 77-HY-120 section of 115 mm chord. Foil support struts, rudder and rear stabiliser are NACA 0012 section of 170 mm chord.

produces a wake and hence added drag, somewhat more than one might expect from such a small and narrow object moving through water. The problem can be reduced by adding fences to the foils, thin fins perpendicular to the long axis of the foil as visible on the leeward foil in fig. 8.28. Although they help in reducing ventilation they add considerably to the drag.

It is thus evident that fully submerged foils which are free of these problems would be more efficient. But now we lose automatic height control and would have to introduce the considerable complication (and

225

weight) of surface sensing and incidence control mechanisms.

Buoyancy in waves is fundamentally different for a hydrofoil than for a displacement boat. The ideal arrangement would be to have the foil lifting surfaces well down where the wave amplitude is less so that the boat would 'platform', i.e. stay at a constant height above the mean surface irrespective of wave action. There is no need to dwell on the obvious advantages of this when sailing to windward in a chop. A surface-piercing hydrofoil, even if it doesn't ventilate, is affected by wave action though the motion is smoother than for a displacement boat. The reason for the fundamental difference in buoyancy characteristics from those of a displacement boat arises because the lift is dynamic in origin rather than static; it therefore depends on the local motion of the water. Fig. 8.29 shows the orbital motion of water particles in a wave. Although waves march in lines across the ocean, the particles of water of which they are composed move only in circles of diameter equal to the wave height. It is only the shape of the sea which moves, not the sea itself. Water motion is forward in the crest, upwards on the face of the wave and backwards in the trough.

When the hydrofoil is near point A (fig. 8.29) the angle of attack of the foil increases giving increased lift, so the boat will have a tendency to shoot out of the face of a wave and dig in to the back of the next wave. This effect can be minimised by using the largest possible aspect ratio, both to maximise the mean depth of immersion of the foil for a given lift area and to minimise the rate of lifting area increase as the foil immerses farther into a wave crest.

In choppy seas the change in immersion depth of one surface-piercing foil can be rapid and different from the others. This tends to produce an uncontrollable yawing of about 10° each way. Wave action also seems to induce ventilation in many instances.

Although there are a few sailing hydrofoil enthusiasts in the world, their number is very small and so this interesting and exciting form of sailing is little developed and still in its infancy. Figs. 8.28 and 8.30 show a successful first generation flying hydrofoil. Foils were added to a standard 5.3 m catamaran; sail area is 9.3 m$^2$ and all-up weight including crew is 198 kg. The main foil dihedral, sweep angle and angle of attack are 35°, 15° and 3.5° respectively. Foil support struts, rudder and rear stabiliser are approximately NACA 0012 section of 170 mm chord. The main lifting foils are Wortman 77-HY-120 section of 115 mm chord.

A major problem with hydrofoils is obtaining a sufficient strength/weight ratio. This is not easy as the entire weight of the boat may end up being supported on a few square centimetres of foil area. The boat featured here used fairly standard glassfibre and resin techniques for foil construction, the total weight of foils plus associated supports being 41 kg. The boat pictured will fly to windward in 12 knots of wind with most of the leeward foil immersed and most of the windward one out of the water. Only in fairly fresh winds is it likely to be faster than a normal catamaran.

# Chapter Nine

# THREE DIMENSIONAL LIFT THEORY

## §9.1 Circulation, Vorticity and Lift

The very earliest attempt to explain lift scientifically as opposed to ascribing it to mysterious psychic or occult powers possessed only by birds, was that of Newton whose model was exactly that of the momentum change theory described in §3.3.2 and §8.4. As we saw, this is a useful qualitative way of visualising the development of lift while sailing, but it does not conform with quantitative reality.

Probably the first person to understand lift properly was the Englishman F. W. Lanchester. A paper describing his theories was rejected by both the Royal Society and the Physical Society in 1897. It was ten years before his ideas were published in a book. Unfortunately his writings were not easy reading; he did not use mathematical description and those not familiar with his line of thought found him impossible to understand. His reputation was rescued by the German scientist Ludwig Prandtl who said: 'Lanchester's treatment is difficult to follow, since it makes a very great demand on the reader's intuitive perceptions, and only because we had been working on similar lines were we able to grasp Lanchester's meaning at once.' The two great ideas conceived by Lanchester were: the idea of circulation as the cause of lift, and the idea of tip vortices as the cause of that part of the drag now known as vortex drag or induced drag.

The purpose of this section is to give a somewhat more quantitative definition of these concepts than was made in §4.1. Vortex flow conjures up in most people's minds a vision of a whirlpool, which is confined mainly to the surface of water. The kind of vortex flow of interest to us is unseen, like the 'invisible and creeping wind' of the Shakespeare quotation at the beginning of chapter 4. A *line vortex* is a line of fluid particles spinning on a common axis and carrying with them a swirl of particles which flow around in circles. A cross-section of such a vortex shows a spinning point outside of which is streamline flow in concentric circles as shown in fig. 9.1. This is of course a mathematical idealisation since points and lines have no extent. Real vortices are common in nature but differ in that they have a core of fluid which is rotating like a solid. This means there is no sliding motion of fluid particles over one another within the core, whereas outside the core the flow is like that of a line vortex. As we shall now see, the speed of flow varies inversely as the distance from the centre so adjacent particles of water or air are sliding over one another.

Since a vortex is a stable rotation of fluid, the centrifugal force experienced by the fluid particles must be balanced by a pressure force in towards the centre. The pressure at the centre of a vortex must therefore be lower than that at the outside. Bernoulli's rule tells us that where the pressure is lower the velocity is higher, so we conclude that for a vortex to have a stable existence the velocity must increase toward the centre. A full mathematical analysis shows that the velocity is inversely proportional to the radius. This is normally written mathematically: $v = K/2\pi r$. $v$ is the tangential flow velocity and $r$ the distance out from the centre of the vortex. K is constant for a given vortex and is known as the circulation. The circulation may be visualised in another way. If one follows the track of a particle of fluid in a vortex it travels in a circle of radius $r$ and the distance covered in one full rotation is therefore

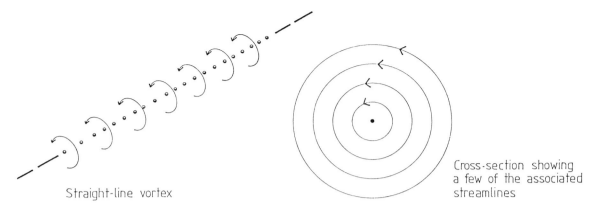

Straight-line vortex

Cross-section showing
a few of the associated
streamlines

*Fig. 9.1* A straight-line vortex consists of a chain of fluid particles spinning on a common axis and carrying around with them a swirl of fluid particles which move in circles. An 'ideal' vortex has a central core of zero extent. A real vortex has a central core of fluid rotating about its own axis like a solid. Outside of this concentric layers of fluid move at

different speeds continually sliding over one another. The tangential speed varies inversely as the distance from the centre. Because of the increased speed near the centre the pressure there is less than at the periphery, according to Bernoulli's rule.

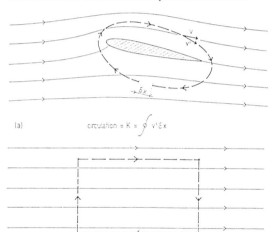

(a)

circulation = K = $\oint v' \delta x$

(b)

*Fig. 9.2* Take any closed path in a fluid flow. For each small element of path $\delta x$, determine the component of the flow velocity parallel to $\delta x$. Multiply this component $v'$ by $\delta x$. If $v'$ is in the same direction as $\delta x$ then the product is taken as positive; if in the opposite direction as negative. Do this for every $\delta x$ making up the complete closed circuit and add up all the results taking due account of sign. This mathematical quality is called a line integral and is here the definition of the circulation K in the fluid flow. In a uniform flow like that of (b) the circulation is zero. This can be seen by taking a rectangular path for the line integral calculation. There is no component of velocity parallel to the ends so they do not contribute to the circulation. The top and bottom make equal and opposite contributions so the circulation is zero. This conclusion is independent of the choice of path used for the calculation since it is a property only of the fluid flow.

$2\pi r$. The product of this distance and the velocity is just the circulation since $2\pi r v = K$ from above. The idea of circulation is not just limited to vortices, however. Take any closed path in a fluid flow and multiply the velocity component parallel to the path by an increment of path length for all such increments around the closed path, add up the results and the answer is the circulation! Fig. 9.2 may make this a little clearer. The circulation is the sum of incremental contributions $v' \delta x$ around a closed path, known mathematically as a *line integral*. It is easy to see that in uniform flow the circulation is zero. Fig. 9.2b shows circulation calculated around a rectangular path in a uniform flow. The contribution to the circulation from the vertical ends of the rectangle is zero since the velocity has no component parallel to them. The contributions from the horizontal sides of the rectangle are equal and opposite since on the lower side the velocity and the path are in the same direction whereas at the top they are in opposite directions. The same conclusion would have been drawn if the rectangle had been narrow and high, or indeed if the path had been a circle. In other words the circulation determined by this procedure does not depend on the shape of the path used to evaluate it, but is a property only of the shape of the fluid flow.

## §9.2 Relationship between Lift and Circulation

Fig. 9.3 shows the flow around a lifting surface of unit

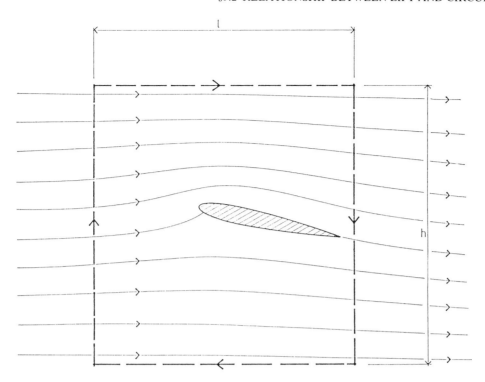

*Fig. 9.3* If the circulation around a lifting surface is calculated using a rectangular path of length l and height h, the only contribution comes from the right hand end. This is wh, where w is the downwash velocity. Downwash is simply the component parallel to h of the fluid flow. The circulation is therefore K = wh. Lift is equal to the rate of change of vertical component of momentum, which is the mass per second of fluid going through end h. If unit length of foil is considered this has the value $\varrho$Vhw. So lift is given by $\varrho$VK. This shows the direct and fundamental relationship between lift of a sail and the circulation in the flow that its shape produces.

length. If the circulation around it is calculated by following the dashed rectangular path of length l and height h, then the only side which makes a contribution is the right hand one of length h. At top and bottom the flow is parallel to the path but of opposite sign so these contributions cancel; at the left hand end the flow has no component parallel to the path. The vertical component of flow at the right hand end is simply the downwash velocity w averaged over the length h. So the circulation is just K = wh.

Now the lift is the rate of change of the vertical component of momentum. A mass of fluid flows in each second at the left and gets pushed downwards with a velocity w. This mass of fluid is equal to $\varrho$Vh, where V is the speed of flow and h is the area (remembering that we are considering unit length of foil). Thus the rate of change of the vertical component of momentum, which is the lift, is given by L = $\varrho$Vhw. But we have just shown that K = hw, so we have finally: L = $\varrho$VK

This is an extremely important result in fluid flow theory and forms the basis for the lifting line theory of sail and wing forces.

Just as lift is measured by the product of the undisturbed flow velocity and the circulation, so drag, which is at right angles to lift, is determined by the product of the downwash velocity and the circulation (fig. 9.4). Thus we have L = $\varrho$VK and D = $\varrho$wK. For an elliptic lift distribution the downwash velocity w is given by w = K/4s where s is the semi-span (§4.1.1). Substituting this value of w into the expression for drag, we have drag proportional to $K^2$. But lift is proportional to circulation K, so we see that *vortex drag is proportional to the square of the lift*. A full mathematical analysis gives $C_{DV} = C_L^2/\pi(AR)$ for an elliptic lift distribution.

229

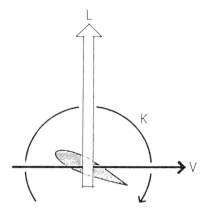

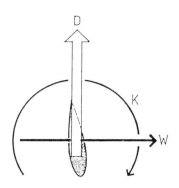

$$l = \rho VK$$

$$L = \int_{-s}^{s} \rho VK \, dy$$

$$d_v = \rho wK$$

$$D_v = \int_{-s}^{s} \rho wK \, dy$$

*Fig. 9.4* Just as lift is related to circulation by an expression proportional to the undisturbed fluid flow velocity V, so is vortex drag related to the downwash velocity w, as can be seen here by simply rotating the diagram through 90°.

## §9.3 Helmholtz's Theorems of Vortex Motion

Before discussing these theorems it is well to point out the surprising fact that the fluid in an ideal vortex flow is *irrotational*. What can this possibly mean? Imagine a large Ferris wheel at a fun fair. People sit in little cars and the whole thing rotates in a vertical plane but the people always remain vertical; they do not rotate about their own axes. They are irrotational but moving in a circle. This is exactly the motion of fluid particles in a vortex. However, the core of a vortex contains fluid particles rotating like a solid, just like the hub of the Ferris wheel. This is rotation about its own axis and is truly called *rotational*.

The first theorem of Helmholtz maintains that the strength of a vortex remains constant along its length. The strength of a vortex is the magnitude of the circulation around it. The second theorem states that a vortex cannot end in the fluid. It must form a closed loop or end on a solid surface. A vortex tube cannot change strength unless it branches (fig. 9.5). This, then, is why a change in circulation up a sail results in the shedding of vortices.

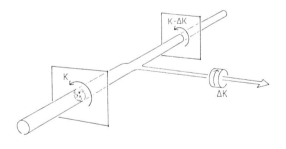

*Fig. 9.5* According to a theorem of Helmholtz the total vorticity or circulation within a fluid must remain fixed and the vortex must be a continuous loop unless there is a solid surface for it to end on. If the circulation decreases in one region it must be compensated by a branching vortex filament. Since circulation decreases with height up a sail, vortices must be shed from the leech to compensate. These combine to form a trailing vortex system which accounts for most of the drag.

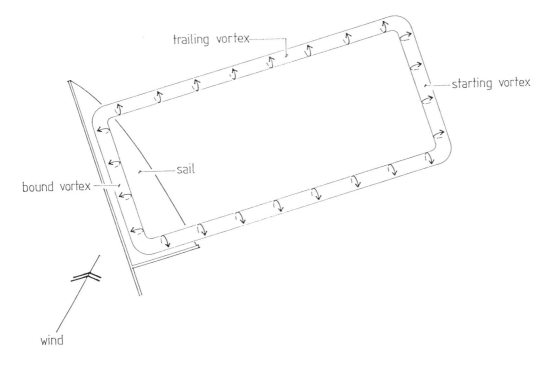

*Fig. 9.6* How the bound vortex and the starting vortex shown in fig. 3.21 are connected by the trailing vortices to form a closed loop, as demanded by a theorem of Helmholtz.

In §3.3.3 we showed the existence of the starting vortex and the bound vortex. We can now see why these must exist and how they are related to the trailing vortices. Fig. 9.6 shows a sail generating lift. It must therefore have circulation or a bound vortex. According to Helmholtz's theorem this vortex cannot end in the air but must make a continuous loop. This is formed by the trailing vortices connecting to the starting vortex.

## §9.4 Elliptic and Non-elliptic Lift Distributions

The relationship between the span-wise load variation and the trailing vortex strength can be readily seen by reference to fig. 9.7. The change in circulation from section to section is equal to the strength of the vorticity shed between these sections. The substitution of an aerofoil or sail by a system of vortices is really a means to an end, a model. It allows us to build up a relationship between the physical

load distribution which depends on sail shape, chord length and angle of attack, and the trailing vortex system which determines the magnitude of the drag.

Although it is a relatively simple matter to produce an airplane wing with an elliptic lift distribution, this is neither possible nor in fact desirable with soft sails. Theory shows that if the lift distribution does not differ too greatly from the elliptic, only simple modifications of the formulae already presented are needed to cover the more general case. The vortex drag coefficient is now given by:

$$C_{DV} = \frac{C_L^2[1 + \delta]}{\pi(AR)}$$

In this formula $\delta$ is always a positive number which is a measure of the amount by which the lift distribution differs from the elliptic. The vortex drag is thus always greater than the value associated with an elliptic lift distribution.

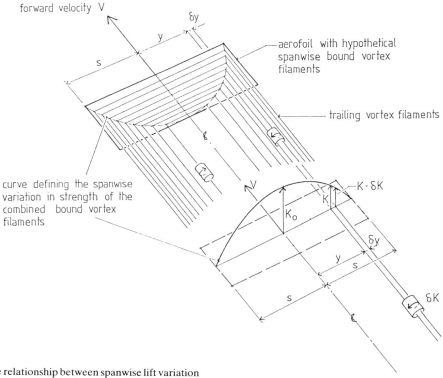

*Fig. 9.7* The relationship between spanwise lift variation and trailing vortex strength. Any change in lift along the span means a change in circulation which is accompanied by a shed vortex of strength equal to the change.

## §9.5   Lifting Line Theory

We have established that a wing or sail producing lift has associated with it a bound vortex or circulation. This is not a bodily motion of the air around the wing but simply a circular tendency superimposed on the general flow. It is as though the air were trying to circulate around the wing but gets dragged off downwind before it can make it. In §9.2 we derived a relationship between lift and circulation. This was for a two-dimensional situation, however. A real three-dimensional sail has a lift per unit area which varies from foot to head; i.e. there is a spanwise variation in the loading. This can only occur if the bound vortex changes in strength with position, which can only be done by shedding some vortex strength. Thus we have the situation depicted in fig. 9.7. Each line represents a vortex filament. If several vortex filaments pass through a certain cross-sectional area, then the total vorticity contained in that area is the sum of the vorticities of the individual filaments.

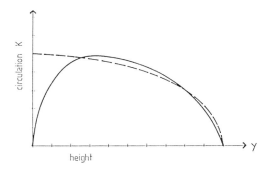

*Fig. 9.8* A plot showing the trend of the variation of circulation with height for a mainsail. Height is labelled y and is plotted horizontally! The foot of the sail corresponds to y = 0. The dashed curve is the elliptic distribution of circulation which gives the maximum lift/drag ratio. If there were a hermetic seal between the sail and the deck the circulation would not fall to zero at the foot. Since the area under this graph represents the total lift, a deck seal both increases the sail force as well as giving a better approximation to an elliptic lift distribution with its higher lift/drag ratio.

Imagine now that we have a circular plane whose diameter is greater than the chord of the aerofoil in fig. 9.7 and that it is oriented perpendicular to the long axis or y direction. When it is at the centre where y = 0, nine vortex filaments pass through it. Adding these up gives the circulation $K_o$ at the centre. As the plane moves out to greater values of y the number of vortex filaments passing through it falls off: thus we get a variation in circulation across the span. The total lift is then determined by using the formula L = $\varrho$VK at each position of the plane and adding up all these contributions.

This kind of calculation can be done mathematically by the process of *integration*. The total lift is written: L = $\varrho$V $\int$ K(y)dy for a uniform air flow. If you ever studied calculus in the past you will remember that the process of integration is equivalent to finding the area under a graph, as in fig. 9.8 where a graph of K(y), that is K as a function of y, is shown. If this graph has the elliptic form shown dotted, then the lift/drag ratio is a maximum. In theoretical calculations the functional relationship between K and y is usually expressed by means of a Fourier series. This has the property of being able to fit any loading distribution.

If the cross-sectional shape of a sail is known at various heights, and if the two-dimensional lift associated with these shapes has been measured in a wind tunnel, the circulation can then be plotted as in fig. 9.8 and the total lift determined.

# Chapter Ten

# FURTHER HULL
# CHARACTERISTICS

## §10.1    Righting Moment Measurements

This section gives some details of how the self-righting index quoted in §5.3.1 is obtained. Also the various terms used in the formula will be justified.

The setup for making the measurements is most quickly grasped by reference to fig. 10.1. The boat must be moored fore and aft as low down as possible so that it can pivot freely on an axis close to the waterline. Suitable points of attachment are the towing eye and the rudder pintles; it is important that these mooring lines be led off horizontally. The boat is heeled by means of a tackle added to the spinnaker halyard. The tension is measured with a spring balance which should be attached as close to the water surface as possible otherwise measurements at 90° become difficult. When measurements are made the boat should be in normal sailing trim: outboard motor, toolbox and fuel tank in normal position, mainsail rolled or flaked on the boom, which needs to be securely held amidships, and the rudder should be down. Precautions should be taken that battery acid or fuel cannot spill or siphon out and that heavy items cannot fall about the cabin.

The angle of heel is measured by an observer on the dock using an inclinometer which can be sighted parallel to the mast. For each angle of heel the angle of pull of the tackle to the horizontal must also be measured. This is because one needs to know the component of tension perpendicular to the mast in order to calculate the heeling moment. This can then be determined geometrically as shown in fig. 10.2. $T$ is the measured tension at an angle $\theta$ to the horizontal. We require the component of tension along the line AB perpendicular to the mast, given by

$T\cos\beta$. We need to relate $\beta$ to the two angles which are measured, namely $\theta$ and $\varphi$. Two facts are evident: the sum $\varphi + \alpha + \beta = 90°$ and also $\theta + \alpha = 90°$. Since both these sums are equal to 90° they can be equated to each other giving $\varphi + \alpha + \beta = \theta + \alpha$, or $\beta = \theta - \varphi$. So the true value of the tension is $T\cos(\theta - \varphi)$, plotted as 'corrected load' in fig. 5.13. A further correction to the spring balance reading which is important is due to the fact that its own weight and that of the tackle is not included in the reading but is applied to the boat. This weight must be added to get the correct applied tension.

A graph of tension against heel angle should be plotted, drawing a smooth curve through the measured points. Any points well away from a smooth curve should be ignored as being probably due to an error. The values for $T_{75}$ and $T_{90}$ as used in the SRI formula given in §5.3.1 and below should then be picked off this curve.

It remains now to give a physical justification for the various terms in the SRI formula:

$$\text{SRI} = \frac{(3T_{90} - T_{75})(I_s + 0.5\,\text{FML})}{6B^2L + 3B^2I_s + (40L \times \text{FML})}$$

It has been found that most boats show a linear decline in their righting moment with angle of heel beyond about 60°. Since measurements cannot be conveniently made at 95°, which is typical of a knockdown, it is a simple matter to extrapolate the graph because of its linearity. It can easily be shown that $T_{95} = (4T_{90} - T_{75})/3$. The first factor in the SRI equation is modified slightly from this to provide an allowance for wave action. $I_s + 0.5\text{FML}$ is just the lever arm, on the assumption that the boat pivots at a

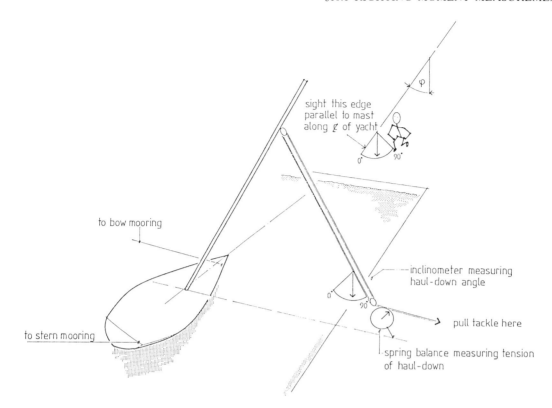

*Fig. 10.1* Procedure for measuring the full stability curve up to 90° for yachts up to about 7 m. The haul-down tension is measured with a spring balance secured as low as possible to facilitate the 90° measurement. The angle of heel and the

angle of the haul-down tackle must be measured simultaneously. The method of converting the raw measurements to righting moment is explained in the text.

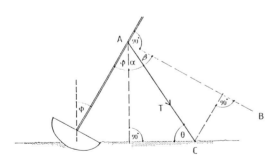

*Fig. 10.2* In measuring the righting moment of a small yacht the heeling force at right angles to the mast is required. What is conveniently measured is the tension T in a line not generally perpendicular to the mast. The component perpendicular to the mast is given by $T\cos\beta$. From the geometry of the right angled triangles involved it is easy to show that $\beta = \theta - \varphi$.

point somewhat above the waterline plane. The top line of the formula is then the righting moment provided by the boat. The bottom line is an estimate of the various overturning moments present when the boat is heeled to 95° in a 30 knot wind.

Photographs of overturned yachts have shown that the projected area of the hull above the water looking downwind in line with the mast is about 0.63LB, where B is the maximum beam and L the overall length. The overturning moment due to wind on the exposed hull is given by wind force × moment arm. The wind force is $\frac{1}{2}\rho AV^2C_D$, where A is the projected area of hull above water, V the wind speed (15 m/sec or 30 knots) and $C_D$ the hull drag coefficient. Since the hull approximates a hemisphere with the round side upwind, the drag coefficient appropriate to such a shape may be used; it has the value $C_D = 0.65$. As can be seen from fig. 10.3 a

235

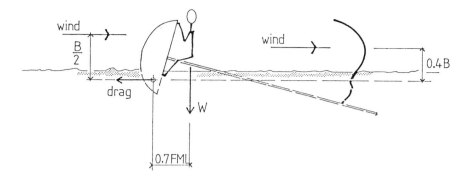

*Fig. 10.3* The overturning moments present in a complete knockdown. There are three such moments: the wind force on the hull with a lever arm B/2 (B is maximum beam), the wind force catching part of the spinnaker with an assumed lever arm of 0.4B and the weight force of the crew with a lever arm of about 0.7FML. (FML is the freeboard at mid-length.) For the boat to be self-righting its righting moment at 95° must exceed the sum of these three overturning moments.

reasonable value for the lever arm is B/2. Multiplying out these factors and putting into units of kg-m gives a value which rounds off to $3B^2L$.

The second term on the bottom line of the SRI formula (see §5.3.1 and above) is due to wind in the spinnaker. As before one must calculate the force from the effective area, drag coefficient and wind speed and decide on a reasonable moment arm. The effective area of the spinnaker is assumed to have a height of about 0.75B above the water and a maximum width of about $0.35I_s$: this gives an area of about $0.26BI_s$. The drag coefficient is now appropriate to that of a sphere with open side upstream: $C_D = 1.3$. The assumed effective lever arm has been taken as 0.4B. Again for a wind velocity of 30 knots the overturning moment becomes $1.5B^2I_s$ kg-m when rounded off.

The final term is the overturning moment due to crew weight. The effective load is assumed to be proportional to the length of the craft with a factor of 30 kg-m. For a 7 m boat this corresponds to three each weighing 70 kg. The effective lever arm with the crew assumed to be draped across the vertical deck and cabin top, is a function of the height of the hull sides: a figure of 0.7 FML has been used. Hence the overturning moment becomes 20L × FML when rounded off. Tests on a number of yachts have shown this figure to be reasonable.

It can be seen from this analysis that the main factors preventing a yacht from righting itself after a knockdown have been included. The fact that they are only rough estimates is unimportant as the formula can have no absolute significance. Its value is

solely in comparing one type of boat with another under the assumed conditions. It must also be remembered that the stiffness of a boat under normal sailing conditions is no indication of its SRI value. This is because the SRI depends on ballast stability whereas stiffness at small angles of heel is more dependent on form stability. (*I would like to thank W. D. Spry for kindly supplying the above material on the development of the SRI formula.*)

If the yacht you own is larger than a trailer type and more heavily ballasted, then a full measured righting moment curve is not a practical reality. In any case a knockdown to 90° is far less likely (though no less dangerous), because the static stability of a boat scales as $L^4$. (Scaling is explained in §5.5.) Thus in going from a 6 m boat to one 10 m long, the righting moment increases by a factor of $(10/6)^4 = 7.7$. Of course overturning moments are increased, but not as much since these scale as $L^3$ approximately, so an index such as SRI will increase as $L^4/L^3 = L$. That is, it increases in direct proportion to the size of the boat.

It is, however, interesting to know the small angle stability of a boat. The following is a simple procedure by which this can be measured.

The boat should be moored or in a marina in calm wind and water. All gear should be stowed in normal positions with the boom secured amidships. As the angle of heel to be measured is a fairly small one, a reasonably accurate method of measurement is required. This is done with a pendulum which should be at least 1.5 m long and consists of a light cord with a weight at the bottom. The upper end is attached

*Fig. 10.4* Measuring the righting moment at 90° for a 6 m boat. From such measurements a self-righting index can be formulated which is useful in comparing the ability of different designs to recover from a knockdown.

inside the yacht and on the centreline. To provide damping for the pendulum the weight is immersed in a bucket of water, and in order to further improve the damping action it is a good idea to add to the weight a scrubbing pad of bronze wool from the galley. The deflection of the pendulum is then measured on a metre stick fastened horizontally just above the bucket (fig. 10.5). The ratio of the pendulum deflection PD to the pendulum length PL is the angle of heel in *radians*, a mathematically natural non-dimensional method of measuring angle. Unfortunately few people are familiar with radians and we must convert to our arbitrary but useful system of degrees of angle. The connection between

the two is made by the fact that a full circle or 360° is equal to $2\pi$ radians, which works out to 1 radian = 57.3°. Thus the angle of heel in degrees is given by $(PD/PL) \times 57.3$.

A simple way to apply a known heeling moment to the boat is to get previously weighed crew members to stand on the toerail at the point of maximum beam. Their weight times the distance to the centreline of the boat is the heeling moment. Possibly a more accurate method is to follow the procedure prescribed by the IOR in which known weights are hung from a spinnaker pole rigged outboard. Whichever method is used, the heeling moments for two angles of heel of approximately 1° and 2° should

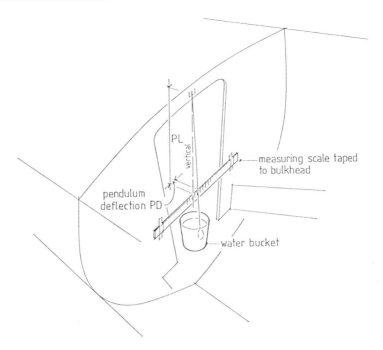

*Fig. 10.5* For larger boats it is only practical to measure the initial stiffness. This is a standard measurement procedure specified by the International Offshore Rule. The angle of heel is measured with a pendulum whose motion is damped with a bucket of water as shown. The heel angle in degrees is 57.3 × (PD/PL). The heeling is produced by calibrated weights on a measured lever arm.

be measured for heel to both port and starboard. Basically what these measurements determine is the quantity GM (fig. 5.7) which is the distance from the centre of gravity of the yacht to the metacentre. GM is thus a parameter of the yacht design.

The equation that makes this connection is: RM = $\Delta$GMsin$\varphi$. RM is the measured righting moment and $\Delta$ is the boat's displacement. Turning the formula around gives GM = RM/$\Delta$ sin$\varphi$. Providing the four values for GM so obtained do not differ from each other by more than a few %, the average value should be used. The righting moment for 1° heel is then given by $\Delta$GMsin (1°) or 0.0175$\Delta$GM.

If the yacht's lines are known so that the position of the metacentre can be determined, then the inclining measurement can be regarded as a determination of the height of the centre of gravity. In addition it can be used to calculate the Dellenbaugh angle and the wind pressure coefficient as described in §5.3.1. For the latter quantity the righting moment at 20° heel is required. Since GM is now known this can be determined from GMsin(20°). This is a rough approximation as we saw in §5.3.1 and will tend to slightly overestimate the WPC.

The IOR method of measuring righting moment for small angles of heel is a basically sound procedure even though the apparatus seems 'crude'. The most critical part of the measurement is the scale deflection, PD, which should be measured to a fraction of a millimetre if possible. In the IOR rule a 'tenderness ratio', TR, is first calculated from the righting moment and then this is used in a formula to calculate the centre of gravity factor, CGF. The form of this formula is such that if the TR is close to the minimum allowed value of 5.15, a large error in CGF will result even for quite accurate heeling angle measurements. Since the final rating is proportional to CGF, a given percentage error in the latter will produce the same percentage error in the rating.

## §10.2   A Crude Estimate of Wave Drag

The energy of waves per unit area of sea surface can be shown to be equal to $\frac{1}{8}\varrho gH^2$, where H is the

*Fig. 10.6* A very rough way of determining the order of magnitude of wave drag is to use the fact that the energy of a wave is proportional to the square of its height, so the energy of waves on an area of sea surface can be calculated. A boat in moving a distance d produces waves in the cross-hatched area of water. Equating the energy of these waves to the work done by the boat against drag enables one to show that wave drag is approximately proportional to the fourth power of the boat speed.

height of the waves. Referring to fig. 10.6, assume the boat moves a distance d in time $t = d/V_B$, where $V_B$ is the boat speed. During this time the boat will produce waves which are initially within the area $Bd$, where B is the beam (fig. 10.6). The work done by the boat in producing the waves is then $\frac{1}{8}\varrho gH^2 \times Bd =$ drag $\times$ d. From this we see that the wave drag is proportional to $H^2$.

For trochoidal waves (an approximate mathematical model of water waves) the height is proportional to the wavelength. So we have H proportional to $\lambda = 2\pi V^2/g$ for waves in deep water. That is H is proportional to $V^2$. But the drag is proportional to $H^2$, which in turn is proportional to $V_B{}^4$.

## §10.3 Summary of Scaling Laws

Have you ever wondered why birds have skinny legs and elephants have thick ones? It is a result of the fact that different physical properties depend on size in different ways. In this context size is represented by a typical linear dimension: for yacht hulls this is usually the waterline length, but it could just as easily be the beam of draft. Any of these dimensions gives us a feel for the size of the boat in an approximate way.

Consider a column such as a mast or a bird's leg. The stress on the structural material is measured by the weight it must support per unit of cross-sectional area. Since weight is proportional to volume, which depends on the cube of the linear dimensions, and area depends on the square of the linear dimension, we have: stress = weight/area = $L^3/L^2 = L$. Here L is a typical linear dimension, i.e. as the size of an object increases the stresses go up in proportion. Since birds' legs and elephants' legs are both made of bone with about the same tensile strength, it is clear that elephants' legs must be thicker to withstand the greater stresses. The same increase in stresses occurs when one scales up from a model yacht to the full size vessel. Drop a model from a height equal to its length and little damage is likely to occur; do the same thing with your 10 m pride and joy and it would be totally destroyed.

Scaling laws are important in yacht design mainly because of their use in tank testing. Improvements in yacht design are basically an evolutionary process. Building models and testing them in a towing tank is expensive but it speeds up the evolutionary process. If the gains, such as winning the America's Cup, are considered worth the expense, and the quantum leap in evolution is incrementally greater than the opposition's, then success might ensue.

To arrive at the scaling laws we concern ourselves only with the dimensions of physical quantities. For instance the drag force on a spinnaker can be represented by $\frac{1}{2}C_D A V^2 \varrho$. $C_D$ is the (dimensionless) drag coefficient, A the area and V the apparent wind speed. We are only interested in the parts of this drag equation which are going to vary when we change the size of the spinnaker or the wind strength. For this purpose it is sufficient to say only that the drag is proportional to $AV^2$. A, being an area, is proportional to the square of the linear dimensions so we say the drag is proportional to $V^2L^2$. Thus if we double the size of the boat we also roughly double the linear dimensions of the spinnaker and hence quadruple ($2^2$) the drag force. Since the drag force also depends on the square of the velocity it increases in the same way.

With this introduction I will now write down the relevant dimensions of various quantities so that you can see how they scale with the size as represented by L.

**Displacement** = weight = density $\times$ volume. Volume is proportional to $L^3$ so doubling the length of a boat will increase the displacement by a factor of $2^3 = 8$. Since the cost of a boat is more or less proportional to its displacement it is easy to see why costs rise so sharply with size.

**Heeling Moment** = sail force $\times$ arm = $L^2V^2 \times L = V^2L^3$

**Righting Moment** = displacement $\times$ arm = $L^3 \times L = L^4$

We see from this that righting moment increases very rapidly indeed with size of boat. Increasing boat length by a factor of 2 produces a factor of $2^4 = 16$ times increase in stability! The ratio of heeling moment to righting moment which is used in one form or another as a measure of relative stability then becomes: HM/RM = $V^2/L$. Thus even though the larger boat has more sail area, the relative stability increases in direct proportion to size.

**Wave Formation:** this is characterised by Froude's number: $F_n = V/\sqrt{gL}$. A model hull and a 'ship' hull of the same shape will have the same wake pattern if their Froude number is the same. This will be true if $V^m/\sqrt{L^m} = V^s\sqrt{L^s}$, where the superscript m refers to the model and s refers to the full size boat. Rearranging the equation gives: $V^m = V^s \times (L^m/L^s)^{1/2}$. This condition ensures a constant wave drag coefficient. The actual drag is in turn subject to a scaling law since it is proportional to the displacement. Wave drag is proportional to the weight of water moved aside by the boat which in turn is proportional to $L^3$.

**Reynolds' Number:** this quantity must be kept constant where similarity of flow is required in situations where viscosity plays an important role. $R_n = VL/v$. Viscosity $v$ is normally fixed so we have $V^mL^m = V^sL^s$. So the relationship between model speed and ship speed that keeps the friction drag coefficient the same is $V^m = (L^s/L^m) \times V^s$. Since surface friction drag is proportional to $AV^2$ it therefore scales as $L^2V^2$.

**Sail Stress:** stress is defined as the ratio of the internal forces acting to the cross-sectional area over which they act. For a sail the cross-sectional area is rather small since it is of the order of the product of the sail chord and the sail thickness. Since the origin of the forces is the wind on the sail the stress is proportional to $L^2V^2/Lt$, where t is the cloth thickness. If the cloth thickness is held constant then the stress in the material scales like $LV^2$.

# Chapter Eleven

# SOME DETAILS OF DYNAMIC MOTION

## §11.1 Yacht Steering Performance

The control which a rudder gives depends almost entirely upon the turning moment that it can produce. This moment results from the lift force of the rudder multiplied by its distance from the centre of gravity of the boat (fig. 11.1). Y is the lift or side force produced by the rudder and d is the distance of this force from the CG. The turning ability of the rudder is measured by the 'leverage' that it can exert about the CG. It is called the turning or yawing moment and defined by $N = Y \times d$.

Very few scientific measurements of yacht controllability appear to have been published. In one interesting study, carried out in the 1960s when keel-hung rudders were still in vogue, models of the three hulls shown in fig. 11.2 were tested. The full size version of A, though it sailed to windward satisfactorily, was completely unmanageable with the wind on the quarter. The yacht was converted to the

profile shown in B and then performed satisfactorily. Tank tests were made of these two and also a further version without the skeg, C.

Fig. 11.3a shows the relative rudder side force as a function of the rudder angle. The most modern looking profile, C, generates the least side force and A the most (fig. 11.2). This may come as a surprise in view of the fact that boats have in the interim evolved away from A and toward C. The evolutionary trend is partly answered when we look at fig. 11.3b which plots the turning moment N against the rudder angle. From this it is clear that B is best and A worst. This, of course, is a result of the increase in lever arm being greater than the decrease in side force.

What is interesting is to enquire why the side force is so much greater for the traditional profile A. Evidently a rudder on the trailing edge of a keel acts like a wing flap: it modifies the smooth flow around the keel and produces a large amount of lift. On the other hand the spade rudder finds itself in a region where flow separation and hence turbulence may be occurring around the hull. This is probably a result of too sharply changing curvature of buttock lines. For the designs shown in fig. 11.2 this effect is partly offset by the use of a skeg which improves the flow around the rudder. Better hull design, on the other hand, could eliminate the need for a skeg which can impair manoeuvrability.

Another artifact of the rear-hung rudder is the possibility of ventilation at large angles of heel (fig. 11.4b). Beyond a rudder angle of 20° the rate of increase of moment falls off when the boat is heeled. In fact the moment actually decreases for increase in rudder angle at 30° heel. The effect is not nearly so great for the traditional keel-hung rudder in fig.

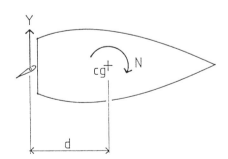

*Fig. 11.1* The turning moment N can be thought of as the leverage supplied by the rudder about the centre of gravity. Y is the lift force generated by the rudder and d is the leverage arm. The turning moment is defined as $N = Y \times d$.

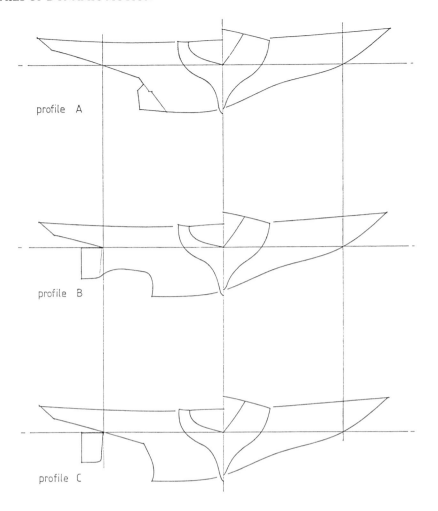

profile   A

profile   B

profile   C

*Fig. 11.2* Three models for which test tank measurements of steering control were made. The full size version of A was unmanageable downwind, whereas the modification B proved satisfactory. The next two diagrams show some of the results of the measurements on these models.

11.4a. An interesting feature of these graphs is that the rudder angle for zero turning moment is about 4° for profile A and about 9° for profile B. That is, a greater rudder angle is needed for the aft-hung rudder. This apparent contradiction is explained by the effects of downwash from the keel when making leeway. This gives a greater effective angle of attack to the rudder mounted close to the keel.

The directional stability of commercial vessels has been studied for many years. The problem is whether or not a ship moving in smooth water, when given a perturbation in yaw or sway, will become steady in time. If the course becomes a straight line after the perturbation, the ship is said to be stable or to have fixed controls stability.

Clearly a ship which is stable must have some characteristics which set it apart from an unstable one. It is not possible by just looking at the lines to make this determination. So how does one specify steering stability? It starts by writing down the equations of motion, basically statements of Newton's laws of motion in mathematical form. Mass and all accelerations along the axes x, y and z (fig. 6.1) are taken into account as well as their rotational

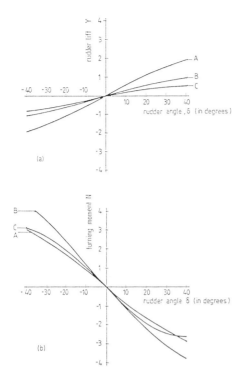

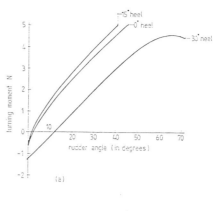

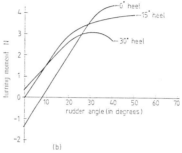

*Fig. 11.3* The results of measurements of the rudder lift Y and the turning moment N for the three profiles A, B and C of fig. 11.2. Although the aft-hung rudders in models B and C provide less rudder side force, they produce a greater turning moment because of the longer arm d.

*Fig. 11.4* The effect of heel on the turning moment for profiles A and B of fig. 11.2. The symmetry of fig. 11.3 is now lost (the curves do not pass through the origin) because these measurements were carried out with leeway. (a) pertains to profile A and (b) to profile B. When heeled the aft-hung rudder rapidly loses efficiency for rudder angles greater than 20°, a result of the disturbed flow near the stern and the effects of air entrainment.

equivalents, moment of inertia and angular acceleration. In their simplest form such equations neglect the roll motion and aerodynamic forces.

These equations have solutions which give the yaw angle as a function of time in the following form: r(t) = $r_1 e^{s_1 t} + r_2 e^{s_2 t}$. In this equation $r_1$ and $r_2$ are constant yaw velocities and represent the initial perturbation in yaw angle; r(t) which is yaw velocity then tells us how the yaw angle of the ship subsequently develops in time t. Clearly the subsequent motion is then determined entirely by the quantities $s_1$ and $s_2$. These quantities are called the *stability roots* of the ship and depend in a complicated way on the forces and moments acting on it. They are usually determined experimentally by rotating arm and dynamic oscillation tests on models. In the former the model is attached to a rotating arm through two dynamometers measuring the side force at two locations along the length of the model. The sum of

the two readings gives the total side force while the difference is a measure of the turning moment about the centre of gravity.

The stability roots $s_1$ and $s_2$ are generally complex numbers, which is mathematical jargon for numbers which have two parts, a real part and an imaginary one. This terminology may mean something to you, but if not it is of no consequence for the subsequent discussion. It turns out that if the stability roots have a negative real part then the ship is stable. If the real part is positive the ship is unstable against an external perturbation in yaw. The imaginary part determines the period of an oscillatory motion which will be damped in the stable case or increase in amplitude in the unstable case. These two possibilities are shown in fig. 11.5.

For a yacht there is normally a relatively large

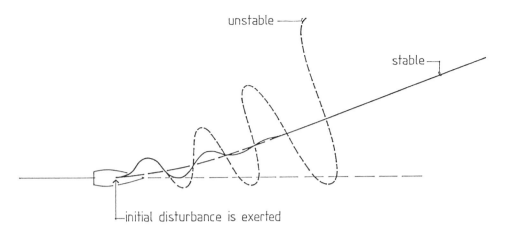

unstable

stable

initial disturbance is exerted

*Fig. 11.5* The yaw stability of a vessel may be characterised by a mathematical quantity called the stability roots, usually numbers with two parts, real and imaginary. A stable ship must have stability roots with a negative real part. The oscillations shown are a result of the imaginary part of the stability roots. A stable ship when perturbed will eventually settle down to a straight-line course which will in general be different from the original. An unstable ship will oscillate about the mean path with ever-increasing amplitude.

vertical distance between the centre of lateral resistance and the centre of gravity. This means that a sway or yaw motion introduces a rolling moment while a roll can give rise to sway or yaw. When this hydrodynamic coupling is included the stability roots are modified. Compared with the roots of the basic sway-yaw system which imply stability for all yachts, the roots of the coupled sway-yaw-roll system show that the coupling with roll has a destabilising influence. This means that the oscillatory motion is damped out more slowly.

For sailing yachts an important ingredient has not yet been added. Whenever there is a course change there is a change in the sail forces: our equations of motion should therefore include also the aerodynamic forces of the sails. When this is done the stability roots are further modified. As with roll, the inclusion of sail forces has a further destabilising influence.

Remember that what we have been speaking of here is *fixed controls stability*. Because of the slow reaction time of a large ship the helmsman will make an adjustment after which it will be some time before another adjustment is made. Under these conditions fixed controls stability is obviously of paramount importance. Yachts, on the other hand, because of their much smaller mass react much faster to the helm so that nearly continuous rudder action is required to offset the effects of disturbances. Thus an analysis of

the steering performance of a yacht should be extended to include the performance of the helmsman. The stability roots determine the physical time constants of the system. As long as these do not differ too much from human time constants or reaction times, the boat can be steered whether or not it has controls fixed stability. Of course it will still be more difficult to steer an unstable boat.

If it can be assumed that the helmsman provides an almost continuous rudder action, it is possible to regard the crew-boat combination as a mechanical or electrical system with feedback. Viewed in this way, standard engineering control systems theory can be applied. We assume that the oscillatory rudder action has sinusoidal Fourier components of various angular frequencies $\omega$ such that the position of the rudder at any time t is given by: $\delta(t) = \delta_a \sin(\omega t)$ where $\delta_a$ is the maximum rudder angle used. This use of the rudder will result in a course deviation given by: $\psi(t) = \psi_a \sin(\omega t + \varepsilon)$ where $\psi_a$ is the maximum course deviation and $\varepsilon$ is the phase difference between rudder angle and course deviation.

This phase difference requires a little explanation although it is common experience, especially when sailing in waves with the wind on the quarter. The boat is usually yawing and rolling and regular tiller action is normally required to keep the boat on a reasonably straight course. The tiller action will require the same repetition rate as the yawing motion

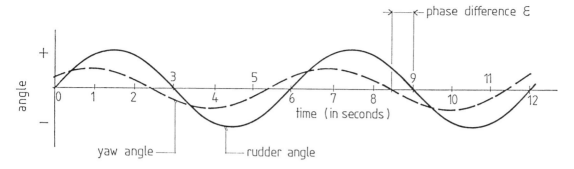

Fig. 11.6 A *periodic* variation of the rudder angle every 6.3 sec gives a periodic variation in the yaw angle (deviation from the mean path) also with a period of 6.3 sec. However there is a time difference, or phase difference, of about 0.6 sec between the two motions.

but will need to be applied at a different time. This is the phase difference shown in fig. 11.6

Viewing the helmsman-boat combination as a feedback system we can follow control engineering and define the ratio $\psi_a/\delta_a$ as the response amplitude operator. It is simply a measure of the effectiveness of the rudder angle in producing a given yaw angle. A graph on which is plotted this ratio against the circular frequency $\omega$ is called a Bode diagram for the system. (Incidentally, circular frequency of a periodic motion is related to ordinary frequency by the expression $\omega = 2\pi f$, where f is frequency in cycles per second.) Fig. 11.7 shows a Bode diagram for an IOR Half-Tonner. The upper half of the diagram gives the response of the boat to rudder actions of various periodic rates. The lower half shows the phase difference between the yaw angle and the rudder angle.

To sum up: any disturbance such as wind or waves will cause a yacht to deviate from its course or even broach. The helmsman reacts to such deviations and tries to neutralise them with the rudder. The response amplitude operator then gives a measure of the ability of the yacht to correct the disturbance and this could be called the steering power of the yacht. The frequency range of greatest interest for yachts is in the range $\omega = 0.5$ to $\omega = 6$. This corresponds to waves from astern of about 4 to 60 m length. Thus the Bode diagram (fig. 11.7) contains all the information to describe the steering qualities of a yacht. Unfortunately the relationship between the shape of this diagram and actual handling qualities is not yet known.

Finally, it is interesting to note that in the few

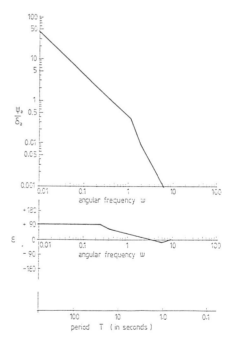

Fig. 11.7 A Bode diagram, used extensively in control system engineering. The ordinate of the upper graph is the ratio of the yaw amplitude produced to the rudder amplitude when it is moved periodically with the angular frequency or period given on the abscissa. The lower graph shows the phase difference between rudder angle and yaw angle. In principle such a diagram gives all the necessary information to describe the steerability of a yacht.

investigations of steering controllability that have so far been made with yacht models, it appears that long keels do not give better fixed controls stability as many sailors commonly believe.

## §11.2   Movement of Hydrodynamic CLRs

The following simple argument shows that the position of the centre of lateral resistance is expected to depend both on the angle of heel and the leeway angle.

We assume that the total yawing moment N is made up of a moment due to side force or lift acting at the upright CLR position $x_o$ together with a moment M proportional to the angle of heel $\varphi$. This latter moment is produced by the asymmetric form of the heeled hull. Thus by taking moments about the bow we may write: $N = x_oL + M$ where L is the side force on the hull. The CLR position x measured from the bow (fig. 11.8) must be such that x = moment/side force, thus: $x = x_o + M/L$. Now the lift force L is directly proportional to the leeway angle $\lambda$ and if we assume that the asymmetric moment M is proportional to angle of heel $\varphi$ then M/L is proportional to $\varphi/\lambda$. Introducing a constant of proportionality $\mu$ allows us to write: $M/L = \mu\varphi/\lambda$. Hence: $x = x_o + \mu\varphi/\lambda$.

As it is more useful to give the position of the CLR as a fraction of the waterline length from the bow, we simply divide both sides of the equation by the LWL and write it: $x' = x'_o + \mu'\varphi/\lambda$. x' is the CLR position as a fraction of the waterline length from the bow, $x'_o$ is the upright position of the CLR as a fraction of the waterline length from the bow, and $\varphi$ and $\lambda$ are the angles of heel and leeway respectively.

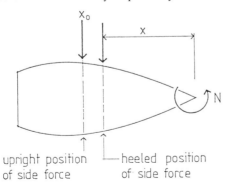

Fig. 11.8 The position $x_o$ of the CLR in the absence of heel. On heeling the CLR moves forward, reducing its moment about the bow. An approximate method for calculating this movement with heel angle is given in the text.

Values of $\mu'$ will depend on hull type but will generally be negative, giving a forward movement of the CLR with heeling. A value suggested by data from Southampton University was used to plot fig. 6.9.

## §11.3   Some Physics of Waves

It has always seemed to me a remarkable thing that most people, aided perhaps by a pint of ale, can hold forth on such topics as politics, modern art, music or sport, yet the conversation dries up if the discussion turns to questions such as why does the wind blow, why do waves break, or why are some waves bigger than others? For sailors these are important topics worthy of conversation. Waves can be used to the yachtsman's advantage or they can be his deadliest enemy, so it behoves everyone to understand their inner workings.

In this section some facts about waves will be presented. Even if they don't help you to sail better, they will certainly make you a better conversationalist. Sea waves have complicated shapes and 'non-linear' characteristics which make an exact description difficult. For that reason scientists often resort to models, imaginary physical systems whose behaviour closely resembles a real wave but which are more easily understood.

The sensible discussion of any subject must begin by defining the technical jargon to be used. Fig. 11.9 does this for waves. The distance from crest to crest (or from trough to trough) is the *wavelength* and is always denoted by the Greek letter $\lambda$ (lambda). The wave *height* is the vertical distance from trough to crest. Although this is a more natural quantity for an observer to use in describing a wave, a more natural quantity for the mathematical description of a wave is the *amplitude* which is just half the height. If fig. 11.9 is thought of as a graph, then what it shows is the height of the wave surface at a fixed time as a function of position, a snapshot if you like.

Waves move also in time so the position of the surface at a fixed place changes with time according to the lower curve of fig. 11.9. This is the graph you would obtain by observing the height of water on a wharf pile as waves go by. The time between the passage of crests is called the wave *period* T. If we count the number of waves that pass the pile per second we get the frequency f, which is simply the inverse of the period: f = 1/T. The speed c of a wave is found from how far a crest moves in a given time. Obviously the crest moves a distance equal to the

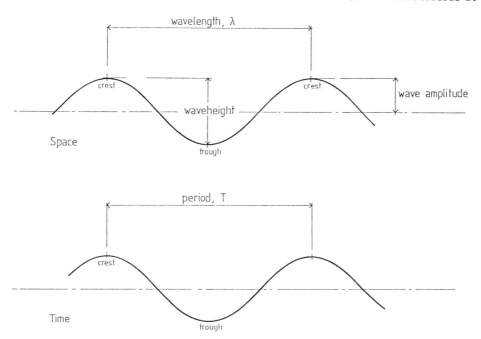

*Fig. 11.9* The upper curve shows the instantaneous appearance of a wave as it would be caught by a photograph. The lower one shows the variation in wave height with time.

This is not something you actually see since only spatial configurations can be seen. Consider it a graph of the height of water on a wharf pile as a function of time.

wavelength in a time equal to the period, so c = λ/T or c = fλ.

The speed of water waves depends on the relationship between wavelength and depth. The full mathematical expression for speed is:

$$c = \sqrt{g\lambda/2\pi \tanh 2\pi d/_\lambda}$$

where d is the water depth and g is the acceleration of gravity. If the depth is more than half the wavelength the expression for the velocity simplifies to: c = $\sqrt{g\lambda/2\pi}$ = 2.5 $\sqrt{\lambda(m)}$ in knots. Such a wave is called a deep-water gravity wave. Note that its speed depends on its wavelength. This is why waves in the ocean seem so jumbled. Longer ones move faster through the others, alternately adding to and subtracting from them as they go by.

If the depth of water underneath a wave is less than 1/25th of its length it is referred to as a shallow water wave and its speed is simply: c = $\sqrt{gd}$ = 6.1 $\sqrt{d(m)}$ knots. Note that now its speed is no longer dependent on wavelength, so all shallow water waves travel at the same speed. This is why waves approaching shore tend to form up in long equally spaced lines with an orderliness never seen in deep water. When the

depth is between 1/25th and ½ the wavelength the waves are called transitional and the full formula for speed must be used.

The period of a wave is always the same no matter what the depth of water, but as a deep water wave enters shallower water its velocity is reduced and its length shortened. Combining the relationship T = λ/c with the expression for the velocity of a deep water wave gives: T = $\sqrt{2\pi\lambda/g}$ = 0.8 $\sqrt{\lambda(m)}$ seconds.

It is important to realise that although we speak of the speed of waves we do not mean that the water itself moves at this speed. What we mean is that the *shape* of the water moves at this speed. Sound waves are movements of the air just as water waves are movements of the water: however when we speak to someone across a room we don't expect the air in our mouth to go whistling across the room into the listener's ear!

Although the water particles do not move with the wave speed, they certainly do move. In fact each particle of water moves in a circle whose centre remains fixed. That this is so is not too easy to visualise; fig. 11.10 may help. Each of the diagrams

247

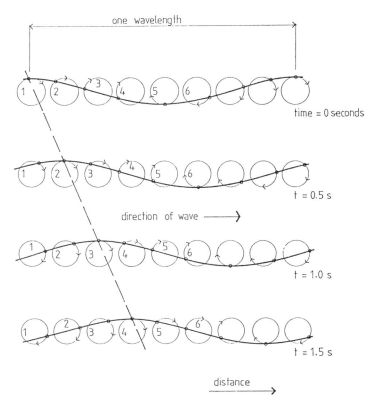

*Fig. 11.10* The water particles in a wave rotate in circles of diameter equal to the wave height. The wave surface is composed of these particles, so adjacent particles are in different positions in their rotation cycle. As they all rotate they form a shape, the wave, which moves forward although the water itself merely moves in circles. This progression can be seen by studying the diagrams for half-second intervals. After 1.5 sec the crest has travelled from particle 1 to particle 4. The period of this wave is 4 seconds, which is the time for its successive crests to pass a fixed point.

is a 'snapshot' taken at half-second intervals. The small dots are water particles; for clarity only six particles per wavelength are shown. Each dot is rotating in a clockwise direction at the same speed. These dots are water particles at the surface so their position forms the contour of the wave.

Follow down the dashed line and you see the position of a water particle at half-second intervals and how it contributes to the wave contour. At time = 0 it is at the crest and has a tangential velocity in the direction of the wave. At time = 0.5 s this particle is starting to move down whereas the adjacent one to the right has reached the top of its trajectory and therefore forms the crest in its new position. After 1.5 sec, which is nearly half the wave period, the particle is almost at the bottom of its trajectory and moving in the opposite direction to the wave motion. The diagonal dashed line traces the progress of the crest

of the wave which has moved from particle 1 to particle 4 in 1.5 sec.

Have you ever thought how it is possible for a wave to have a sloping face? Put water in a bucket and the surface is horizontal, so why doesn't gravity just flatten the sea out instantly as it does in a bucket? There is, however, a way that you can tilt the water in a bucket from the horizontal. Hold the bucket by the handle and rotate your body, observing the water surface at the same time. Its angle to the horizontal will depend on your speed of rotation. Motion in a circle can only occur if there is an attendant force and hence acceleration: this is the familiar centrifugal force. So the net force acting on the water in the bucket is the vector sum of this and gravity. Clearly this will no longer be vertically downward. As far as the water is concerned 'vertical' is always perpendicular to the water surface. Fig. 11.11 shows

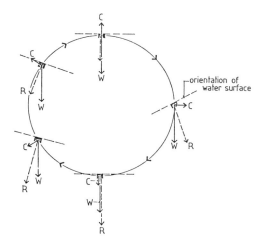

Fig. 11.11 The forces on a water particle in a wave as it moves in a circle. The weight W of the particle always acts downwards. Centrifugal force C always acts out from the centre of the circle. The resultant of these two forces is R which swings from side to side as the particle rotates. This is the 'local' gravity. The local water surface must always be perpendicular to R. Conversely the local gravity of an object stationary with respect to the water is always perpendicular to the wave slope.

in detail the forces acting on an object moving in a vertical circle. The weight W always acts vertically downward and the centrifugal force C always acts radially outward from the centre of the circle. From the point of view of the object moving in the circle, the resultant or vector sum of these is the local direction of gravity.

Compare this diagram now with fig. 11.10. The direction of local gravity for the particle of water in each circle swings from right to left as it rotates, creating a local water surface always perpendicular to the local gravity. Thus a wave has a sloping surface

because the resultant gravitational force is not down but perpendicular to its slope.

This local gravity gives rise to some interesting effects, one of which is shown in fig. 6.21. A raft in a large ocean swell will essentially follow the motion of one of the water particles in the wave. Local gravity for the occupants will therefore always be perpendicular to the wave slope, as the diagram shows. The situation for a surfer is different. He is moving *through* the water but *with* the wave, his motion is unaccelerated, so gravity for him is still vertically down.

Another effect of the local gravity is shown in fig. 11.12. A shipwrecked sailor in a liferaft judges the height of an approaching wave from its angular height above the horizontal. But 'horizontal' for him is the dashed line so he greatly overestimates the height of the approaching wave. The same effect can also occur in yachts in big seas.

A boat which is lying hove-to and therefore not moving through the water experiences less gravity on the crest of a wave than in the trough. Assuming the heeling forces are the same in both cases, the boat will heel more on the crest because the weight of its ballast in this low-g region is less. It is also likely that the wind speed will be greater on the exposed crest thereby accentuating the effect.

On the other hand, if the boat is moving through the water with the waves on an almost unaccelerated path it will sink slightly into the low-g water near the crest and float higher in the high-g water of the trough. When a boat starts surfing down the face of a wave near the crest, the water beneath it is effectively lighter, the boat sinks lower and much 'light weight' spray is thrown up.

Waves are produced by normal pressure forces and surface friction forces caused by the wind. Their size and shape depend on the wind's strength, fetch and

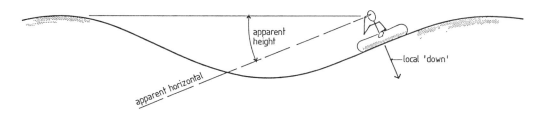

Fig. 11.12 A man in a raft which is stationary with respect to the water in a wave senses horizontal as being parallel to the local wave slope. The apparent height of an approaching wave can therefore be easily overestimated.

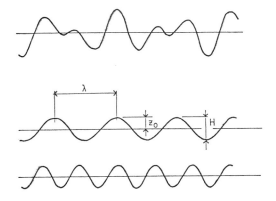

*Fig. 11.13* Waves at sea have a complicated shape because they result from the interference of many waves of different length moving at different speeds. This figure shows the effect of superimposing just two waves. The upper curve is the resultant of the two lower ones added together.

duration. Fetch is the length of open water upwind from the observation point. Duration is the time for which wind conditions have been constant prior to a measurement. The wave patterns produced are complex in appearance with waves travelling in different directions and having a range of heights and lengths. Such a wave pattern is called an 'irregular short-crested sea'; the crests are short because the waves are travelling in slightly different directions and one wave, crossing another diagonally, can cancel out the crest.

So far our model of a wave has been a 'regular long-crested sea'. This is no disadvantage as it is possible to represent the real sea state by a superposition of ideal waves. Fig. 11.14a shows the time history of the height of water at a fixed point, not to be confused with the appearance of the waves seen in space. Fourier's theorem (see §5.4.1) tells us that such a complicated wave train consists of a combination of ideal sinusoidal waves of different frequencies. Fig. 11.13 shows how the addition of just

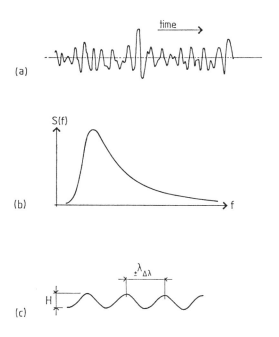

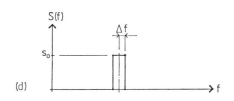

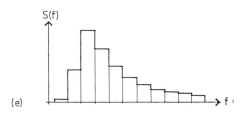

*Fig. 11.14* (a) shows variation in height of a wave as a function of time. Such a variation can be considered as being made up of a large number of sinusoidal components as shown in (c). Each of these components has associated with it an energy density s(f). For waves occupying a narrow band of frequencies f, its contribution to the overall energy density is shown in (d). A large number of contributing waves of different frequencies will each contribute to the energy density in a way which depends upon their amplitude, giving a graph like (e). If the bandwidth f is made small enough the histogram (e) becomes a smooth curve like (b). This is called the wave energy spectrum.

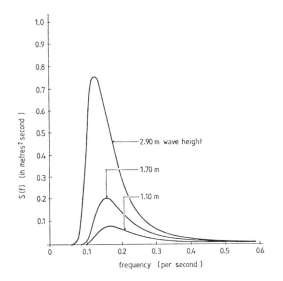

two waves of the same amplitude can give rise to a quite complex irregular wave. A sea state is defined as the sum of the energies of its component waves. The energy of a wave, per square metre of water surface is:
$E = \varrho gH^2/8$ where H is the wave height.

A plot of the energy contained at each frequency in the component waves is called a sea state spectrum (fig. 11.14b). The vertical axis, the energy density s(f), represents the energy of waves having a frequency f. To see this more clearly, consider a single sinusoidal wave as in (c): its spectrum will look like (d). If a large number of regular waves having different heights and therefore different energies are superimposed a spectrum like that in (e) will result. If the component frequency intervals are made narrow enough it is easy to see how (e) could eventually look like (b). The actual spectrum which was used for calculating added resistance in waves in §6.4 is shown in fig. 11.15.

The wave heights shown on this diagram are *significant* wave heights, defined as the average of the heights of the highest third of the waves. This definition is used because it has been found to be the

*Fig. 11.15* Standard wave spectra for significant wave heights of 2.9, 1.7 and 1.1 m. Significant wave height refers to the average height of the highest one-third of the waves.

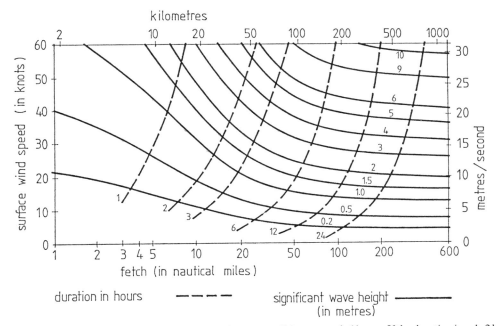

*Fig. 11.16* The empirical relationship between wind speed, fetch, duration and significant wave height. The graph is strictly applicable to coastal waters but is approximately valid also for oceanic conditions. To determine the significant wave height for a wind speed of say, 20 knots, lay a ruler through the 20 knot point parallel to the horizontal axis. If the fetch is 50 nautical miles the significant wave height will be 1.5 m providing the duration of these wind

conditions exceeds 6 hours. If the duration is only 2 hours, for instance, the wave height will be just over 0.5 m. In other words, fetch or duration can be the limiting factor on wave height. *(Adapted from Darbyshire and Draper, Engineering, 1963, **195**, 482–4. Reproduced by permission of the Institute of Oceanographic Sciences, Wormley, England)*

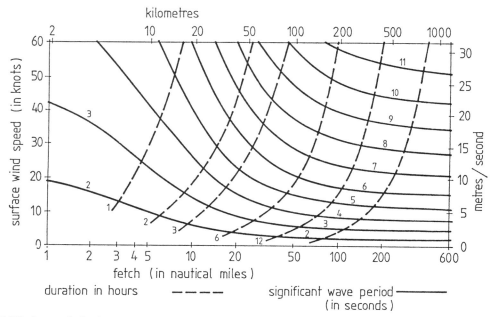

*Fig. 11.17* Wind strength, fetch and duration related to significant wave period. The graph is used in the same way as fig. 11.16. *(Adapted from Darbyshire and Draper, ibid.)*

*Fig. 11.18* Waves in deep water always seem confused and short-crested. This is because the speed of deep water waves depends on their length, so that many waves are passing through each other alternately reinforcing and cancelling each other. The waves in this picture are shallow water waves whose speed does not depend on wavelength but on depth. All the waves therefore travel at the same speed giving rise to the long even crests so characteristic of waves on a gently shoaling shore.

252

wave height that an observer will report as being typical of the sea state. Comparisons between measured observations and estimations by skilled observers in weather ships have shown that the agreement is close, so significant wave height is the most useful way of characterising a sea state.

You may have noticed that the linear theory of waves gave formulae connecting the quantities $\lambda$, $T$ and $c$ but not the wave height. Wave height depends mainly on wind conditions and water depth. If the ratio of wave height to length exceeds 1/7, the waves become unstable. Only in this sense is wave height connected to wavelength. To predict wave height an empirical approach is necessary. Figs. 11.16 and 11.17, which were based on measurements of wave heights and periods in Lough Neagh in Northern Ireland and at lightships in the North Sea, allow the relationship between height, wind speed, fetch, duration, period and depth to be determined.

# BIBLIOGRAPHY

Ira H. Abbott and Albert E. von Doenhoff, *Theory of Wing Sections*. Dover Publications, New York, 1959

H. M. Barkla, 'The behaviour of the sailing yacht'. *Transactions* of the Royal Institution of Naval Architects 1961, **103**, 1

P. R. Crewe, 'Estimation of effect of sail performance on yacht closehauled behaviour'. *Transactions* of the Royal Institution of Naval Architects 1964, **106**, 287

Darbyshire and Draper, in *Engineering* 1963, **195**, 482–4

K. S. M. Davidson, 'Some experimental studies of the sailing yacht'. *Transactions* of the Society of Naval Architects and Marine Engineers 1936, **44**, 288

J. Gerritsma, J. E. Kerwin and G. Moeyes, 'Determination of sail forces based on full scale measurements and model tests'. Symposium on Yacht Architecture, 1975 (*Standfast* data paper)

H. F. Kay, *The Science of Yachts, Wind and Water*. G. T. Foulis and Co. Ltd, Henley-on-Thames, Oxfordshire

J. S. Letcher, 'Surfing: motions of a vessel running in large waves'. Society of Naval Arcnitects and Marine Engineers, Chesapeake Sailing Yacht Symposium, Annapolis, 1977

C. A. Marchaj, 'Instability of sailing craft – rolling'. Second Symposium on Yacht Architecture, edited by I. J. Gerritsma, 1971

C. A. Marchaj, *Sailing Theory and Practice*. Adlard Coles Ltd, London, and Dodd, Mead, New York, 1982

T. Tanner, 'Full scale tank tests of an International 10 Square Metre class canoe'. *Transactions* of the Royal Institution of Naval Architects 1961, **103**, 25

C. J. Wood and S. H. Tan, 'Towards an optimum yacht sail'. *Journal of Fluid Mechanics* 1978, **85**, 459

# Appendix A

# NOMENCLATURE

## SUBSCRIPTS
a    air *or* apparent wind
h    hydrodynamic *or* hull
w    wind
s    sails *or* full size ship or yacht
t    true wind
m    model
r    rudder
T    total

## CENTRES
CE    centre of effort (aerodynamic) of sails
CLR   centre of lateral resistance (hydrodynamic) of hull
CG    centre of gravity
CB    centre of buoyancy
GM    height from centre of gravity to metacentre
BM    height from centre of buoyancy to metacentre

## GEOMETRY
L    length
LOA or $L_{oa}$    length overall
LWL or $L_{wl}$    waterline length
B    beam *or* buoyancy
A    projected area of a body or lifting surface (one side only)
$A_o$    maximum section area
S    surface area (both sides)
b    span of a lifting surface
c    chord length
I    foretriangle height
J    foretriangle base
P    mainsail height
E    mainsail base

$\nabla$    immersed volume
$\Delta$    displacement (in units of weight)

## ANGLES
$\beta$    angle between apparent wind direction and course
$\beta'$    measured apparent wind angle between apparent wind and heading $= \beta - \lambda$
$\gamma$    angle between true wind and course
$\gamma'$    angle between true wind and heading $= \gamma - \lambda$
$\alpha$    geometric angle of attack
$\alpha_\infty$    effective two-dimensional angle of attack
$\varphi$    heel angle
$\theta$    pitch angle
$\psi$    yaw angle (a dynamic quantity)
$\lambda$    leeway angle (a steady state quantity)
$\varepsilon$    drag angle $= \tan^{-1}(D/L)$
$\delta$    angle between boom and centreline
$\delta_r$    rudder angle

## MOTIONS
u, v, w    velocities along x, y, z axes respectively (see fig. 6.1)
p,q,r    angular velocities of roll, pitch and yaw respectively
$V_a$    apparent wind velocity
$V_t$    true wind velocity
$V_{mg}$    component of boat speed contrary to wind direction (speed made good)
$V_s$    ship speed *or* boat speed

g    acceleration of gravity $= 9.8 \text{m s}^{-2}$ (metres per second per second)

## FORCES
X,Y,Z    force components along the x, y, z axes (fig. 6.1) respectively
D    drag force along fluid flow direction
L    lift force perpendicular to fluid flow direction
H    side force in horizontal plane
R    resistance *or* drag
$C_D, C_L$    force coefficients where force $=$ coefficient $\times AV^2 \varrho/2$
$F_R$    driving force of sails (along course direction)
$F_H$    heeling force of sails (H $= F_H \cos \varphi$)
$F_V$    vertical force of sails ($F_V = F_H \sin \varphi$)
$C_F$    friction resistance coefficient
$C_T$    total resistance coefficient
$F_T$    total force when either hull or sails are implied
$F_S$    total sail force
$F_h$    total hull force
$F_r$    rudder force
W    weight or displacement in units of weight
B    buoyancy force

## MOMENTS
K, M, N    rolling, pitching, yawing moments respectively
$I_x, I_y, I_z$    moments of inertia about x, y, z axes respectively (fig. 6.1)
k    radius of gyration (k $= \sqrt{(I/m)}$ )

## SYMBOLS FOR CONSTANTS

g   acceleration of gravity = $9.8\,\mathrm{ms^{-2}}$ (m/sec/sec)

$\varrho_a$   density of air = $1.20\,\mathrm{kg/m^3}$ (dry air at 20°C and one atmos. press.)

$\varrho_w$   density of water = $1000\,\mathrm{kg/m^3}$ (fresh); $1030\,\mathrm{kg/m^3}$ (salt)

$\nu_a$   kinematic viscosity of air = $1.46 \times 10^{-5}\,\mathrm{m^2s^{-1}}$ (sq m/sec)

$\nu_w$   kinematic viscosity of water = $1.14 \times 10^{-6}\,\mathrm{m^2s^{-1}}$ (fresh); $1.11 \times 10^{-6}\,\mathrm{m^2s^{-1}}$ (salt)

$F_n$   Froude's number ($F_n = V/\sqrt{gL}$)

$R_n$   Reynolds' number ($R_n = VL/\nu$)

$\Delta$   mass displacement

$\nabla$   volume displacement

## RATIOS

$\Delta/(0.01L)^3$   displacement/length ratio (depends on units used)

AR   aspect ratio = (span)$^2$/area (the span used depends on the physical situation, see for instance fig. 5.22)

$C_P$   prismatic coefficient = $\nabla/A_oL$ (dimensionless)

# Appendix B

# DIMENSIONS AND CONVERSION FACTORS

This book has been written using primarily the SI (Système International) units. Although these are basically simpler, they are not so familiar to some. For this reason many diagrams are labelled with both the SI units and those traditional in English speaking countries.

Despite the trend to SI units, knots and nautical miles are preferred to metres per second and kilometres in marine applications, basically because systems of navigation are based on nautical miles. Because of this inevitable mixture of units, the following tables should assist in converting one to the other.

Another point which is worth making here in the Appendix is the difference between mass and weight. In the chapter on dynamic motion it was essential to use the term 'mass'. To most people a pound of sugar or a kilogramme of potatoes has a well defined meaning: it is the *force* one feels when these objects are lifted. It is the result of gravity acting on a *mass* of 1 lb or 1 kg. Weight is therefore a force. When you pick up the bag of potatoes your greengrocer has just sold you, the force you exert on them is equal to their mass times the acceleration of gravity which is 9.8 newtons for each kg of mass. The mass of potatoes is what you are interested in. The fact that their weight on the moon, for instance, is less is of no concern.

A heavy horizontal flywheel is difficult to set in motion and just as difficult to stop, but since its motion is perpendicular to gravity it has nothing to do with weight. The flywheel's properties are due to its large moment of inertia which is determined by its distribution of mass. It behaves the same on the moon as it does on the earth.

Because of this confusion between mass and weight, pounds and kilogrammes are often used incorrectly when force is implied. The SI unit of force is the newton. The simplest way of getting a 'feel' for newtons is to remember that when you pick up a 1 kg block of cheese you are feeling a force of about 10 newtons.

## CONVERSION FACTORS

| Length | $m$ | $km$ | $i$ | $f$ | $nm$ |
|---|---|---|---|---|---|
| 1 metre = | 1 | $10^{-3}$ | 39.37 | 3.281 | $5.4 \times 10^{-4}$ |
| 1 kilometre = | 1000 | 1 | $3.9 \times 10^4$ | 3281 | 0.54 |
| 1 inch = | 0.0254 | $2.54 \times 10^{-5}$ | 1 | 0.0833 | $1.37 \times 10^{-5}$ |
| 1 foot = | 0.3048 | $3.048 \times 10^{-4}$ | 12 | 1 | $1.65 \times 10^{-4}$ |
| 1 naut. mile = | 1852 | 1.852 | $7.29 \times 10^4$ | 6076.1 | 1 |

1 league = 3 nautical miles          1 nautical mile = 10 cables

**Area**

| | $m^2$ | $cm^2$ | $f^2$ | $i^2$ |
|---|---|---|---|---|
| 1 square metre = | 1 | $10^4$ | 10.76 | 1550 |
| 1 square centimetre = | $10^{-4}$ | 1 | $1.076 \times 10^{-3}$ | 0.1550 |
| 1 square foot = | $9.29 \times 10^{-2}$ | 929.0 | 1 | 144 |
| 1 square inch = | $6.45 \times 10^{-4}$ | 6.452 | $6.94 \times 10^{-3}$ | 1 |

**Volume**

| | $m^3$ | $cm^2$ | $f^3$ | $i^3$ |
|---|---|---|---|---|
| 1 cubic metre = | 1 | $10^6$ | 35.31 | $6.1 \times 10^4$ |
| 1 cubic centimetre = | $10^{-6}$ | 1 | $3.53 \times 10^{-5}$ | 0.061 |
| 1 cubic foot =e | $2.832 \times 10^{-2}$ | 28,320 | 1 | 1728 |
| 1 cubic inch = | $1.64 \times 10^{-5}$ | 16.39 | $5.79 \times 10^{-4}$ | 1 |

1 Imp. gallon = the volume of 10 lb of water at 62°F = 277.42 $i^3$

1 litre = the volume of 1 kg of water at its maximum density = 1000.028 cm$^3$

**Mass**

| | $g$ | $kg$ | $lb$ |
|---|---|---|---|
| 1 gramme = | 1 | 0.001 | 0.002205 |
| 1 kilogramme = | 1000 | 1 | 2.205 |
| 1 pound = | 453.6 | 0.4536 | 1 |

1 metric tonne = 1000 kg = 2205 lb

1 long tonne = 2240 lb

**Speed**

| | $f/s$ | $km/h$ | $m/s$ | $mile/h$ | $knot$ |
|---|---|---|---|---|---|
| 1 foot/second = | 1 | 1.097 | 0.3048 | 0.6818 | 0.5925 |
| 1 kilometre/hour = | 0.9113 | 1 | 0.2778 | 0.6214 | 0.5400 |
| metre/second = | 3.281 | 3.6 | 1 | 2.237 | 1.944 |
| 1 statute mile/hour = | 1.467 | 1.609 | 0.4470 | 1 | 0.869 |
| 1 knot | 1.688 | 1.852 | 0.5144 | 1.151 | 1 |

1 knot = 1 nautical mile per hour

**Force**

| | $kgf$ | $N$ | $p$ |
|---|---|---|---|
| 1 kilogramme-force = | 1 | 9.8 | 2.205 |
| 1 newton = | 0.102 | 1 | 0.2248 |
| 1 pound-force | 0.4536 | 4.448 | 1 |

**Pressure**

| | atm | $N/m^2$ | $p/i^2$ |
|---|---|---|---|
| 1 atmosphere = | 1 | $1.013 \times 10^5$ | 14.7 |
| 1 newton/sq metre = | $9.87 \times 10^{-6}$ | 1 | $1.45 \times 10^{-4}$ |
| 1 pound/sq inch = | $6.805 \times 10^{-2}$ | $6.895 \times 10^3$ | 1 |

1 millibar = 100 $N/m^2$

**Energy**

| | fp | J | kWh |
|---|---|---|---|
| 1 foot-pound = | 1 | 1.356 | $3.8 \times 10^{-7}$ |
| 1 joule = | 0.7376 | 1 | $2.8 \times 10^{-7}$ |
| 1 kilowatt-hour = | $2.66 \times 10^6$ | $3.6 \times 10^6$ | 1 |

**Power**

| | fp/s | hp | kW |
|---|---|---|---|
| 1 foot-pound/second = | 1 | $1.82 \times 10^{-3}$ | $1.36 \times 10^{-3}$ |
| 1 horsepower = | 550 | 1 | 0.7457 |
| 1 kilowatt = | 737.6 | 1.341 | 1 |

# INDEX

# INDEX